VOLUME ONE HUNDRED AND FIVE

ADVANCES IN VIRUS RESEARCH

Complementary Strategies to Understand Virus Structure and Function

VOLUME ONE HUNDRED AND FIVE

ADVANCES IN VIRUS RESEARCH

Complementary Strategies to Understand Virus Structure and Function

Edited by

FÉLIX A. REY
Structural Virology Unit
Virology Department and CNRS UMR3569
Institut Pasteur, Paris, France

ELSEVIER
ACADEMIC PRESS
An imprint of Elsevier

Academic Press is an imprint of Elsevier
50 Hampshire Street, 5th Floor, Cambridge, MA 02139, United States
525 B Street, Suite 1650, San Diego, CA 92101, United States
The Boulevard, Langford Lane, Kidlington, Oxford OX5 1GB, United Kingdom
125 London Wall, London, EC2Y 5AS, United Kingdom

First edition 2019

ISBN: 978-0-12-818456-1
ISSN: 0065-3527

For information on all Academic Press publications
visit our website at https://www.elsevier.com/books-and-journals

Publisher: Zoe Kruze
Acquisition Editor: Ashlie M. Jackman
Editorial Project Manager: Peter Llewellyn
Production Project Manager: James Selvam
Cover Designer: Miles Hitchen

Typeset by SPi Global, India

Contents

Contributors

Fasséli Coulibaly
Infection & Immunity Program, Department of Biochemistry and Molecular Biology, Biomedicine Discovery Institute, Monash University, Clayton, VIC, Australia

Isabel Fernández de Castro
Cell Structure Laboratory, National Center for Biotechnology, CSIC, Madrid, Spain

Pedro J. de Pablo
Departamento de Física de la Materia Condensada and Instituto de Física de la Materia Condensada (IFIMAC), Universidad Autónoma de Madrid, Madrid, Spain

Jasmin Dülfer
Heinrich Pette Institute, Leibniz Institute for Experimental Virology, Hamburg, Germany

Daniel J. Goetschius
Department of Biochemistry and Molecular Biology, The Pennsylvania State University, University Park; Department of Medicine, College of Medicine, The Pennsylvania State University, Hershey, PA, United States

Susan Hafenstein
Department of Biochemistry and Molecular Biology; Huck Institutes of the Life Sciences, The Pennsylvania State University, University Park; Department of Medicine, College of Medicine, The Pennsylvania State University, Hershey, PA, United States

Juha T. Huiskonen
Division of Structural Biology, Wellcome Centre for Human Genetics, University of Oxford, Oxford, United Kingdom; Helsinki Institute of Life Science HiLIFE and Research Programme in Molecular and Integrative Biosciences, Faculty of Biological and Environmental Sciences, University of Helsinki, Helsinki, Finland

Alan Kadek
Heinrich Pette Institute, Leibniz Institute for Experimental Virology, Hamburg; European XFEL GmbH, Schenefeld, Germany

Young Chan Kim
Division of Structural Biology, Wellcome Centre for Human Genetics, University of Oxford, Oxford, United Kingdom

Janine-Denise Kopicki
Heinrich Pette Institute, Leibniz Institute for Experimental Virology, Hamburg, Germany

Boris Krichel
Heinrich Pette Institute, Leibniz Institute for Experimental Virology, Hamburg, Germany

Hyunwook Lee
Department of Biochemistry and Molecular Biology; Huck Institutes of the Life Sciences, The Pennsylvania State University, University Park, PA, United States

Maolin Lu
Department of Microbial Pathogenesis, Yale University School of Medicine, New Haven, CT, United States

Xiaochu Ma
Department of Microbial Pathogenesis, Yale University School of Medicine, New Haven, CT, United States

Walther Mothes
Department of Microbial Pathogenesis, Yale University School of Medicine, New Haven, CT, United States

Weng M. Ng
Division of Structural Biology, Wellcome Centre for Human Genetics, University of Oxford, Oxford, United Kingdom

Martin Obr
Institute of Science and Technology Austria (IST Austria), Klosterneuburg, Austria

Cristina Risco
Cell Structure Laboratory, National Center for Biotechnology, CSIC, Madrid, Spain

Martin Sachse
Unité Technologie et service BioImagerie Ultrastructurale, Institut Pasteur, Paris, France

Florian K.M. Schur
Institute of Science and Technology Austria (IST Austria), Klosterneuburg, Austria

Robert Stass
Division of Structural Biology, Wellcome Centre for Human Genetics, University of Oxford, Oxford, United Kingdom

Raquel Tenorio
Cell Structure Laboratory, National Center for Biotechnology, CSIC, Madrid, Spain

M. Alejandra Tortorici
Department of Biochemistry, University of Washington, Seattle, WA, United States; Institut Pasteur; CNRS UMR 3569, Unité de Virologie Structurale, Paris, France

Charlotte Uetrecht
Heinrich Pette Institute, Leibniz Institute for Experimental Virology, Hamburg; European XFEL GmbH, Schenefeld, Germany

David Veesler
Department of Biochemistry, University of Washington, Seattle, WA, United States

Preface

Structural biology has had—and will continue to have—a profound impact in virology, as it provides formidable tools to understand the molecular mechanisms deployed by a virus to disseminate in the environment, to enter cells and to replicate within them. It also provides crucial insight into virus evolution, as structure conservation is maintained even when all sequence similarity has been lost. This aspect is especially true given the poor fidelity of viral polymerases in general, and the very fast generation time of viruses. Structural insights have allowed to understand the mechanism of antibody cross-neutralization of highly variable pathogens, such as the human immunodeficiency viruses (HIV) or the influenza viruses. They have also led to the development of specific strategies to interfere with the replication of certain viruses within cells. The most conspicuous example is the case of the hepatitis C virus (HCV), which causes chronic disease and leads over the years to liver cirrhosis and cancer. HCV was discovered only 30 years ago, yet it is possible today to cure patients with new drugs inhibiting specific steps of its replication cycle. The detailed molecular understanding of some of the key steps of the HCV replication cycle, illuminated by the structures of the viral proteins and complexes involved, was crucial for these developments.

Structural virology began with pioneering studies made by Wendell Stanley and by John Bernal on crystalline tobacco mosaic virus back in the 1930s, and was accelerated enormously in the 1980s, when the use of synchrotron radiation became widespread in X-ray crystallography. It was further boosted in the 1990s when it was shown that nuclear magnetic resonance (NMR) was an important additional tool. Today, recent breakthroughs such as cryoelectron microscopy (cEM) breaking the resolution barrier, atomic force microscopy (AFM) or Förster resonance energy transfer (FRET) methods allowing to track single particle dynamics, or mass spectrometry providing multiple levels of information, including the unambiguous determination of subunit stoichiometries in macromolecular complexes, have opened a new era. The integration of these methods with further developments in crystallography and in NMR allow now to go beyond static images and incorporate the time dimension. Indeed, understanding the dynamics of macromolecular interactions is essential. And in addition to the dynamics, the new methods make it possible to analyze viruses in high detail even beyond the special case of highly symmetric particles (as in the previous century), thereby extending the scope of structural virology to study particles that are only partially symmetric, and even to particles that are pleomorphic altogether.

The complementary biophysical approaches available today thus allow the study of complex molecular components in their natural context—viral or cellular—whereas previously they were studied in isolation. To state it differently, the end of the previous century saw the success of a "divide and conquer" approach in structural biology, which had led to understanding in detail isolated parts of viruses, or static assemblies of these parts. The novel methodologies today provide the opportunity to put these parts back together and understand their function as a whole. This volume attempts a recapitulation of many of these new developments, and their key contributions to understand virus function.

In Chapter 1, Sachse et al. describe the current arsenal available to study viral replication complexes within an infected cell, the so-called "viral factories." They include a combination of light microscopy with electron microscopy, and allow to follow the evolution in time of these complexes. In Chapter 2, Stass et al. describe the use of electron cryomicroscopy and tomography procedures to study enveloped viruses, whether symmetric enveloped viruses that can be studied by single particle analyses, or pleomorphic particles that require tomographic procedures. In Chapter 3, Goetschius et al. provide examples of symmetric particles bound asymmetrically to a receptor or to antibodies, illustrating how the details of such nonsymmetric interactions can be extracted. In Chapter 4, Tortorici and Veesler provide an illustration of the power of single particle approaches in cryo-EM to obtain a high-resolution picture of the coronavirus spike—a very large, heavily glycosylated protein. They analyze its interactions with receptor for entry, as well as the way the spike is engaged by neutralizing antibodies. Obr and Schur show in Chapter 5 how subtomogram averaging approaches in cryo-ET can be used to provide structures to near atomic resolution, using the example of the retroviral core particles. Pedro de Pablo explores in Chapter 6 the use of atomic force microscopy to probe mechanical properties of virus particles, while Dülfer et al. discuss the use of mass spectrometry to investigate the structure and dynamics of viral particles or their complexes under conditions very close to their native environment in Chapter 7. Similarly, in Chapter 8, Lu et al. analyze the use of single-molecule FRET to probe the conformational landscape of viral proteins in their natural environment.

Finally, in Chapter 9 Fasséli Coulibaly discusses recent advances in synchrotron radiation to determine structures using in-cell grown crystals, allowing the investigation of virus-induced structures that are key for their life cycle. The volume of such crystals is in the order of $\sim 1\ \mu^3$, requiring the use of highly collimated optics and the full power of the novel X-ray free-electron lasers.

Structural biology has gone a full circle: the crystals of tobacco mosaic virus discovered by Wendell Meredith Stanley—a discovery that led to Stanley being awarded the Nobel Prize for chemistry in 1946—were natural crystals that grew in infected plant leaves. Those crystals were very large, however. The recent developments in structural biology are allowing now the study of micron sized natural structures directly appearing within infected cells. The new exciting era we are seeing today will very likely appear pioneering when future scientists look back to the exceptional progress made in the first two decades of the XXIst century. Present-day structural biologists are thus earning a special place in history.

Félix A. Rey
Structural Virology Unit
Virology Department and CNRS UMR3569
Institut Pasteur, Paris, France

CHAPTER ONE

The viral replication organelles within cells studied by electron microscopy

Martin Sachse[a,*], Isabel Fernández de Castro[b], Raquel Tenorio[b], Cristina Risco[b,*]

[a]Unité Technologie et service BioImagerie Ultrastructurale, Institut Pasteur, Paris, France
[b]Cell Structure Laboratory, National Center for Biotechnology, CSIC, Madrid, Spain
*Corresponding authors: e-mail address: martin.sachse@pasteur.fr; crisco@cnb.csic.es

Contents

Abstract

Transmission electron microscopy (TEM) has been crucial to study viral infections. As a result of recent advances in light and electron microscopy, we are starting to be aware of the variety of structures that viruses assemble inside cells. Viruses often remodel cellular compartments to build their replication factories. Remarkably, viruses are also able to induce new membranes and new organelles. Here we revise the most relevant imaging technologies to study the biogenesis of viral replication organelles. Live cell microscopy, correlative light and electron microscopy, cryo-TEM, and three-dimensional imaging methods are unveiling how viruses manipulate cell organization. In particular, methods for molecular mapping in situ in two and three dimensions are revealing how macromolecular complexes build functional replication complexes inside infected cells. The combination of all these imaging approaches is uncovering the viral life cycle events with a detail never seen before.

Advances in Virus Research, Volume 105
ISSN 0065-3527
https://doi.org/10.1016/bs.aivir.2019.07.005

Abbreviations

CEMOVIS	cryo-electron microscopy of vitreous sections
CLEM	correlative light and electron microscopy
EM	electron microscopy
ET	electron tomography
FIB	focused-ion beam
FS	freeze-substitution
GFP	green fluorescence protein
HPF	high-pressure freezing
LM	light microscopy
METTEM	metal-tagging TEM
MT	metallothionein
RO	replication organelle
RT	room temperature
SEM	scanning electron microscopy
STEM	scanning-transmission electron microscopy
3D	three-dimensional
TEM	transmission electron microscopy
2D	two-dimensional
VF	virus factory
VRC	viral replication complex
vRNP	viral ribonucleoprotein

Virus names

BUNV	Bunyamwera virus
DENV	dengue virus
FHV	flock house virus
FMDV	foot-and-mouth disease virus
HCV	hepatitis C virus
HIV	human immunodeficiency virus
SARS CoV	severe acute respiratory syndrome coronavirus
SFV	Semliki forest virus
TBSV	tomato bushy stunt virus
VV	vaccinia virus

1. Introduction

The common hallmark of all viruses is their dependency on a host cell to infect and multiply. During this cycle the virus utilizes and transforms the machinery of the cell for the biosynthesis of the own genetic material, proteins, and lipids. This can lead to vast reorganization in the subcellular organization inside the cell (Risco et al., 2014; Romero-Brey and Bartenschlager, 2014). The size of most viruses is beyond the resolution limit of optical microscopy, and only electron microscopy (EM) provides the resolution required to study viruses and their life cycles (Pelchen-Matthews and Marsh, 2007).

The technical specifications of an electron microscope are, however, only one part of the equation to deliver insight into the viral life cycle within the host cell. The first obstacle comes from the dynamic nature of the virus cycle. Infection, replication, virus particle morphogenesis and egress are different, consecutive steps. But in contrast to optical microscopy, EM does not permit the observation of the same object over time. This aspect can be solved with the study of different samples after finding the right time post-infection for each step. Next to the dynamics of these different steps during viral infection, another challenge arises from their topology: only entry and release of the viral particle take place at the surface of the host cell. All other steps utilize machineries inside the cell and their study requires the analysis of the morphology of the cell interior. For this reason, transmission electron microscopy (TEM) combined with thin sections of infected cells has been for a long time the standard approach to study the intracellular aspects of virus life cycles. However, the morphology of a cell is three-dimensional and thus random thin sections provide a limited view. As pointed out more than 60 years ago, the impact of EM for the morphological analysis of viruses depends highly on technological progress on sample preparation (Hillier, 1950). Since this time, sample preparation as well as the sensitivity of electron microscopes and their detectors has evolved to a great extent and different approaches for tomography were developed to study samples in 3D. But the key point raised at that time has remained: each sample preparation/imaging technique produces artifacts and it is important to be aware of them and be critical about the amount of information that can be extracted. Simultaneously with the technological development, our knowledge about the life cycle of viruses has increased, and the diversity of the questions addressed has increased concomitantly. It is often required to combine different approaches to answer the question at hand. Therefore, this review will focus on the different preparation and imaging techniques that are used to visualize the cell biology of viral infections (Fig. 1). We will discuss the technological basis of the different approaches and the findings they made possible. In particular, we will focus on the characterization of virus factories and virus replication organelles. Viruses express their fundamental biological properties during the intracellular stage of their life cycle,

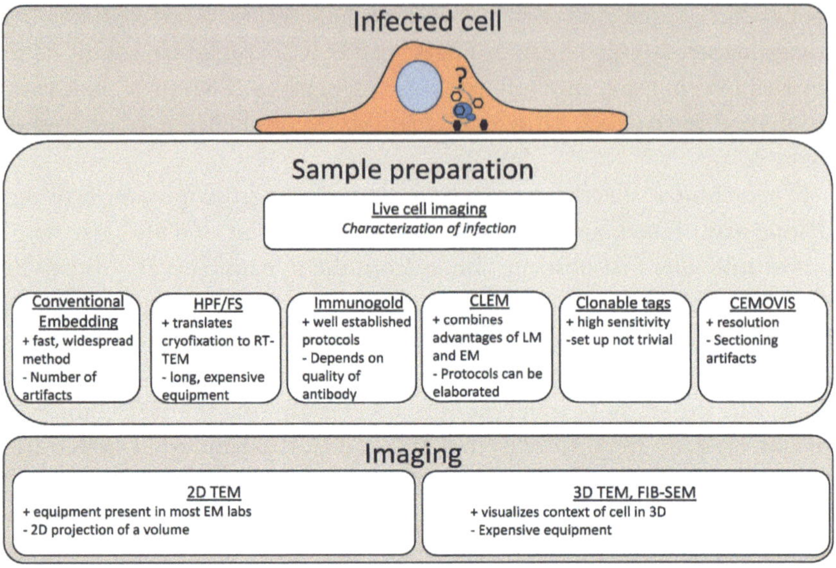

Fig. 1 Scheme summarizing the different approaches for sample preparation and imaging, which are discussed in this review. Life cell imaging provides the best approach to characterize the dynamics of infection and the assembly of replication organelles. Based on these data the EM experiments can be designed. The choice of sample preparation protocols will depend on the questions of interest and the available possibilities. Main advantages (+) and limitations (−) of each method are indicated.

which in most cases takes place within virus factories (Fernández de Castro et al., 2013). Viral factories (VF) are intracellular membranous compartments harboring viral replication organelles (RO) that contain viral replication complexes (VRC) and the sites of virus particle assembly (Harak and Lohmann, 2015; Novoa et al., 2005). VRCs are often located in single-membrane vesicles known as spherules, but also in double-membrane vesicles (DMVs), flat membranes and cubic/convoluted membranes (CM) (Fernández de Castro et al., 2013). To build these neoorganelles, viruses manipulate lipid synthesis and flows to create platforms with particular lipid composition and biophysical properties that facilitate viral genome replication (Altan-Bonnet, 2017; Fernández de Castro et al., 2016; Fernández-Oliva et al., 2019). Due to their complex composition and functions, virus factories could be contemplated as the living state of viruses (Fridmann-Sirkis et al., 2016). Two main pathways have been identified so far as being responsible for viral factory biogenesis: aggresome formation for DNA viruses and autophagy for RNA viruses (Abernathy et al., 2019; Jackson et al., 2005; Kajitani et al., 2013; Wileman, 2007).

Mitochondria and cytoskeleton elements are often found around replication organelles building a large structure that usually transforms the whole cell content (Fernández de Castro et al., 2013). Viral factories are dynamic structures that change over time and maintain communication with intracellular transport pathways such as endocytosis and exocytosis, which are necessary in order to incorporate essential factors into the factories and to facilitate virus egress (Altan-Bonnet, 2017; Fontana et al., 2010). Some viruses build factories inside the nucleus (Erickson et al., 2012) but most of the structural information available to date comes from cytoplasmic viral factories (Risco et al., 2014).

2. Live cell imaging

Despite the focus on EM of this review we must highlight light microscopy (LM) as a great ally of electron microscopists. This is due to the ability of LM to characterize dynamic changes in the cell associated to viral infection, albeit with less resolution but with shorter times for sample preparation. Furthermore, new advances in fluorescent probes and LM methods are providing new insights into the cell biology of viruses. Very interesting recent reviews deal with light microscopy of viral infections (Alonas et al., 2016; Dirk et al., 2016; Hanne et al., 2016; Witte et al., 2018). However, microscopy of fixed cells provides static views of complex, dynamic events. To study multi-step processes such as a virus infection, time-course assays are useful and provide data that we try to organize sequentially. Nevertheless, during viral infections what we perceive as a logical sequence of events might not correspond to reality, and many findings would have never been uncovered without dynamic, real time, live cell microscopy. Live-cell imaging of viral infections has been mainly focused on processes of entry and release of viral particles (Baktash et al., 2018; Francis and Melikyan, 2018; Jouvenet et al., 2008) but has shown great potential to study all steps of the virus replication cycle (Campbell and Hope, 2008). Live cell microscopy can also reveal a heterogeneous response of cells to infection (Movie 1 in the online version at https://doi.org/10.1016/bs.aivir.2019.07.005, Fig. 2). When mammalian cells infected with a fluorescent bunyavirus (Shi et al., 2010) are studied by live cell imaging, it can be observed that some cells rapidly build a virus factory and produce many progeny viruses (Fig. 2, cell 1), a few cells die soon (Fig. 2, cell 2) and some others apparently do not get infected (Fig. 2, cell 3). Indeed, some cells receive the fluorescent virus and pass it to neighboring cells while remaining

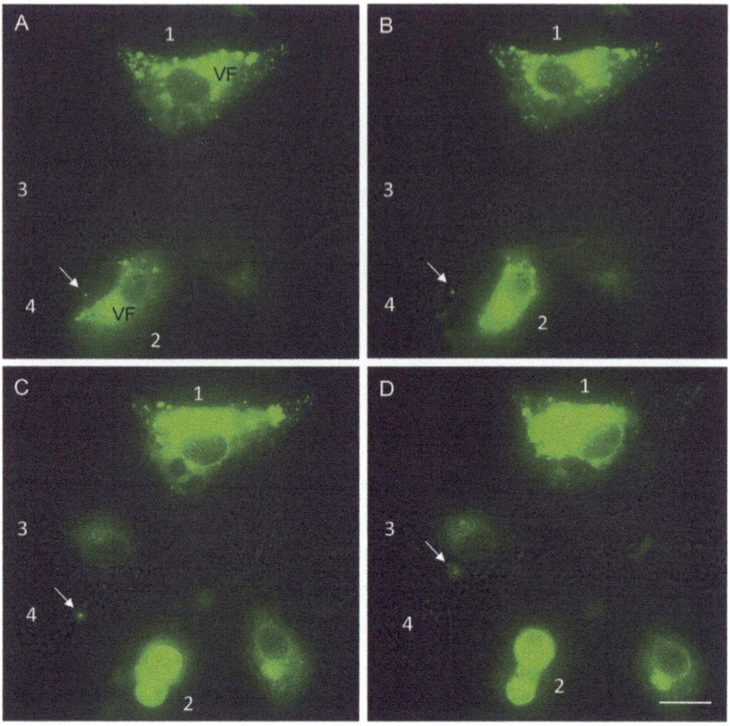

Fig. 2 Snapshots from Movie 1 in the online version at https://doi.org/10.1016/bs.aivir.2019.07.005 showing different stages of viral factory biogenesis and a heterogeneous response of cells to the virus. BHK-21 cells infected with the fluorescent bunyavirus eGFP-Gc-BUNV at a multiplicity of infection (MOI) of 5 PFU/cell, were transferred to a TCS-SP5 microscope (Leica Microsystems) and images recorded every 15 min from 1 h post-infection (hpi) to 18 hpi. Images correspond to 8 h and 15 min post-infection (A), 8 h and 45 min pi (B), 13 h and 45 min pi (C), and 15 h and 30 min pi (D). In cell #1, a large viral factory (VF) assembles and grows over time. Cell #2 assembles a VF and dies soon. Cell #3 remains apparently non-infected and cell #4 receives fluorescent viruses (arrows in A–C), passes them to another cell (arrow in D) and remains non-infected. Scale bar: 25 μm. Sanz-Sánchez and Risco, unpublished results.

apparently non-infected (Fig. 2, cell 4). These findings demonstrate that cells in the cultured cell monolayer are not identical and respond to virus infection heterogeneously.

An interesting case was reported in a study of mosquito cells infected with an arbovirus (López-Montero and Risco, 2011). Video-microscopy showed that all infected cells went through three different stages of infection, early, acute and persistent, something that can be fully appreciated only when following each single cell in real time (López-Montero and Risco, 2011). Live-cell imaging has been used to study the biogenesis, dynamics and trafficking of viral replicative structures. Remarkably, these studies have shown that unrelated viruses such as Vaccinia Virus, Hepatitis C Virus, rotaviruses and the mouse hepatitis coronavirus, assemble small replication organelles that move through the cytoplasm in a microtubule-dependent manner, together with larger replication platforms that are rather immobile (Baktash and Randall, 2019; Eichwald et al., 2012; Eyre et al., 2014; Hagemeijer et al., 2010; Schramm et al., 2006; Wölk et al., 2008). Live-cell imaging has also shown the transformation of cell compartments induced by viruses to build their replication organelles. This is the case of the enterovirus coxsackievirus B3 that disassembles the Golgi complex (Van der Schaar et al., 2016) or the human reovirus that remodels the endoplasmic reticulum to build viral inclusions (Tenorio et al., 2018). New probes for imaging viral RNA will be powerful tools to study viral replication and assembly in live cells (Alonas et al., 2016).

For obtaining reproducible results by live cell imaging, the experiment must be carefully designed and adequate controls have to be included. Before starting, a number of important factors should be taken into account. Fluorescent probes must be compatible with in vivo conditions (Galas et al., 2018). Cells have to be grown on special plates with a clear bottom made of plastic or glass and plated with an appropriate confluence, usually 70% or less. A previous time-course study by confocal microscopy of fixed cells is important to determine the best time post-infection for launching the live cell imaging recording. To limit potential effects caused by phototoxicity, the right choice of fluorescent proteins, light intensity, exposure time and time interval between frame acquisitions will be critical (Magidson and Khodjakov, 2013; Icha et al., 2017).

For a detailed understanding of events, either expected or rare, live cell imaging is a powerful partner of EM through the methods we know as Correlative Light and Electron Microscopy (CLEM) (see Section 4).

3. General aspects of sample preparation for transmission electron microscopy

For a morphological study by EM the goal should be to preserve the samples as close as possible to the condition of interest and keep all cellular components in place. The electron microscope adds restrains to this goal because the electrons used for imaging interact with matter, requiring samples under high vacuum during imaging. Except for cryo-TEM, this excludes the observation of samples in their natural, hydrated state.

The second, main restrain introduced by electrons is the thickness of the sample: only thin samples can be observed by TEM, whereas most biological samples are soft, which makes the preparation of thin slices impossible.

Therefore, samples have to be prepared adequately for TEM. In general, this involves the arrest of any biological activity by fixation. For morphological studies, conventional TEM preparation at room temperature (RT) relies on aldehyde fixation (Sabatini et al., 1963), followed by a post-fixation with osmium to stabilize lipids (White et al., 1976), and dehydration to replace gradually the water of the sample by an organic solvent. In the final step, the sample is infiltrated with a plastic resin, which becomes hard after polymerization and allows the preparation of thin sections for TEM. In this procedure, each step has well documented artifacts (Kellenberger, 1991), which have to be taken into account. But it should also not be forgotten that these sample preparation protocols have contributed to a vast basis of knowledge before alternative sample preparation methods became available. Characteristic viral factories as seen by TEM of ultrathin sections after conventional embedding are shown in Fig. 3. In mammalian BHK-21 cells,

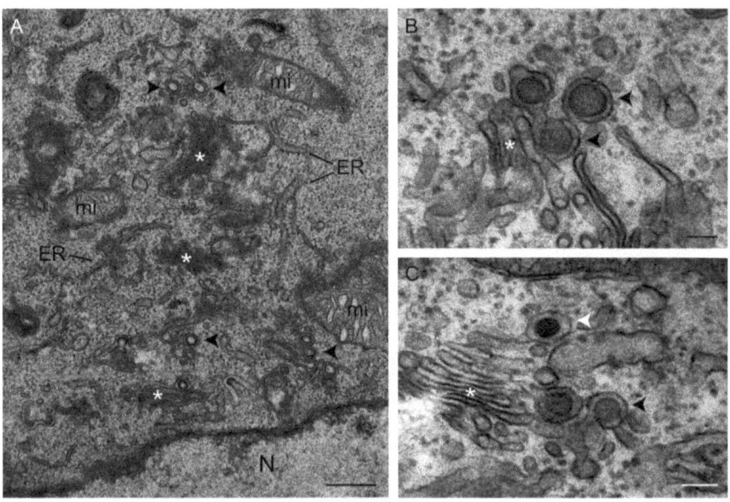

Fig. 3 Viral factories as seen by TEM of ultrathin sections after conventional embedding. BHK-21 cells were adsorbed with Bunyamwera virus (BUNV) at a MOI of 1 PFU/cell and imaged by TEM at 10 hpi. (A) Low magnification image of a BUNV factory that is a juxtanuclear structure made of Golgi stacks (white asterisks) with spherules (black arrowheads) that contain the VRCs (Fontana et al., 2008), mitochondria (mi) and ER cisternae. N: nucleus. (B and C) High magnification images of BUNV factories showing spherules (black arrowheads) in Golgi stacks (asterisks). A viral particle (white arrowhead) in the lumen of a Golgi sacculus is also seen (white arrowhead in C). Scale bars: 500 nm in A; 100 nm in B and C. Fernández de Castro, unpublished results.

Bunyamwera virus assembles a large cytoplasmic factory using Golgi complex stacks, ER and mitochondria (Fig. 3A). Characteristic spherules with VRCs are formed in Golgi membranes (Fontana et al., 2008) where progeny viruses are assembled (Fig. 3B and C).

4. Correlative light and electron microscopy

The term correlative and light electron microscopy (CLEM) covers an enormous area of approaches, different combinations of light microscopy techniques with EM strategies. Two extensive handbooks give an idea about this field, which showed over the last years a very dynamic development and continues to do so (Müller-Reichert and Verkade, 2012, 2014). Therefore, this section does not aim to be exhaustive but should outline examples of different strategies. In general, the different approaches can be defined by the complementing aspects that light microscopy (LM) and EM have in their combination. The light microscope offers an unmatched field of view compared to EM, together with the above discussed possibility of imaging the dynamics of the living sample. On the other side, EM allows to visualize the sample at a higher resolution and in all the structural context of the cell and not only the structures labeled for the light microscope. The different approaches can be grouped: (1) To use the field of view from the light microscope and find a rare event by fluorescent labeling in a population of cells or tissue. After imaging by LM, the sample can be prepared by different methods for EM. (2) Characterize a dynamic process by LM and fix the sample at the event of interest to study it in the context of the cell. (3) Visualize your protein of interest in the context of the cell. Often this approach is done with LM and EM on sections or with cryo-fluorescence followed by cryo-TEM on plunge frozen cells. The latter is reviewed elsewhere (Wolff et al., 2016).

To identify a rare fluorescently labeled event in a culture of infected cells the most straightforward approach is to grow them on a support that has coordinates and can be imaged by LM, allowing to identify the location of the cell. This support can be photoetched glass coverslips or gridded plastic dishes for conventional preparation, sapphire discs with a carbon pattern for high pressure freezing (HPF), or a finder grid for plunge freezing (Madela et al., 2014; Santarella-Mellwig et al., 2018; Tenorio et al., 2018). The pattern on the substrate allows to localize the cell of interest by phase contrast. After embedding in resin the pattern is present on the surface of the resin and will allow to locate the area with the cell of interest and prepare serial sections. A visual example of correlation is shown in Fig. 4. For tissue,

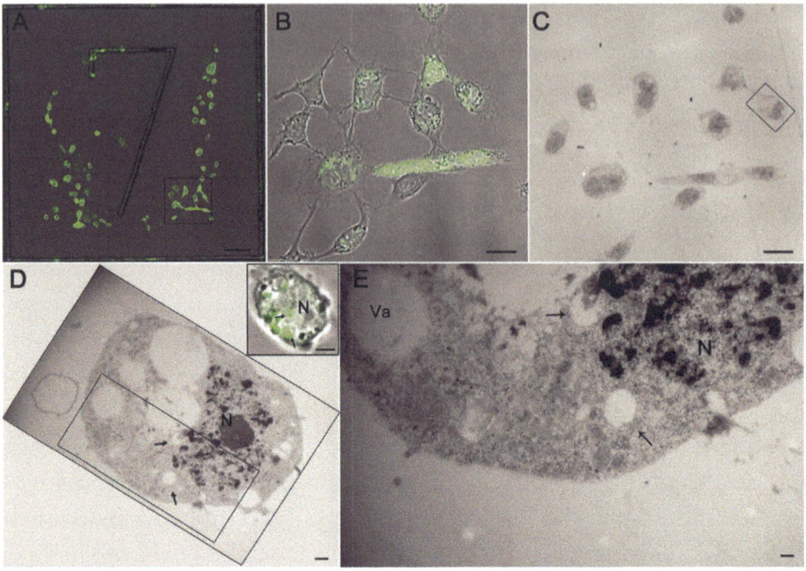

Fig. 4 Correlative light electron microscopy (CLEM) of viral factories. C6/36 mosquito cells grown on gridded plates were adsorbed with the fluorescent bunyavirus eGFP-Gc-BUNV for 24 h and imaged using phase contrast and fluorescence microscopy at 200 × and 630 × magnification (A and B, respectively). Gc is one of the glycoproteins of the virus and a marker of viral factories. Ultrathin sections of selected cells (square in A) were imaged by TEM at low magnification (C) and at higher magnification (D and E). (D) TEM image (mainfield) and confocal image (inset) of the cell selected in (C). The arrows point to fluorescent zones (inset) and the approximate corresponding areas in the TEM image. (E) High magnification image shows that the marked fluorescent zones (inset in D) contain vacuole–like structures (arrows). N, nucleus; Va, vacuole. Scale bars: 50 μm in A; 10 μm in B and C; 0.5 μm in D; 5 μm in inset in D; 200 nm in E. Tenorio, unpublished results.

transfer of coordinates provides often a challenge. Fluorescent dyes can be used to highlight specific structures of tissues and create internal landmarks to position the region of interest and embed the samples for EM in such a way that the fluorescence is retained, and either the block or the sections can be imaged by LM (Lucas et al., 2012). Another approach introduced landmarks on the surface of the tissue by an infrared laser (Karreman et al., 2014).

CLEM on sections after sample preparation for EM has the advantage to minimize changes between the LM and EM image, which result from sample preparation. It can be done on Tokuyasu sections (Oorschot et al., 2014), or on methacrylate resins (Kukulski et al., 2011). For both methods, protocols are available that can preserve the fluorescent signal of fluorescent fusion proteins. Alternatively, antibody labeling can be done on sections (Oorschot et al., 2014; Schwarz and Humbel, 2014). The fluorescent signal can be observed in the LM and afterward the grids are contrasted and observed by EM. Because in the EM the section itself is devoid of landmarks, a first correlation can be obtained with a finder grid. Finer correlation is possible when the DNA inside the nuclei of cells is stained with DAPI or Hoechst for LM. Higher precision can be obtained when fluorescent spheres are applied as landmarks to the section, though this approach requires TEM tomography since the spheres are not visible in a 2D projection image. With this approach it was possible to detect a single HIV-eGFP-delEnv particle at the plasma membrane of a MDCK cell (Kukulski et al., 2011).

Live cell imaging combined with EM allows to study the dynamics of a process and arrest it at the point of interest with the aim of visualization within the context of the cell. The method of live cell imaging depends on the dynamics of the studied process and the resolution required. As mentioned above for imaging live cells, care needs to be taken to avoid phototoxicity. The ideal arrest of the cell after imaging would be cryofixation due to its rapidity. However, this is often limited by practical issues. Often a cryoplunge freezer or a high-pressure freezer are not localized next to a light microscope. And even if they are, the transfer has to be fast enough not to lose the benefit of rapid immobilization. Transfer should also avoid changes in the environment (temperature, physical bending on grid, etc.), which can influence the sample's morphology. Currently there are two ways to combine fast transfer of sapphire discs from the light microscope to a high-pressure freezer (Heiligenstein et al., 2014; Verkade, 2008), both of which allow a short time frame of live cell imaging before high pressure freezing. If biosafety or the length of imaging does not permit this, chemical fixation is mandatory. The delay of the fixation depends on the localization of the process inside the sample. For rat liver, a 4% Paraformaldehyde solution penetrated the tissue to a depth of 2.5 mm within 4 h, which corresponds on average to a penetration of about 170 nm/s (Dempster, 1960). Thus, for events near the plasma membrane like the transport of viral particles for release, chemical fixation will work faster than for the fixation of a viral factory in the centre of the cell. However, it might not be fast enough to visualize fusion or budding events at the plasma membrane. It is still recommendable to control the localization of the event of interest after chemical fixation by light microscopy before processing for EM, either by cryofixation (plunge freezing or high-pressure freezing/HPF), or conventional preparation.

CLEM approaches are advancing our understanding of the biogenesis of viral replication organelles (Bykov et al., 2016). Some interesting examples include the flaviviruses hepatitis C virus (HCV) and Dengue virus (DENV), the alphavirus Semliki Forest Virus (SFV), the enterovirus coxsackievirus B3, and the human reovirus. The genome replication sites of HCV have been located in a complex vesicular network interconnected with the ER known as the "membranous web" (Paul et al., 2013). This structure originates by accumulation of ER-derived double-membrane vesicles (DMVs). CLEM showed that NS5A, a phosphoprotein that facilitates viral replication by interacting with other viral non-structural proteins and cell factors, accumulates in double-membrane vesicles that locate close to lipid droplets, a key organelle for HCV morphogenesis (Filipe and McLauchlan, 2015; Romero-Brey et al., 2012). These results provided evidence of the spatial connection between HCV genome replication and viral particle morphogenesis inside viral factories. In the case of DENV, CLEM is assisting in the characterization of the biogenesis of the replication factories, in particular in the role of viral proteins in membrane remodeling and virion morphogenesis (Bykov et al., 2016; Scaturro et al., 2015). CLEM allowed the unambiguous identification of the enterovirus coxsackievirus B3 replication organelle morphologies as a first step to understand the requirements for their biogenesis and their evolution over time (Van der Schaar et al., 2016). For the alphavirus SFV, CLEM demonstrated that the spherules with VRC are first formed on the plasma membrane to end up in cytoplasmic vacuoles upon internalization (Spuul et al., 2010). Viral and cell factors required for spherule biogenesis were also studied by CLEM (Kallio et al., 2013; Spuul et al., 2011). A combination of live cell imaging and CLEM unveiled that nascent reovirus replication organelles, known as viral inclusions (VI) are made of aggregates of tubules and vesicles originated by remodeling of the peripheral endoplasmic reticulum and that two reovirus non-structural proteins, σNS and μNS, are responsible for ER remodeling (Tenorio et al., 2018). Besides, CLEM has great potential for evaluating compounds to treat viral infections (Berger et al., 2014; García-Serradilla et al., 2019; Lowen et al., 2018; Martínez et al., 2014). For those interested in performing CLEM of virus-infected cells, detailed, step-by-step CLEM workflows describing sample preparation, imaging and correlation, have been published recently (Hellström et al., 2015; Santarella-Mellwig et al., 2018).

5. High-pressure freezing and freeze-substitution

The arrest of cells by chemical fixation is determined by the diffusion of the fixative into the cell and might take seconds. This results in a gradient: first proteins at the plasma membrane are inactivated, whereas intracellular organelles are reached by the fixative at a later time point. To circumvent this problem, an alternative is to freeze cells fast, which arrests them in their state in milliseconds. However, to keep all cellular structures in place, it is mandatory to vitrify the water in the sample and avoid formation of ice crystals during freezing. At atmospheric pressure this requires high cooling rates and allows only the vitrification up to a thickness of a few micrometers (Escaig, 1982). Bulk samples can be vitrified by high pressure freezing. In this approach, for a few milliseconds a pressure of more than 2000 bars is applied to the sample before cooling is applied (Riehle and Hoechli, 1973). Since water in the liquid state is almost incompressible, the application of the high pressure has no influence on the sample's morphology. But the pressure changes the physical properties of water by lowering the melting point and the temperature for homogenous ice nucleation (Kanno et al., 1975). Depending on the water content of the sample, this allows the vitrification of samples up to 200 μm thickness. After HPF, the sample can be processed and imaged in the frozen state by cryo-FIB (Focused Ion Beam)/cryo-TEM or by CEMOVIS (Cryo-Electron Microscopy of Vitreous Sections) (Hagen et al., 2015; Hayles et al., 2010). Most often the samples are processed by freeze substitution (FS). During FS, the water of the sample is replaced by an organic solvent that contains chemical fixatives to stabilize the sample during warm up. Here, the order of events is the reverse to the above described conventional preparation at RT: during FS the sample is first dehydrated (usually at −90 °C, where chemical fixatives are not active) and then fixed when the temperature is raised (Hurbain and Sachse, 2011). Dehydration in the cold avoids most artifacts of dehydration at RT (Kellenberger, 1991). Since the solvent is also the carrier of the fixatives, these are in place and ready to react with the sample and the gradient of fixation at RT is omitted as well. After dehydration, samples can be infiltrated with a resin that permits immunolabeling, such as Lowicryl- or LR-resins, or in epoxy resin for pure morphological studies.

The advantage of HPF is the fast immobilization of the sample from the native, unfixed state. The fast immobilization permits to catch high dynamic processes that are difficult to visualize by slow chemical fixation, such as cytoskeleton changes, budding or fusion processes (Bilý et al., 2015; Fontana et al., 2010; Wan et al., 2015). However, for virus-infected cells this raises the issue of biosafety. With one exception, most HPF machines are not mobile and it is impossible to move them into a BSL-2 or BSL-3 laboratory. In this case the sample should be inactivated by aldehyde fixation. The advantage of a fast immobilization is lost but a superior morphology can be achieved even with prefixed samples by HPF/FS compared to conventional preparation because the major harm to morphological preservation is caused by the dehydration step at RT in the latter (Kellenberger, 1991; Sosinsky et al., 2008). However, depending on the virus and the stress the viral infection causes to the host cell, HPF/FS might not be suitable. It has been reported that stressed cells can freeze poorly with ice crystal formation (McDonald, 2007).

For some viruses, HPF/FS allows visualization of viral structures that remain invisible after conventional preparation protocols. This is the case of Foot-and-mouth disease virus (FMDV). The intracellular particles of this virus are not visible by conventional preparation techniques but in cells processed by HPF/FS, newly formed virions are easily seen (Monaghan et al., 2003). This is apparently due to a high sensitivity of viral particles to low pH during conventional sample processing. An example of viral replication organelles visualized by TEM after conventional embedding or HPF/FS is shown in Fig. 5. Viral inclusions (VIs) that are the replication neoorganelles of the human reovirus (Desmet et al., 2014; Rhim et al., 1962), present a more homogeneous content after HPF/FS (compare Fig. 5A and B). Fine structural details such as the morphology of viral particles inside VIs are better preserved after HPF/FS (Fig. 5A and B, insets).

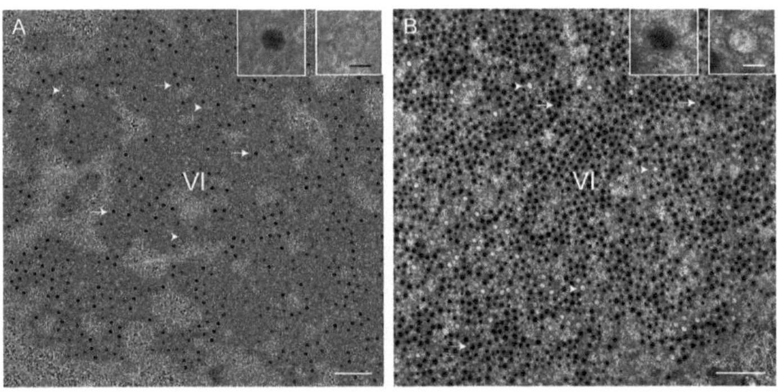

Fig. 5 Conventional embedding vs HPF/FS of human reovirus inclusions. Human brain microvascular endothelial cells (HBMECs) were infected with reovirus at a MOI of 1 PFU/cell. At 18 hpi, cells were fixed with aldehydes and either dehydrated with ethanol at 4 °C and embedded in EML-812 epoxy resin (A) or processed by high-pressure freezing, freeze-substitution and embedding in EML-812 (B). Ultrathin sections show that the reovirus replication organelles known as viral inclusions (VIs) have empty areas after conventional embedding (A) but a more homogeneous content after HPF/FS (B). Both mature virions (arrows) and empty viral particles (arrowheads) are easily distinguished in VIs after either conventional preparation or HPF/FS. Higher magnification views of individual viral particles (insets) show that fine details in both mature virions and empty viral particles are better appreciated after HPF/FS. Scale bars: 500 nm in main fields, 50 nm in insets. Fernández de Castro, unpublished results.

Another viral RO studied by HPF/FS is that assembled by Rubella virus (RUBV) in modified lysosomes that are surrounded by RER cisternae and mitochondria (Fontana et al., 2007; Lee et al., 1994; Magliano et al., 1998). HPF/FS has shown fine details of the complex organization of these structures such as arrays of filaments surrounding the ROs, macromolecular complexes connecting the RO with RER and mitochondria, spherules connected with flat membranes that contain macromolecules organized in oligomeric arrays inside the RO, and pores opened to the cytosol on the periphery of the RO (Fontana et al., 2010). Furthermore, HPF/FS is helping to understand how remodeled membranes build very large viruses such as vaccinia virus and the giant Mimivirus (Milrot et al., 2016; Mutsafi et al., 2013; Suárez et al., 2013).

6. Frozen-hydrated sections: Cemovis and cryo-lamella

HPF increases the depth of vitrification in the sample but in a TEM only thin specimens can be visualized. After the discovery of vitrified water and the onset of cryo-EM (Brüggeller and Mayer, 1980; Dubochet and McDowall, 1981) the transition from thin films of vitrified small biological specimens to vitrified sections of bulk specimen was fast (Lepault et al., 1983; McDowall et al., 1983). To prepare a sample for cryo-electron microscopy of vitreous sections (CEMOVIS), first the sample and the surrounding medium is vitrified. Next, it is sectioned in a cryo-microtome below the devitrification temperature of pure water and the sections are transferred to a grid. The sections are observed in the frozen state with a cryo-TEM using low dose conditions. For a long time, CEMOVIS was the only approach for cryo-TEM of bulk specimens. Compared to the thin film method of plunge frozen material, a frozen hydrated section has artifacts, mainly knife marks, compression, and crevasses of which the last two can be controlled (Al-Amoudi et al., 2005). CEMOVIS offers to visualize at high resolution a section of a bulk sample close to the native state. As for all cryo-EM approaches, also for CEMOVIS it is important to minimize the radiation damage of the beam during imaging. Though beam damage takes place also for resin embedded samples and results in shrinkage and mass loss (Luther et al., 1988). Nevertheless, this is mainly a concern for tomography of resin embedded material because the implications for 2D projection images of thin sections are negligible. By contrast, in cryo-EM of hydrated specimens beam damage is also visible on 2D projection images, which is impressively seen by the phenomenon of bubbling of organic material after a certain amount of electron dose (Dubochet et al., 1982). Bubbling is caused by H_2 formation due to radiolysis of the biological specimen and the onset of it depends on the composition of the sample. On CEMOVIS sections it often starts on membranes or in protein rich regions. However, the structural damage starts already at low doses of electrons (Lepault et al., 1983). In our hands a dose between 5 and 8 $e/Å^2$ per image allows at 120 kV to take images at several defocus values. For TEM tomography of CEMOVIS sections the cumulative dose should be kept below 80 $e/Å^2$ (Hagen and Grünewald, 2008). However, for TEM tomography the use of CEMOVIS sections has remained a challenge, which is mainly due to the difficulties to attach the whole section flat on the grid and the pronounced formation of crevasses on sections thicker than 60 nm but it is not impossible (Bouchet-Marquis and Hoenger, 2011; Hagen et al., 2015). To avoid these problems for TEM tomography of vitreous sections, the preparation of cryo-lamella by FIB-SEM (Scanning Electron Microscopy) has been developed during the last years (Hagen et al., 2015; Hayles et al., 2010). Infected cells are plunge frozen on a TEM grid and are thinned in the FIB. With the absence of compression and especially crevasses, the sample is better suited for TEM tomography than a CEMOVIS section. However, care needs to be taken to thin the area of interest since a big part of the cell is removed by the FIB (Arnold et al., 2016). A current drawback of this approach is the cost of the equipment.

Not used yet to study viral replication organelles and VRCs, CEMOVIS has shown the structures associated to assembly and movement of large viruses within cells (Hagen and Grünewald, 2008; Suárez et al., 2013) and has a great potential to visualize the native organization of viral and cellular macromolecular complexes that build and work inside viral factories.

7. Immunogold labelling

For all immunolabeling approaches a high affinity probe binds specifically to its target. The probe is either directly linked to an electron dense marker that can be directly visualized in the TEM or detected indirectly with a secondary probe, which carries an electron dense marker. Ferritin was used in the past to be later substituted by colloidal gold probes (Singer and Schick, 1961; Slot and Geuze, 1981). The majority of high affinity probes used today are antibodies against proteins, but also lectins can be used and in situ hybridization for RNA localization is possible (Gershon et al., 1994; Herpers et al., 2010). Different strategies exist to label the target of interest: they can be divided in general between pre- and post-embedding labeling. For pre-embedding labeling the whole cell is incubated with the affinity probe, followed by the electron dense marker. However, without permeabilization only epitopes exposed to the exterior of the cell are accessible (Gelderblom et al., 1987; Hinners et al., 1999). Permeabilization will provide access to intracellular epitopes, but will extract also cellular material and result in a compromised morphology. For post-embedding labeling, the sample is embedded in a way that preserves the structure of the epitope so that the affinity probe can bind to it when exposed on the surface of the section. This can be done when the samples are embedded in an acrylic or methacrylic resin, as LR White or Lowicryl (Armbruster et al., 1982; Ellinger and Pavelka, 1985). Methacrylic resins were developed by the group of Villiger-Kellenberger. They do not chemically react with the sample, and exist with a versatile range of polarities, which makes them compatible with immuno labeling (Carlemalm et al., 1982). An alternative for resin embedding is labeling on thawed cryosections, an approach which is also called the Tokuyasu technique, after its pioneer Kiyoteru Tokuyasu (Möbius and Posthuma, 2019). In this approach, the sample is arrested by a mild chemical fixation with aldehydes. Since chemical fixation is mandatory, in this approach it can be combined with the often necessary inactivation of viruses presenting a biosafety risk. Chemical fixation makes the plasma membrane of cells permeable for small solutes and the sample can be infiltrated with a sucrose solution (Penttila et al., 1974; Tokuyasu, 1973). The sucrose has a dual role: first, it works as a cryo-protectant for the subsequent freezing step, and second, it influences the elasticity of the sample to facilitate cryo-sectioning. After cryo-protection, the sample can be frozen in liquid nitrogen and becomes hard enough to be sectioned in a cryo-microtome. Frozen sections are thawed and transferred to a grid. Once attached to a grid they can be immunolabeled. The mild fixation of the sample combined with the absence of dehydration results in a good preservation of antigenicity for proteins. Together with the secondary probes conjugated to colloidal gold with different, defined size, this technique allows to study the distribution of two up to three different proteins in one labeling experiment (Slot and Geuze, 1981). But it is important to have the right controls for the specificity of the labeling (Möbius and Posthuma, 2019). After the labeling, the sections are embedded in a thin film of a mixture of uranylacetate and methylcellulose (Griffith et al., 1982). The uranylacetate works as a contrast agent, whereas the methylcellulose provides support to avoid too much shrinkage during the air-drying of the—until this moment—hydrated sections. With uranylacetate as the only contrasting agent the image of the sample from a Tokuyasu section gives a mixture of positive (e.g., nucleus) and negative contrast (e.g., membranes) with certain cellular structures (microtubules, ribosomes) unstained. The selective staining is more than compensated by the good preservation of antigens. The possibility of double or triple labeling makes it possible to study the distribution and sorting of viral proteins during their synthesis, and identify their pathways by cellular markers. This was elegantly demonstrated in the studies of the morphogenesis of the human immunodeficiency virus (HIV) in macrophages (Deneka et al., 2007). With the help of immunogold labeling on Tokuyasu cryosections, the authors demonstrated that the virus assembles in an intracellular membranous compartment that contains several tetraspanins and that is connected with the plasma membrane. Immunogold labeling contributed also to identify the cytoplasmic replication organelles of Vaccinia Virus (VV) that are endoplasmic reticulum-enclosed mini-nuclei (Tolonen et al., 2001) and to visualize the connections of nascent VV particles with ER/ERGIC cellular membranes that participate in the assembly of immature viruses (Moss, 2015; Salmons et al., 1997). Localization of viral proteins and cell markers helps to explore the origin of viral replication organelles. Moreover, assays of brome-uridine or brome-deoxiuridine incorporation for RNA or DNA viruses, respectively, in combination with immunogold detection on Tokuyasu cryosections, have been very useful to identify active replication organelles (Besse and Puvion-Dutilleul, 1994; Fontana et al., 2008; Tenorio et al., 2018). An example is shown in Fig. 6. Reovirus VIs are labeled with antibodies specific for ER markers, such as the protein disulfide isomerase (PDI) (Fig. 6A). These organelles support viral genome replication because in the presence of inhibitors of cellular RNA synthesis, they incorporate BrU, as confirmed with specific antibodies in immunogold labeling assays (Fig. 6B). For RNA viruses, antibodies specific for double-stranded RNA, an intermediate formed during RNA replication, are also used to label active VRCs. Anti-dsRNA antibodies work particularly well for positive-sense RNA viruses (Cotton et al., 2017; Fontana et al., 2007; Welsch et al., 2009) and some negative-sense RNA viruses (Fontana et al., 2008; Mateer et al., 2019). For bunyaviruses and flaviviruses it has been demonstrated that the anti-dsRNA signal co-localizes with that of BrU incorporation sites (Fontana et al., 2008; Westaway et al., 1999).

An important advantage of immunogold labeling is that quantitative analysis of the distribution or the labeling densities of the protein of interest is easy to perform (Mayhew, 2005).

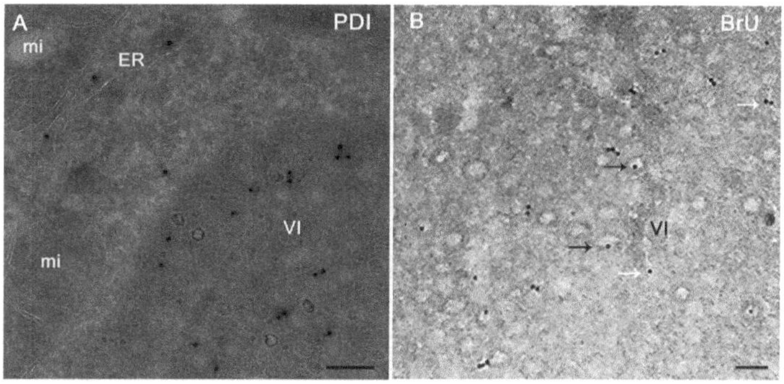

Fig. 6 Immunogold labeling in Tokuyasu cryosections of reovirus inclusions. HeLa cells were adsorbed with reovirus and incubated for 24 h (A) or 14 h (B), fixed with aldehydes, cryoprotected, frozen in liquid nitrogen, and sectioned at −120 °C. (A) Thawed cryosections were processed for immunogold labeling using an anti-PDI antiserum and a secondary antibody bound to 10-nm colloidal gold particles. Anti-PDI antibody labels RER cisternae in the cytosol and membranes inside inclusions. (B) Cells were processed to localize viral RNA synthesis with a bromouridine (BrU) incorporation assay. The thawed cryosection was labeled with an anti-BrU antibody and a secondary antibody bound to 10-nm colloidal gold particles. Inside the VI, the anti-BrU antibody labels viral particles (black arrows) and membrane fragments (white arrows) VI, viral inclusion; RER, rough endoplasmic reticulum; mi, mitochondrion. Scale bars: 200 nm in A; 100 nm in B. *Modified from Tenorio, R., Fernández de Castro, I., Knowlton, J.J., Zamora, P.F., Lee, C.H., Mainou, B.A., Dermody, T.S., Risco, C. 2018. Reovirus σNS and μNS proteins remodel the endoplasmic reticulum to build replication neo-organelles. mBio 9, e01253-18.*

8. 3D imaging/tomography

The images that a transmission electron microscope generates from a thin section are 2D projections of the 3D volume of the section. In this projection, the spatial information within the volume of the thin section is lost. Over the last decades different 3D imaging techniques for biological samples have evolved. In general, for the different approaches there is an inverse relation between the resolution that can be obtained and the volume that can be imaged. TEM electron tomography (TEM-ET) started with the groundbreaking work of Hoppe et al. (1974). For ET, a series of 2D projection images of a sample is acquired over a large tilt series (+70° to −70°) in small increments of 1° (Frank, 1995) and the 3D model of the sample is computationally reconstructed. Due to the geometry of the sample holder, images at tilts above 70° are impossible to acquire and a part of the sample will be lost in the reconstruction (missing wedge). For plastic sections, this wedge can be reduced and the resolution of the final tomogram increased by the acquisition of a double tilt series (Mastronarde, 1997). TEM-ET offers a good resolution but is limited by the thickness of the section that can be visualized (maximum 300–500 nm), and the size of the area that can be analyzed. STEM (Scanning Transmission Electron Microscopy) tomography allows the acquisition of tilt series from thicker sections up to 700 nm (Yakushevska et al., 2007).

FIB-SEM tomography allows the study of larger volumes than TEM tomography (De Winter et al., 2009). The sample is inside an SEM, which is equipped with a second, ion beam column. The blockface of a resin embedded sample is imaged by the backscattered electrons or the secondary electrons of the beam. After imaging the ion beam removes a layer of 5–10 nm of the sample and prepares a new blockface for imaging. Compared to TEM tomography, the resolution is lower but volumes up to $20 \times 20 \times 20\,\mu m^3$ can be analyzed (Titze and Genoud, 2016).

For serial sectioning tomography, serial sections are collected on slot grids and the 3D volume of the sample is combined by the stacking of the 2D projections data from each single section (Harris et al., 2006). Therefore, the resolution in X and Y is down to few nanometers while in Z it is determined by the section thickness. Compared to FIB-SEM, serial sections are a non-destructive approach because the sections can be stored and re-evaluated. It also allows the reconstruction of large volumes and even whole cells can be studied. To avoid the difficulties of serial section tomography that come with varying compression and folds, Denk and Horstmann (2004) placed the sample into an SEM that contains a microtome. The image of the blockface is generated by the backscattered electrons of the SEM beam and after imaging, the microtome removes 30–50 nm of the sample surface to prepare a new blockface. Due to the imaging in the SEM, the resolution is lower compared to TEM serial section tomography, but larger volumes can be analyzed.

Three-dimensional imaging of viral infections is showing how much viruses modify the cell architecture to build their own structures for viral genome replication, morphogenesis, egress and propagation (Bilý et al., 2015; Chlanda et al., 2009; Earl et al., 2013; Felts et al., 2010; Gómez-Aix et al., 2015; Ibiricu et al., 2011; Milrot et al., 2016; Mutsafi et al., 2010, 2013, 2014; Villinger et al., 2015; Wan et al., 2015; Welsch et al., 2009). The first study of a viral replication organelle visualized in 3D by electron tomography showed that flock house virus (FHV) in Drosophila cells assembles spherules with VRCs in mitochondria, spherules that connect with the cytosol through pores (Kopek et al., 2007). The 3D characterization of SARS CoV and DENV replication platforms had a significant impact in the field (Knoops et al., 2008; Welsch et al., 2009). Since then, many studies have demonstrated the power of these methods. Structures harboring the VRCs such as single-membrane spherules and double membrane vesicles have been studied in detail within the replication organelles of West Nile Virus (Gillespie et al., 2010), HCV (Romero-Brey et al., 2012), coxsackievirus B3 (Limpens et al., 2011), Bunyamwera virus (Fontana et al., 2008), Rubella virus (Fontana et al., 2010), Tombusvirus (Fernández de Castro et al., 2017a), a cardiovirus (Melia et al., 2018) and the human reovirus (Fernández de Castro et al., 2014a; Tenorio et al., 2018). The reovirus replication neoorganelle visualized in 3D with two different methods is shown in Fig. 7. With 3D reconstruction of serial sections large volumes that contain whole viral factories were

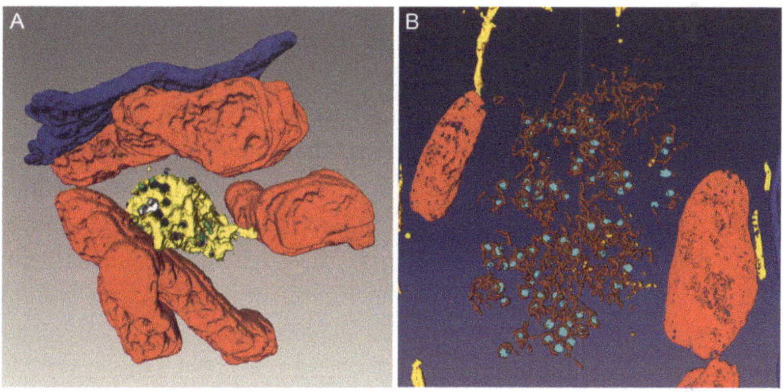

Fig. 7 3D models of reovirus inclusions. (A) HeLa cells were adsorbed with reovirus and fixed at 12 hpi. The inclusion was visualized by TEM of serial sections, 3D volume reconstruction, and image processing. The VI contains a network of membranes (yellow) surrounded by mitochondria (red). The inclusion is adjacent to the nuclear envelope (blue). Inside the VI, mature virions (black), empty viral particles (white) and microtubules (green) are integrated in the membranous network. (B) Reovirus-infected cells were fixed at 14 hpi, frozen in liquid nitrogen, and sectioned at −120°C following the Tokuyasu method. Thawed cryosections were processed by electron tomography. RER cisternae (yellow), mitochondria (red), nuclear envelope (dark blue) are close to the viral inclusion. The VI is a collection of tubules (brown) and vesicles (orange) with viral particles (light blue) attached to membranes. Although reoviruses are non-enveloped particles, we hypothesize that VI membranes might provide physical support for viral genome replication or particle assembly. *Panel (A): Reproduced from Fernández de Castro, I., Zamora, P.F., Ooms, L., Fernández, J.J., Lai, C.M., Mainou, B.A., Dermody, T.S., Risco, C., 2014a. Reovirus forms neo-organelles for progeny particle assembly within reorganized cell membranes. mBio 5, e00931-13 and Panel (B): Tenorio, R., Fernández de Castro, I., Knowlton, J.J., Zamora, P.F., Lee, C.H., Mainou, B.A., Dermody, T.S., Risco, C. 2018. Reovirus σNS and μNS proteins remodel the endoplasmic reticulum to build replication neo-organelles. mBio 9, e01253-18.*

generated (Fig. 7A). With ET of Tokuyasu cryosections smaller volumes with better definition of membranes and higher resolution were obtained (Fig. 7B).

3D imaging has revealed that inside viral factories there are inter-organelle contacts not observed in non-infected cells (Fontana et al., 2010; Risco et al., 2014). Some reports show that structures with VRCs localize close to virus assembly sites, something expected but difficult to detect in two dimensions (Fontana et al., 2008; Welsch et al., 2009). One of the current technical challenges is the development of labeling methods for localizing viral and cellular macromolecules in three dimensions.

9. Clonable tags for electron microscopy

Electron microscopists have been working for decades to localize proteins in cells in two and three dimensions. Together with the fundamental contribution of antibodies and immunogold labeling assays described in Section 7, a number of approaches have been reported to label proteins in cells using clonable tags. Most of them are based on the generation of an electron-dense precipitate of diaminobenzidine as a result of photoactivation of a fluorophore (GFP or Mini-SOG) or an enzymatic reaction (HRP or APEX2) (Connolly et al., 1994; Grabenbauer et al., 2005; Martell et al., 2012). An alternative approach that uses the small metal binding protein metallothionein (MT) as a clonable tag for electron microscopy was proposed by two scientists working at Brandeis University (Mercogliano and DeRosier, 2006, 2007). The idea of MT-tagging was the basis of METTEM (Metal-Tagging TEM), a highly sensitive labeling technique that was validated in bacteria (Diestra et al., 2009a, 2009b), yeast (Barajas et al., 2014; Morphew et al., 2015), and mammalian cells (Fernández de Castro et al., 2017b; Risco et al., 2012; Sachse et al., 2018; Tenorio et al., 2018) and the only one used so far to label intracellular viral structures. In live cells incubated briefly with gold salts, proteins bearing MT will form ~1 nm-gold nano-clusters, readily detectable by electron microscopy (Risco et al., 2012). If the protein of interest is fused with MT and a fluorescent tag such as GFP, the same molecules will be visualized by light and electron microscopy (Fernández de Castro et al., 2014b). Combining METTEM with immunogold allows double and multiple labeling approaches (Fernández de Castro et al., 2014b). MT-tagging was used to detect one of the proteins of the Rubella virus VRC within replication organelles (ROs) (Risco et al., 2012). The study revealed that the viral polymerase molecules first localized in the plasma membrane to be later transported to lysosomes/ROs, where they aggregate into

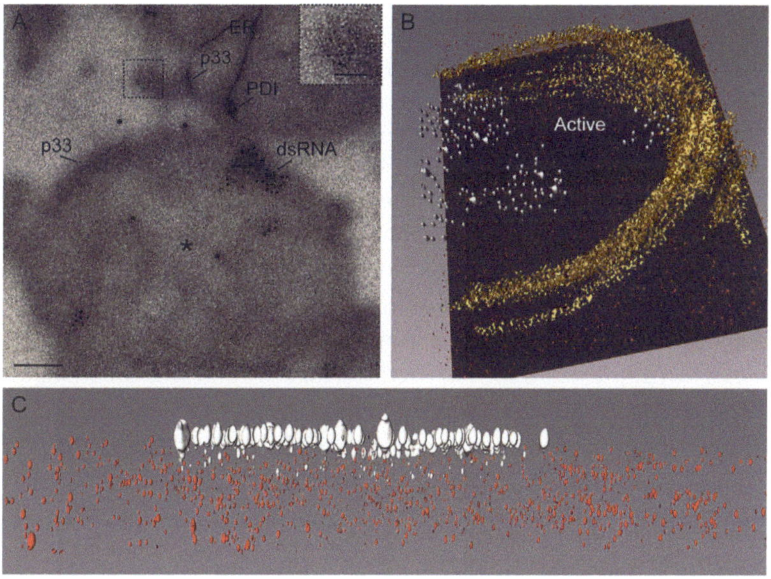

Fig. 8 METTEM of viral replication organelles. *Saccharomyces cerevisiae* Δpah1 cells expressing p33, the most abundant VRC component of the plant virus TBSV (Tomato Bushy Stunt Virus), fused with MT tag as a viral replicon component were incubated with gold salts, embedded in the resin LR White, sectioned, immunogold labeled and visualized by 2D TEM (A) or electron tomography (B and C). (A) METTEM visualization of p33-MT-gold nanoparticles (~1 nm) in an unstained ultrathin section. p33-MT-gold was detected inside the replication organelle (black asterisk) and in the surrounding ER membranes, that were labeled with an anti-PDI antibody and a secondary antibody conjugated with 10 nm colloidal gold particles. Replicating TBSV RNA was labeled with an antibody against dsRNA and a secondary antibody conjugated to 5-nm gold particles. The inset is a high magnification of a group of p33-MT-gold nanoclusters. (B) Electron tomography of active VRCs within the replication organelle. Before ET, sections were processed by immunogold labeling with an antibody against dsRNA and a colloidal gold conjugate (white). Molecules of p33-metallothionein-gold are represented in red and ER membranes in yellow. The active domain labeled with antibodies (black asterisk) within the replication organelle is surrounded by ER membranes, whereas the domain beyond the ER boundaries is not labeled. (C) Lateral view of the tomogram in B, showing the anti-dsRNA antibodies bound to their antigens on the section surface (white), whereas p33 molecules (red) were detected inside the section. Scale bar: 100 nm in mainfield in A; 50 nm in inset in A. *Modified from Fernández de Castro, I., Fernández, J.J., Barajas, D., Nagy, P.D., Risco, C., 2017a. Three-dimensional imaging of the intracelular assembly of a functional viral RNA replicase complex. J. Cell Sci. 130, 260–268, with permission.*

densely packed oligomeric arrays. Studies with the Tombusvirus Tomato Bushy Stunt Virus (TBSV) showed the distribution of the viral replicase molecules in modified ER membranes within the replication factory (Barajas et al., 2014) (Fig. 8). Molecules of the p33-MT-gold subunit of the viral polymerase labeled in live cells were visualized as ~1 nm nano-clusters in ultrathin sections (Fig. 8A). Immunogold labeling on sections confirmed the association of p33-MT-gold molecules with ER membranes labeled with anti-PDI antibodies and with active replication sites labeled with anti-dsRNA antibodies (Fig. 8A). 3D maps of the replicase molecules inside replication organelles were obtained by electron tomography (Fernández de Castro et al., 2017a) (Fig. 8B). These volumes showed RO active domains labeled with anti-dsRNA antibodies together with RO domains devoid of labeling. A lateral view of the tomogram is shown in Fig. 8C. This volume shows why METTEM is a labeling method of high sensitivity: with antibodies only the epitopes exposed on the surface of the section are detected; however, with METTEM all protein molecules carrying an MT-gold-nanocluster inside the section are visualized in 2D and 3D. With METTEM it was discovered that influenza virus uses a modified ER to transport the vRNPs from the nucleus to the plasma membrane in a process that depends on Rab11 (Fernández de Castro et al., 2017b). Early in infection, the reovirus non-structural protein μNS fused with MT was found in strangled ER cisternae and small vesicles near the nucleus. This result suggests that μNS triggers the massive ER fragmentation that generates replication organelles by direct binding to ER membranes (Tenorio et al., 2018).

New developments and improvements for METTEM are in progress. Combining METTEM with CEMOVIS will allow ultrastructural imaging of viral macromolecular complexes in intracellular viral factories in their native state and at molecular scale resolution.

10. Conclusions and future perspectives

Recent advances in sample preparation and imaging by light and electron microscopy are having considerable impact in virology. The powerful combination of different imaging techniques with various levels of resolution shows the viral life cycle from the early steps of virus entry and replication to the final stages of viral particle morphogenesis and egress. All these processes take place at different locations in the infected cell, and after the initial study of infection in live cells, scientists can focus on the subsequent characterization of particular structures at high resolution by electron microscopy. All techniques have advantages and limitations, as summarized in Fig. 1, and the choice of methods depends on the complexity of the study.

Some technical challenges for the future include developing super-resolution light microscopy for live cells, improve CLEM protocols, new image processing tools for 3D TEM, and combining molecular mapping methods such as METTEM with Tokuyasu cryosections and CEMOVIS. The myriad of structural data to be obtained will help to interfere with the biogenesis and function of viral structures in cells. Drugs that block the construction of functional replication organelles, assembly sites and virus egress machineries will be the basis of new antiviral strategies.

Acknowledgments

The authors acknowledge Prof. Terence S. Dermody and his team at the University of Pittsburgh for many useful discussions and Dr. Laura Sanz-Sánchez for data in Movie 1 in the online version at https://doi.org/10.1016/bs.aivir.2019.07.005 and Fig. 2. This work has been supported by grant BIO2015-68758-R (AEI/FEDER, EU) from the Ministry of Science, Innovation and Universities of Spain.

References

Abernathy, E., Mateo, R., Majzoub, K., van Buuren, N., Bird, S.W., Carette, J.E., Kirkegaard, K., 2019. Differential and convergent utilization of autophagy components by positive-strand RNA viruses. PLoS Biol. 17 (1), e2006926.

Al-Amoudi, A., Studer, D., Dubochet, J., 2005. Cutting artefacts and cutting process in vitreous sections for cryo-electron microscopy. J. Struct. Biol. 150, 109–121.

Alonas, E., Vanover, D., Blanchard, E., Zurla, C., Santangelo, P.J., 2016. Imaging viral RNA using multiply labeled tetravalent RNA imaging probes in live cells. Methods 98, 91–98.

Altan-Bonnet, N., 2017. Lipid tales of viral replication and transmission. Trends Cell Biol. 27, 201–213.

Armbruster, B.L., Carlemalm, E., Chiovetti, R., Garavito, R.M., Hobot, J.A., Kellenberger, E., Villiger, W., 1982. Specimen preparation for electron microscopy using low temperature embedding resins. J. Microsc. 126, 77–85.

Arnold, J., Mahamid, J., Lucic, V., de Marco, A., Fernandez, J.J., Laugks, T., Mayer, T., Hyman, A.A., Baumeister, W., Plitzko, J.M., 2016. Site-specific cryo-focused Ion beam sample preparation guided by 3D correlative microscopy. Biophys. J. 110, 860–869.

Baktash, Y., Randall, G., 2019. Live cell imaging of hepatitis C virus trafficking in hepatocytes. Methods Mol. Biol. 1911, 263–274.

Baktash, Y., Madhav, A., Coller, K.E., Randall, G., 2018. Single particle imaging of polarized hepatoma organoids upon hepatitis C virus infection reveals an ordered and sequential entry process. Cell Host Microbe 23, 382–394.

Barajas, D., Fernández de Castro, I., Pogany, J., Risco, C., Nagy, P.D., 2014. Noncanonical role for the host Vps4 AAA+ ATPase ESCRT protein in the formation of Tomato bushy stunt virus replicase. PLoS Pathog. 10, e1004087.

Berger, C., Romero-Brey, I., Radujkovic, D., Terreux, R., Zayas, M., Paul, D., Harak, C., Hoppe, S., Gao, M., Penin, F., Lohmann, V., Bartenschlager, R., 2014. Daclatasvir-like inhibitors of NS5A block early biogenesis of hepatitis C virus-induced membranous replication factories, independent of RNA replication. Gastroenterology 147 (1094–1105) e1025.

Besse, S., Puvion-Dutilleul, F., 1994. High resolution localization of replicating viral genome in adenovirus-infected HeLa cells. Eur. J. Cell Biol. 63, 269–279.

Bilý, T., Palus, M., Eyer, L., Elsterová, J., Vancová, M., Růžek, D., 2015. Electron tomography analysis of tick-borne encephalitis virus infection in human neurons. Sci. Rep. 5, 10745.
Bouchet-Marquis, C., Hoenger, A., 2011. Cryo-electron tomography on vitrified sections: a critical analysis of benefits and limitations for structural cell biology. Micron 42, 152–162.
Brüggeller, P., Mayer, E., 1980. Complete vitrification in pure liquid water and dilute aqueous solutions. Nature 288, 569–571.
Bykov, Y.S., Cortese, M., Briggs, J.A.G., Bartenschlager, R., 2016. Correlative light and electron microscopy methods for the study of virus-cell interactions. FEBS Lett. 590, 1877–1895.
Campbell, E.M., Hope, T.J., 2008. Live cell imaging of the HIV-1 life cycle. Trends Microbiol. 16, 580–587.
Carlemalm, E., Garavito, R.M., Villiger, W., 1982. Resin development for electron microscopy and an analysis of embedding at low temperature. J. Microsc. 126, 123–143.
Chlanda, P., Carbajal, M.A., Cyrklaff, M., Griffiths, G., Krijnse-Locker, J., 2009. Membrane rupture generates single open sheets during vaccinia virus assembly. Cell Host Microbe 6, 81–90.
Connolly, C.N., Futter, C.E., Gibson, A., Hopkins, C.R., Cutler, D.F., 1994. Transport into and out of the Golgi complex studied by transfecting cells with cDNAs encoding horseradish peroxidase. J. Cell Biol. 127, 641–652.
Cotton, B.T., Hyde, J.L., Sarvestani, S.T., Sosnovtsev, S.V., Gree, K.Y., White, P.A., Mackenzie, J.M., 2017. The norovirus NS3 protein is a dynamic lipid- and microtubule-associates protein involved in viral RNA replication. J. Virol. 91, e02138-16.
De Winter, D.A., Schneijdenberg, C.T., Lebbink, M.N., Lich, B., Verkleij, A.J., Drury, M.R., Humbel, B.M., 2009. Tomography of insulating biological and geological materials using focused ion beam (FIB) sectioning and low-kV BSE imaging. J. Microsc. 233, 372–383.
Dempster, W.T., 1960. Rates of penetration of fixing fluids. Am. J. Anat. 107, 59–72.
Deneka, M., Pelchen-Matthews, A., Byland, R., Ruiz-Mateos, E., Marsh, M., 2007. In macrophages, HIV-1 assembles into an intracellular plasma membrane domain containing the tetraspanins CD81, CD9, and CD53. J. Cell Biol. 177, 329–341.
Denk, W., Horstmann, H., 2004. Serial block-face scanning electron microscopy to reconstruct three-dimensional tissue nanostructure. PLoS Biol. 2, e329.
Desmet, E.A., Anguish, L.J., Parker, J.S.L., 2014. Virus-mediated compartimentalization of the host translational machinery. mBio 5, e01463-14.
Diestra, E., Fontana, J., Guichard, P., Marco, S., Risco, C., 2009a. Visualization of proteins in intact cells with a clonable tag for electron microscopy. J. Struct. Biol. 165, 157–168.
Diestra, E., Cayrol, B., Arluison, V., Risco, C., 2009b. Cellular electron microscopy imaging reveals the localization of the Hfq protein close to the bacterial membrane. PLoS One 4, e8301.
Dirk, B.S., Van Nynatten, L.R., Dikeakos, J.D., 2016. Where in the cell are you? Probing HIV-host interactions through advanced imaging techniques. Viruses 8, 200. https://doi.org/10.3390/v8100288.
Dubochet, J., McDowall, A.W., 1981. Vitrification of pure water for electron microscopy. J. Microsc. 124, RP3–RP4.
Dubochet, J., Lepault, J., Fremman, R., Berriman, J.A., Homo, J.C., 1982. Electron microscopy of frozen water and aqueous solutions. J. Microsc. 128, 219–237.
Earl, L.A., Lifson, J.D., Subramaniam, S., 2013. Catching HIV "in the act" with 3D electron microscopy. Trends Microbiol. 21, 397–404.
Eichwald, C., Arnoldi, F., Laimbacher, A.S., Schraner, E.M., Fraefel, C., Wild, P., Burrone, O.R., Ackermann, M., 2012. Rotavirus viroplasm fusion and perinuclear localization are dynamic processes requiring stabilized microtubules. PLoS One 7 (10), e47947.

Ellinger, A., Pavelka, M., 1985. Post-embedding localization of glycoconjugates by means of lectins on thin sections of tissues embedded in LR white. Histochem. J. 17, 1321–1336.
Erickson, K.D., Bouchet-Marquis, C., Heiser, K., Szomolanyi-Tsuda, E., Mishra, R., Lamothe, B., Hoenger, A., Garcea, R.L., 2012. Virion assembly factories in the nucleus of polyomavirus-infected cells. PLoS Pathog. 8 (4), e1002630.
Escaig, J., 1982. New instruments which facilitate rapid freezing at 83 K and 6 K. J. Microsc. 126, 221–229.
Eyre, N.S., Fiches, G.N., Aloia, A.L., Helbig, K.J., McCartney, E.M., McErlean, C.S.P., Li, K., Aggarwal, A., Turvilee, D.G., Beard, M.R., 2014. Dynamic imaging of the hepatitis C virus NS5A protein during productive infection. J. Virol. 88, 3636–3652.
Felts, R.L., Narayan, K., Estes, J.D., Shi, D., Trubey, C.M., Fu, J., Hartnell, L.M., Ruthel, G.T., Schneider, D.K., Nagashima, K., Bess Jr., J.W., Bavari, S., Lowekamp, B.C., Bliss, D., Lifson, J.D., Subramaniam, S., 2010. 3D visualization of HIV transfer at the virological synapse between dendritic cells and T cells. Proc. Natl. Acad. Sci. U. S. A. 107, 13336–13341.
Fernández de Castro, I., Volonté, L., Risco, C., 2013. Virus factories: biogenesis and structural design. Cell. Microbiol. 15, 24–34.
Fernández de Castro, I., Zamora, P.F., Ooms, L., Fernández, J.J., Lai, C.M., Mainou, B.A., Dermody, T.S., Risco, C., 2014a. Reovirus forms neo-organelles for progeny particle assembly within reorganized cell membranes. mBio 5, e00931-13.
Fernández de Castro, I., Sanz-Sánchez, L., Risco, C., 2014b. Metallothioneins for correlative light and electron microscopy. Methods Cell Biol., 124, Elsevier, pp. 55–70.
Fernández de Castro, I., Tenorio, R., Risco, C., 2016. Virus assembly factories in a lipid world. Curr. Opin. Virol. 18, 20–26.
Fernández de Castro, I., Fernández, J.J., Barajas, D., Nagy, P.D., Risco, C., 2017a. Three-dimensional imaging of the intracelular assembly of a functional viral RNA replicase complex. J. Cell Sci. 130, 260–268.
Fernández de Castro, I., Fournier, G., Sachse, M., Pizarro-Cerdá, J., Risco, C., Naffakh, N., 2017b. Influenza virus genome reaches the plasma membrane via a modified endoplasmic reticulum and Rab11-dependent vesicles. Nat. Commun. 8 (1), 1396.
Fernández-Oliva, A., Ortega-Gónzalez, P., Risco, C., 2019. Targeting host lipid flows: exploring new antiviral and antibiotic strategies. Cell. Microbiol. 21 (3), e12996.
Filipe, A., McLauchlan, J., 2015. Hepatitis C virus and lipid droplets: finding a niche. Trends Mol. Med. 21, 34–42.
Fontana, J., Tzeng, W.-P., Calderita, G., Fraile-Ramos, A., Frey, T.K., Risco, C., 2007. Novel replication complex architecture in rubella replicon-transfected cells. Cell. Microbiol. 9, 875–890.
Fontana, J., López-Montero, N., Elliott, R.M., Fernández, J.J., Risco, C., 2008. The unique architecture of Bunyamwera virus factories around the Golgi complex. Cell. Microbiol. 10, 2012–2028.
Fontana, J., López-Iglesias, C., Tzeng, W.-P., Frey, T.K., Fernández, J.J., Risco, C., 2010. Three-dimensional structure of Rubella virus factories. Virology 405, 579–591.
Francis, A.C., Melikyan, G.B., 2018. Live-cell imaging of early steps of single HIV-1 infection. Viruses 10, 275. https://doi.org/10.3390/v10050275.
Frank, J., 1995. Approaches to large-scale structures. Curr. Opin. Struct. Biol. 5, 194–201.
Fridmann-Sirkis, Y., Milrot, E., Mutsafi, Y., Ben-Dor, S., Levin, Y., Savidor, A., Kartvelishvily, E., Minsky, A., 2016. Efficiency in complexity: composition and dynamic nature of mimivirus replication factories. J. Virol. 90, 10039–10047.
Galas, L., Gallavardin, T., Bénard, M., Lehner, A., Schapman, D., Lebon, A., Komuro, H., Lerouge, P., Leleu, S., Franck, X., 2018. Probe, sample and instrument (PSI): the hat-trick for fluorescence live cell imaging. Chemosensors 6, 21.

García-Serradilla, M., Risco, C., Pacheco, B., 2019. Drug repurposing for new, efficient, broad spectrum antivirals. Virus Res. 264, 22–31.

Gelderblom, H.R., Hausmann, E.H., Ozel, M., Pauli, G., Koch, M.A., 1987. Fine structure of human immunodeficiency virus (HIV) and immunolocalization of structural proteins. Virology 156, 171–176.

Gershon, A.A., Sherman, D.L., Zhu, Z., Gabel, C.A., Ambron, R.T., Gershon, M.D., 1994. Intracellular transport of newly synthesized varicella-zoster virus: final envelopment in the trans-Golgi network. J. Virol. 68, 6372–6390.

Gillespie, L.K., Hoenen, A., Morgan, G., Mackenzie, J.M., 2010. The endoplasmic reticulum provides the membrane platform for biogenesis of the flavivirus replication complex. J. Virol. 84, 10438–10447.

Gómez-Aix, C., García-García, M., Aranda, M.A., Sánchez-Pina, M.A., 2015. Melon necrotic spot virus replication occurs in association with altered mitocondria. Mol. Plant Microbe Interact. 28, 387–397.

Grabenbauer, M., Geerts, W.J., Fernández-Rodríguez, J., Hoenger, A., Koster, A.J., Nilsson, T., 2005. Correlative microscopy and electron tomography of GFP through photooxidation. Nat. Methods 2, 857–862.

Griffith, G., Brands, R., Burke, B., Louvard, D., Warren, G., 1982. Viral membrane proteins acquire galactose in trans Golgi cisternae during intracellular transport. J. Cell Biol. 95, 781–792.

Hagemeijer, M.C., Verheije, M.H., Ulasli, M., Shaltiël, I.A., de Vries, L.A., Reggiori, F., Rottier, P.J.M., de Haan, C.A.M., 2010. Dynamics of coronavirus replication-transcription complexes. J. Virol. 84, 2134–2149.

Hagen, C., Grünewald, K., 2008. Microcarriers for high-pressure freezing and cryosectioning of adherent cells. J. Microsc. 230, 288–296.

Hagen, C., Dent, K.C., Zeev-Ben-Mordehai, T., Grange, M., Bosse, J.B., Whittle, C., Klupp, B.G., Siebert, C.A., Vasishtan, D., Bäuerlein, F.J., Cheleski, J., Werner, S., Guttmann, P., Rehbein, S., Henzler, K., Demmerle, J., Adler, B., Koszinowski, U., Schermelleh, L., Schneider, G., Enquist, L.W., Plitzko, J.M., Mettenleiter, T.C., Grünewald, K., 2015. Structural basis of vesicle formation at the inner nuclear membrane. Cell 163, 1692–1701.

Hanne, J., Zila, V., Heilemann, M., Müller, B., Kräusslich, H.-G., 2016. Super-resolved insights into human immunodeficiency virus biology. FEBS Lett. 590, 1858–1876.

Harak, C., Lohmann, V., 2015. Ultrastructure of the replication sites of positive-strand RNA viruses. Virology 479-480, 418–433.

Harris, K.M., Perry, E., Bourne, J., Feinberg, M., Ostroff, L., Hurlburt, J., 2006. Uniform serial sectioning for transmission electron microscopy. J. Neurosci. 26, 12101–12103.

Hayles, M.F., de Winter, D.A., Schneijdenberg, C.T., Meeldijk, J.D., Luecken, U., Persoon, H., de Water, J., de Jong, F., Humbel, B.M., Verkleij, A.J., 2010. The making of frozen-hydrated, vitreous lamellas from cells for cryo-electron microscopy. J. Struct. Biol. 172, 180–190.

Heiligenstein, X., Heiligenstein, J., Delevoye, C., Hurbain, I., Bardin, S., Paul-Gilloteaux, P., Sengmanivong, L., Régnier, G., Salamero, J., Antony, C., Raposo, G., 2014. The CryoCapsule: simplifying correlative light to electron microscopy. Traffic 15, 700–716.

Hellström, K., Vihinen, H., Kallio, K., Jokitalo, E., Ahola, T., 2015. Correlative light and electron microscopy enables viral replication studies at the ultrastructural level. Methods 90, 49–56.

Herpers, B., Xanthakis, D., Rabouille, C., 2010. ISH-IEM: a sensitive method to detect endogenous mRNAs at the ultrastructural level. Nat. Protoc. 5, 678–687.

Hillier, J., 1950. Electron microscopy of microorganisms and viruses. Annu. Rev. Microbiol. 4, 1–20.

Hinners, I., Moschner, J., Nolte, N., Hille-Rehfeld, A., 1999. The orientation of membrane proteins determined in situ by immunofluorescence staining. Anal. Biochem. 276, 1–7.
Hoppe, W., Gassmann, J., Hunsmann, N., Schramm, H.J., Sturm, M., 1974. Three-dimensional reconstruction of individual negatively stained yeast fatty-acid synthetase molecules from tilt series in the electron microscope. Hoppe Seylers Z. Physiol. Chem. 355, 1483–1487.
Hurbain, I., Sachse, M., 2011. The future is cold: cryo-preparation methods for transmission electron microscopy of cells. Biol. Cell 103, 405–420.
Ibiricu, I., Huiskonen, J.T., Döhner, K., Bradke, F., Sodeik, B., Grünewald, K., 2011. Cryo electron tomography of herpes simplex virus during axonal transport and secondary envelopment in primary neurons. PLoS Pathog. 7, e1002406.
Icha, J., Weber, M., Waters, J.C., Norden, C., 2017. Phototoxicity in live fluorescence microscopy, and how to avoid it. Bioessays 39 (8), 1700003.
Jackson, W.T., Giddings Jr., T.H., Taylor, M.P., Mulinyawe, S., Ravinovitch, M., Kopito, R.R., Kirkegaard, K., 2005. Subversion of cellular autophagosomal machinery by RNA viruses. PLoS Biol. 3 (5) e156.
Jouvenet, N., Bieniasz, P.D., Simon, S.M., 2008. Imaging the biogenesis of individual HIV-1 virions in live cells. Nature 454, 236–240.
Kajitani, N., Satsuka, A., Yoshida, S., Sakai, H., 2013. HPV18 E1^E4 is assembled into aggresome-like compartment and involved in sequestration of viral oncoproteins. Front. Microbiol. 4 (251), 1–11.
Kallio, K., Hellström, K., Balistreri, G., Jokitalo, E., Ahola, T., 2013. Template RNA length determines the size of replication complex spherules for Semliki Forest Virus. J. Virol. 87, 9125–9134.
Kanno, H., Speedy, R.J., Angell, C.A., 1975. Supercooling of water to −92°C under pressure. Science 189, 880–881.
Karreman, M.A., Mercier, L., Schieber, N.L., Shibue, T., Schwab, Y., Goetz, J.G., 2014. Correlating intravital multi-photon microscopy to 3D electron microscopy of invading tumor cells using anatomical reference points. PLoS One 9, e114448. https://doi.org/10.1371/journal.pone.0114448. eCollection 2014.
Kellenberger, E., 1991. The potential of cryofixation and freeze substitution: observations and theoretical considerations. J. Microsc. 161, 183–203.
Knoops, K., Kikkert, M., Worm, S.H., Zevenhoven-Dobbe, J.C., van der Meer, Y., Koster, A.J., Mommaas, A.M., Snijder, E.J., 2008. SARS-coronavirus replication is supported by a reticulovesicular network of modified endoplasmic reticulum. PLoS Biol. 6, e226.
Kopek, B.G., Perkins, G., Miller, D.J., Ellisman, M.H., Ahlquist, P., 2007. Three-dimensional analysis of a viral RNA replication complex reveals a virus-induced mini-organelle. PLoS Biol. 5, e220.
Kukulski, W., Schorb, M., Welsch, S., Picco, A., Kaksonen, M., Briggs, J.A., 2011. Correlated fluorescence and 3D electron microscopy with high sensitivity and spatial precision. J. Cell Biol. 192, 111–119.
Lee, J.-Y., Marshall, J.A., Bowden, D.S., 1994. Characterization of Rubella virus replication complexes using antibodies to double-stranded RNA. Virology 200, 307–312.
Lepault, J., Booy, F.P., Dubochet, J., 1983. Electron microscopy of frozen biological suspensions. J. Microsc. 129, 89–102.
Limpens, R.W., van der Schaar, H.M., Kumar, D., Koster, A.J., Snijder, E.J., van Kuppeveld, F.J., Bárcena, M., 2011. The transformation of enterovirus replication structures: a three-dimensional study of single- and double-membrane compartments. mBio 2, e00166-11.
López-Montero, N., Risco, C., 2011. Self-protection and survival of arbovirus-infected mosquito cells. Cell. Microbiol. 13, 300–315.

Lowen, R.G., Bocan, T.M., Kane, C.D., Cazares, L.H., Kota, K.P., Ladner, J.T., Nasar, F., Pitt, L., Smith, D.R., Sloveva, V., Sun, M.G., Zeng, X., Bavari, S., 2018. Countering Zika virus: the USAMRIID response. Adv. Exp. Med. Biol. 1062, 303–318.

Lucas, M.S., Günthert, M., Gasser, P., Lucas, F., Wepf, R., 2012. Bridging microscopes: 3D correlative light and scanning electron microscopy of complex biological structures. In: - Müller-Reichert, T., Verkade, P. (Eds.), Correlative Light and Electron Microscopy. In: Methods in Cell biology, 111, pp. 325–356.

Luther, P.K., Lawrence, M.C., Crowther, R.A., 1988. A method for monitoring the collapse of plastic sections as a function of electron dose. Ultramicroscopy 24, 7–18.

Madela, K., Banhart, S., Zimmermann, A., Piesker, J., Bannert, N., Laue, M., 2014. A simple procedure to analyze positions of interest in infectious cell cultures by correlative light and electron microscopy. In: Müller-Reichert, T., Verkade, P. (Eds.), Correlative Light and Electron Microscopy II. Methods in Cell biology, 124, pp. 93–110.

Magidson, V., Khodjakov, A., 2013. Circumventing photodamage in live-cell microscopy. Methods Cell Biol. 114, 545–560.

Magliano, D., Marshall, J.A., Bowden, D.S., Vardaxis, S., Meanger, J., Lee, J.Y., 1998. Rubella virus replication complexes are virus modified lysosomes. Virology 240, 57–63.

Martell, J.D., Deerinck, T.J., Sancak, Y., Poulos, T.L., Mootha, V.K., Sosinsky, G.E., Ellisman, M.H., Ting, A.Y., 2012. Engineered ascorbate peroxidase as a genetically encoded reporter for electron microscopy. Nat. Biotechnol. 30, 1143–1148.

Martínez, M.G., Snapp, E.-L., Perumal, G.S., Macaluso, F.P., Kielian, M., 2014. Imaging the alphavirus exit pathway. J. Virol. 88, 6922–6933.

Mastronarde, D.N., 1997. Dual-axis tomography: an approach with alignment methods that preserve resolution. J. Struct. Biol. 120, 343–352.

Mateer, E., Paessler, S., Huang, C., 2019. Confocal imaging of double-stranded RNA and pattern recognition receptors in negative-sense RNA virus infection. J. Vis. Exp. 143, e59095.

Mayhew, T., 2005. How to count your gold: a tutorial on TEM immunogold label quantification. Microsc. Anal. 12, 9–12.

McDonald, K., 2007. Cryopreparation methods for electron microscopy of selected model systems. In: McIntosh, J.R. (Ed.), Cellular Electron microscopy. In: Methods in Cell Biology, 79, pp. 23–56.

McDowall, A.W., Chang, J.J., Freeman, R., Lepault, J., Walte, R.C.A., Dubochet, J., 1983. Electron microscopy of frozen hydrated sections of vitreous ice and vitrified biological samples. J. Microsc. 131, 1–9.

Melia, C.E., van der Schaar, H.M., de Jong, A.W.M., Lyoo, H.R., Snijder, E.J., Koster, A.J., van Kuppeveld, F.J.M., Bárcena, M., 2018. The origin, dynamic morphology, and PI4P-independent formation of encephalomyocarditis virus replication organelles. mBio 9 e00420-18.

Mercogliano, C.P., DeRosier, D.J., 2006. Gold nanocluster formation using metallothionein: mass spectrometry and electron microscopy. J. Mol. Biol. 355, 211–223.

Mercogliano, C.P., DeRosier, D.J., 2007. Concatenated metallothionein as a clonable gold label for electron microscopy. J. Struct. Biol. 160, 70–82.

Milrot, E., Mutsafi, Y., Fridmann-Sirkis, Y., Shimoni, E., Rechav, K., Gurnon, J.R., Van Etten, J.L., Minsky, A., 2016. Virus-host interactions: insights from the replication cycle of the large *Paramecium bursaria chlorella virus*. Cell. Microbiol. 18, 3–16.

Möbius, W., Posthuma, G., 2019. Sugar and ice: immunoelectron microscopy using cryosections according to the Tokuyasu method. Tissue Cell 57, 90–102. pii: S0040-8166(18)30214-3. https://doi.org/10.1016/j.tice.2018.08.010.

Monaghan, P., Cook, H., Hawes, P., Simpson, J., Tomley, F., 2003. High-pressure freezing in the study of animal pathogens. J. Microsc. 212, 62–70.

Morphew, M.K., O'Toole, E.T., Page, C.L., Pagratis, M., Meehl, J., Giddings, T., Gardner, J.M., Ackerson, C., Jaspersen, S.L., Winey, M., Hoenger, A., McIntosh, J.R., 2015. Metallothionein as a clonable tag for protein localization by electron microscopy of cells. J. Microsc. 260, 20–29.

Moss, B., 2015. Poxvirus membrane biogenesis. Virology 479-480, 619–626.

Müller-Reichert, T., Verkade, P., 2012. Correlative light and electron microscopy. Methods in Cell Biology, 111, Elsevier.

Müller-Reichert, T., Verkade, P., 2014. Correlative light and electron microscopy II. Methods in Cell Biology, 124, Elsevier.

Mutsafi, Y., Zauberman, N., Sabanay, I., Minsky, A., 2010. Vaccinia-like cytoplasmic replication of the giant mimivirus. Proc. Natl. Acad. Sci. U. S. A. 107, 5978–5982.

Mutsafi, Y., Shimoni, E., Shimon, A., Minsky, A., 2013. Membrane assembly during the infection cycle of the giant Mimivirus. PLoS Pathog. 9 e1003367.

Mutsafi, Y., Fridmann-Sirkis, Y., Milrot, E., Hevroni, L., Minsky, A., 2014. Infection cycles of large DNA viruses: emerging themes and underlying questions. Virology 466-467, 3–14.

Novoa, R.R., Calderita, G., Arranz, R., Fontana, J., Granzow, H., Risco, C., 2005. Virus factories: associations of cell organelles for viral replication and morphogenesis. Biol. Cell 97, 147–172.

Oorschot, V.M., Sztal, T.E., Bryson-Richardson, R.J., Ramm, G., 2014. Immuno correlative light and electron microscopy on Tokuyasu cryosections. 2014, In: Müller-Reichert, T., Verkade, P. (Eds.), Correlative Light and Electron Microscopy II. Methods Cell Biol, 124, pp. 241–258.

Paul, D., Hoppe, S., Saher, G., Krijnse-Locker, J., Bartenschlager, R., 2013. Morphological and biochemical characterization of the membranous hepatitis C virus replication compartment. J. Virol. 87, 10612–10627.

Pelchen-Matthews, A., Marsh, M., 2007. Electron microscopy analysis of viral morphogenesis. Methods Cell Biol. 79, 515–542.

Penttila, A., Kalimo, H., Trump, B.F., 1974. Influence of glutaraldehyde and/or osmium tetroxide on cell volume, ion content, mechanical stability, and membrane permeability of Ehrlich ascites tumor cells. J. Cell Biol. 63, 197–214.

Rhim, J.S., Jordan, L.E., Mayor, H.D., 1962. Cytochemical, fluorescent-antibody and electron microscopic studies on the growth of reovirus (ECHO 10) in tissue culture. Virology 17, 342–355.

Riehle, U., Hoechli, M., 1973. The theory and technique of highpressure freezing. In: Benedetti, E.L., Favard, P. (Eds.), Freeze-Etching Technique and Application. Société Francaise de Microscopie Électronique, pp. 31–61.

Risco, C., Sanmartín-Conesa, E., Tzeng, W.P., Frey, T.K., Seybold, V., de Groot, R.J., 2012. Specific, sensitive, high-resolution detection of protein molecules in eukaryotic cells using metal-tagging transmission electron microscopy. Structure 20, 759–766.

Risco, C., Fernández de Castro, I., Sanz-Sánchez, L., Narayan, K., Grandinetti, G., Subramaniam, S., 2014. Three-dimensional imaging of viral infections Annu. Rev. Virol. 1, 453–473.

Romero-Brey, I., Bartenschlager, R., 2014. Membranous replication factories induced by plus-strand RNA viruses. Viruses 6, 2826–2857.

Romero-Brey, I., Merz, A., Chiramel, A., Lee, J.Y., Chlanda, P., Haselman, U., Santarella-Mellwig, R., Habermann, A., Hoppe, S., Kallis, S., Antony, C., Krijnse-Locker, J., Bartenschlager, R., 2012. Three-dimensional architecture and biogenesis of membrane structures associated with hepatitis C virus replication. PLoS Pathog. 8 (12), e1003056.

Sabatini, D.D., Bensch, K., Barnett, R.J., 1963. Cytochemistry and electron microscopy. The preservation of cellular ultrastructure and enzymatic activity by aldehyde fixation. J. Cell Biol. 17, 19–58.

Sachse, M., Fernández de Castro, I., Fournier, G., Naffakh, N., Risco, C., 2018. Metal-tagging electron microscopy and immunogold labeling on Tokuyasu cryosections to image influenza A virus ribonucleoprotein transport and packaging. Methods Mol. Biol. 1836, pp. 281–301, Springer.

Salmons, T., Kuhn, A., Wylie, F., Schleich, S., Rodríguez, J.R., Rodríguez, D., Esteban, M., Griffiths, G., Krijnse Locker, J., 1997. Vaccinia virus membrane proteins p8 and p16 are cotranslationally inserted into the rough endoplasmic reticulum and retained in the intermediate compartment. J. Virol. 71, 7404–7420.

Santarella-Mellwig, R., Haselmann, U., Schieber, N.L., Walther, P., Schwab, Y., Antony, C., Bartenschlager, R., Romero-Brey, I., 2018. Correlative light electron microscopy (CLEM) for tracking and imaging viral protein associated structures in cryo-immobilized cells. J. Vis. Exp. 2018, 139. https://doi.org/10.3791/58154.

Scaturro, P., Cortese, M., Chatel-Chaix, L., Fischk, W., Bartenschlager, R., 2015. Dengue virus non-structural protein 1 modulates infectious particle production via interaction with the structural proteins. PLoS Pathog. 11, e1005277.

Schramm, B., de Haan, C.A.M., Young, J., Doglio, L., Schleich, S., Reese, C., Popov, A.V., Steffen, W., Schroer, T., Krijnse Locker, J., 2006. Vaccinia-virus-induced cellular contractility facilitates the subcellular localization of the viral replication sites. Traffic 7, 1352–1367.

Schwarz, H., Humbel, B.M., 2014. Correlative light and electron microscopy using immunolabeled sections. In: Kuo, J. (Ed.), Electron Microscopy. Methods Mol Biol, 1117, pp. 559–592.

Shi, X., van Mierlo, J.T., French, A., Elliott, R.M., 2010. Visualizing the replication cycle of Bunyamwera orthobunyavirus expressing fluorescent protein-tagged Gc glycoprotein. J. Virol. 84, 8460–9469.

Singer, S.J., Schick, A.F., 1961. The properties of specific stains for electron microscopy prepared by the conjugation of antibody molecules with ferritin. J. Biophys. Biochem. Cytol. 9, 519–537.

Slot, J.W., Geuze, H.J., 1981. Sizing of protein A-colloidal gold probes for immunoelectron microscopy. J. Cell Biol. 90, 533–536.

Sosinsky, G.E., Crum, J., Jone, S.Y.Z., Lanman, J., Smarr, B., Terada, M., Martone, M.E., Deerinck, T.J., Johnson, J.E., Ellisman, M.H., 2008. The combination of chemical fixation procedures with high pressure freezing and freeze substitution preserves highly labile tissue ultrastructure for electron tomography applications. J. Struct. Biol. 161, 359–371.

Spuul, P., Balistreri, G., Kääriäinen, L., Ahola, T., 2010. Phosphatidylinositol 3-kinase-, actin-, and microtubule-dependent transport of Semliki Forest Virus replication complexes from the plasma membrane to modified lysosomes. J. Virol. 84, 7543–7557.

Spuul, P., Balistreri, G., Hellström, K., Golubtsov, A.V., Jokitalo, E., Ahola, T., 2011. Assembly of alphavirus replication complexes from RNA and protein components in a novel trans-replication system in mammalian cells. J. Virol. 85, 4739–4751.

Suárez, C., Welsch, S., Chlanda, P., Hagen, W., Hoppe, S., Kolovou, A., Pagnier, I., Raoult, D., Krijnse Locker, J., 2013. Open membranes are the precursors for assembly of large DNA viruses. Cell. Microbiol. 15, 1883–1895.

Tenorio, R., Fernández de Castro, I., Knowlton, J.J., Zamora, P.F., Lee, C.H., Mainou, B.A., Dermody, T.S., Risco, C., 2018. Reovirus σNS and μNS proteins remodel the endoplasmic reticulum to build replication neo-organelles. mBio 9, e01253-18.

Titze, B., Genoud, C., 2016. Volume scanning electron microscopy for imaging biological ultrastructure. Biol. Cell 108, 307–323.

Tokuyasu, K.T., 1973. A technique for ultracryotomy of cell suspensions and tissues. J. Cell Biol. 57, 551–565.

Tolonen, N., Doglio, L., Schleich, S., Krijnse Locker, J., 2001. Vaccinia virus DNA replication occurs in endoplasmic reticulum-enclosed cytoplasmic mini-nuclei. Mol. Biol. Cell 12, 2031–2046.

Van der Schaar, H.M., Melia, C.E., van Bruggen, J.A.C., Strating, J.R.P.M., van Geenen, M.E.D., Koster, A.J., Bárcena, M., van Kuppeveld, F.J.M., 2016. Illuminating the sites of enterovirus replication in living cells by using a split-GFP-tagged viral protein. mSphere 1 (4), e00104-16.

Verkade, P., 2008. Moving EM: the rapid transfer system as a new tool for correlative light and electron microscopy and high throughput for high-pressure freezing. J. Microsc. 230, 317–328.

Villinger, C., Neusser, G., Kranz, C., Walther, P., Mertens, T., 2015. 3D analysis of HCMV induced-nuclear membrane structures by FIB/SEM tomography: insight into an unprecedented membrane morphology. Viruses 7, 5686–5704.

Wan, J., Basu, K., Miu, J., Vali, H., Zheng, H., Laliberté, J.-F., 2015. Ultrastructural characterization of turnip mosaic virus-induced celular rearrangements reveals membrane-bound viral particles accumulating in vacuoles. J. Virol. 89, 12441–12456.

Welsch, S., Miller, S., Romero-Brey, I., Merz, A., Bleck, C.K.E., Walther, P., Fuller, S.D., Antony, C., Krijnse-Locker, J., Bartenschlager, R., 2009. Composition and three-dimensional architecture of the Dengue virus replication and assembly sites. Cell Host Microbe 5, 365–375.

Westaway, E.G., Khromykh, A.A., Mackenzie, J.M., 1999. Nascent flavivirus RNA colocalized in situ with double-stranded RNA in stable replication complexes. Virology 258, 108–117.

White, D.L., Andrews, S.B., Faller, J.W., Barrnett, R.J., 1976. The chemical nature of osmium tetroxide fixation and staining of membranes by x-ray photoelectron spectroscopy. Biochim. Biophys. Acta 436, 577–592.

Wileman, T., 2007. Aggresomes and pericentriolar sites of virus assembly: cellular defense or viral design? Annu. Rev. Microbiol. 61, 7905–7912.

Witte, R., Andriasyan, V., Georgi, F., Yakimovich, A., Greber, U.F., 2018. Concepts in light microscopy of viruses. Viruses 10, 202. https://doi.org/10.3390/v10040202.

Wolff, G., Hagen, C., Grünewald, K., Kaufmann, R., 2016. Towards correlative super-resolution fluorescence and electron cryo-microscopy. Biol. Cell 108, 245–258.

Wölk, B., Büchele, B., Moradpour, D., Rice, C.M., 2008. A dynamic view of hepatitis C virus replication complexes. J. Virol. 82, 10519–10531.

Yakushevska, A.E., Lebbink, M.N., Geerts, W.J., Spek, L., van Donselaar, E.G., Jansen, K.A., Humbel, B.M., Post, J.A., Verkleij, A.J., Koster, A.J., 2007. STEM tomography in cell biology. J. Struct. Biol. 159, 381–391.

CHAPTER TWO

Structures of enveloped virions determined by cryogenic electron microscopy and tomography

Robert Stass[a,†], Weng M. Ng[a,†], Young Chan Kim[a,†], Juha T. Huiskonen[a,b,*]

[a]Division of Structural Biology, Wellcome Centre for Human Genetics, University of Oxford, Oxford, United Kingdom

[b]Helsinki Institute of Life Science HiLIFE and Research Programme in Molecular and Integrative Biosciences, Faculty of Biological and Environmental Sciences, University of Helsinki, Helsinki, Finland

*Corresponding author: e-mail address: juha.huiskonen@helsinki.fi

Contents

Abstract

Enveloped viruses enclose their genomes inside a lipid bilayer which is decorated by membrane proteins *that mediate* virus entry. These viruses display a *wide range* of sizes, morphologies and symmetries. Spherical viruses are often isometric and their envelope proteins follow icosahedral symmetry. Filamentous and pleomorphic viruses lack such global symmetry but their surface proteins may display *locally ordered assemblies*.

† These authors contributed equally.

Advances in Virus Research, Volume 105
ISSN 0065-3527
https://doi.org/10.1016/bs.aivir.2019.07.009

Determining the structures of enveloped viruses, including the envelope proteins and their protein-protein interactions on the viral surface, is of paramount importance. These structures can reveal how the virions are assembled and released by budding from the infected host cell, how the progeny virions infect new cells by membrane fusion, and how antibodies bind surface epitopes to block infection. In this chapter, we discuss the uses of cryogenic electron microscopy (cryo-EM) in *elucidating* structures of enveloped virions. Starting from a detailed outline of data collection and processing strategies, we highlight how cryo-EM has been successfully utilized to provide unique insights into enveloped virus entry, assembly, and neutralization.

Abbreviations

BDBV	Bundibugyo virus
CCD	charge coupled device
CHIKV	Chikungunya virus
cryo-EM	cryogenic electron microscopy
CTF	contrast transfer function
DED	direct electron detector
DENV	Dengue virus
DNA	deoxyribonucleic acid
EBOV	Ebola virus
EEEV	eastern equine encephalitis virus
ESCRT	endosomal sorting complex required for transport
EV71	enterovirus 71
FIB	focused ion beam
GP	glycoprotein
HA	hemagglutinin
HAZV	Hazara virus
HBV	hepatitis B virus
HIV	human immunodeficiency virus
HPeV3	human parechovirus 3
HPIV3	human parainfluenza virus 3
JEV	Japanese encephalitis virus
LASV	Lassa virus
MARV	Marburg virus
MCP	major capsid protein
MeV	measles virus
MHV	mouse hepatitis virus
MLD	mucin-like domain
NA	neuraminidase
NC	nucleocapsid
NDV	Newcastle disease virus
RNA	ribonucleic acid
RVFV	Rift Valley fever virus
SARS-CoV	severe acute respiratory syndrome-related coronavirus
SeV	Sendai virus

SFV	Semliki Forest virus
SINV	Sindbis virus
SPA	single particle averaging
SSP	stable signal peptide
STA	subtomogram averaging
TBEV	tick borne encephalitis virus
TULV	Tula virus
UHV	University of Helsinki virus
VACV	Vaccinia virus
VEEV	Venezuelan equine encephalitis virus
VSV	vesicular stomatitis virus
VLP	virus-like particle
VPP	volta phase plate
WNV	West Nile virus
ZIKV	Zika virus

1. Introduction

Enveloped viruses encompass a large group of viruses with different morphologies and genome types, spanning across different virus families and infecting host cells from all three domains of life (Eukaryota, Bacteria and Archaea). Common to all of these viruses is that their virions harbor a lipid bilayer which associates with integral and peripheral membrane proteins. Together, these structural components create a viral envelope that encloses the genome, or an internal protein–genome complex called the nucleocapsid (NC). The virion envelope is derived from the host cell membrane by fission in the budding process of progeny virions. During entry into a new host cell, the envelope must fuse to a host cell membrane for the enclosed genome or the NC to enter the cytoplasm. This process is catalyzed by membrane fusion proteins (Harrison, 2015; Kielian, 2014).

Understanding the structures of enveloped viruses is important for understanding their infection mechanisms, especially entry by membrane fusion in addition to assembly and budding by membrane fission. Various types of enveloped virus morphologies are schematized in Fig. 1. The shapes of the enveloped virions range from spherical or icosahedral to ellipsoidal to pleomorphic and filamentous. The virion envelope surface is covered to varying degrees by membrane proteins that often form multimeric assemblies. Such morphological units on the virion surface are referred to as

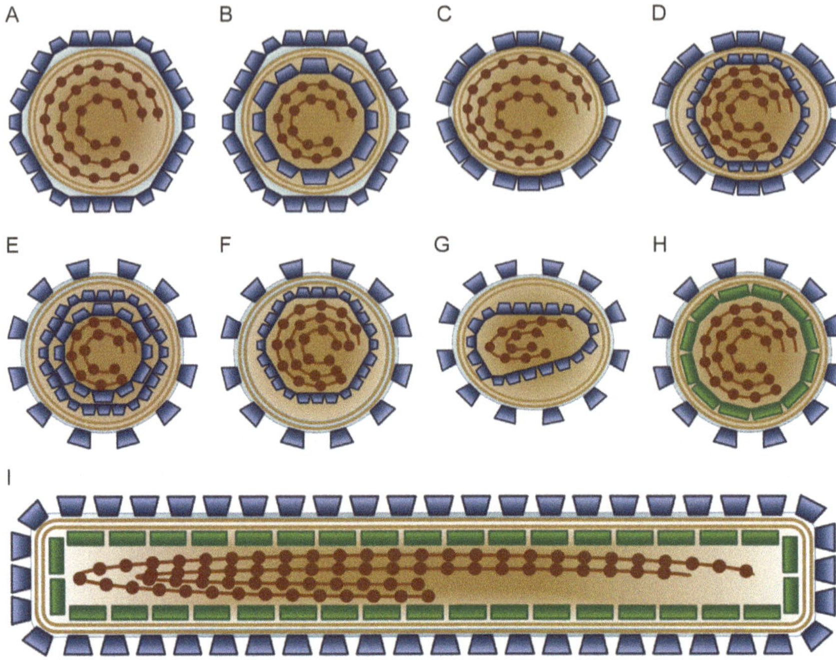

Fig. 1 Approximate schematic presentations of different types of enveloped virion morphologies. (A) Virion with an icosahedrally symmetric outer protein shell covering a lipid bilayer. Example structures include members of *Flavivirus* such as dengue virus (DENV) (Kuhn et al., 2002) and *Phenuiviridae* such as Rift Valley fever virus (RVFV) (Huiskonen et al., 2009) in addition to many membrane-containing prokaryotic viruses including members of *Tectiviridae* such as bacteriophage PRD1 (San Martín et al., 2002). (B) Virion with an icosahedrally symmetric outer protein shell covering a lipid bilayer and an additional icosahedrally symmetric inner protein shell. Example structures include members of *Togaviridae* such as SFV (Mancini et al., 2000). (C) A virion with a nonicosahedrally symmetric, but locally ordered outer protein shell covering most of the lipid bilayer and lacking a matrix layer. Example structures include members of order *Bunyavirales* such as Tula virus (TULV) (*Hantaviridae*) (Huiskonen et al., 2010), Hazara virus (HAZV) (*Nairoviridae*) (Punch et al., 2018) and Bunyamwera virus (BUNV) (*Orthobunyaviridae*) (Bowden et al., 2013). (D) A virion with an icosahedral inner capsid and a nonicosahedrally symmetric, but locally ordered outer protein shell. Example structures include members of *Hepadnaviridae* such as hepatitis B virus (HBV) (Dryden et al., 2006). (E) A virion with two internal icosahedrally symmetric protein shells surrounded by a lipid envelope with surface spikes. Example structures include members of *Cystoviridae* such as bacteriophage Φ6 (Jäälinoja et al., 2007a). (F) Members of *Herpesviridae*, such as HSV-1 (Grünewald et al., 2003), have an icosahedrally symmetric protein shell, enclosed by a tegument layer (not shown) and an external lipid envelope with surface glycoproteins. (G) Some members of *Retroviridae*, such as human immunodeficiency virus 1 (HIV-1) (Briggs et al., 2006), have mature virions with relatively few GPs on the virion envelope. (H) A virion with a lipid bilayer, decorated by glycoprotein spikes and an internal matrix protein layer is shown. Examples include members

capsomers or spikes, as they often protrude from the surface and have pointed appearance. In animal viruses, these surface proteins are typically glycosylated (glycoproteins; GPs). In a seminal study on herpes simplex virus 1 (HSV-1), the GPs were visualized on the virion surface by cryo-electron tomography (cryo-ET) (Grünewald et al., 2003). The capsomers may further form higher order assemblies, which may have local symmetry. In many enveloped viruses the lipid bilayer is almost entirely covered by surface proteins, leaving hardly any naked membrane accessible from the virion exterior. This is the case for example in the members of *Flaviviridae*, such as dengue virus (DENV), where GPs form a continuous icosahedral protein shell on the envelope (Kuhn et al., 2002), and members of *Hantaviridae*, such as Tula virus (TULV), where GPs form locally ordered patches on the envelope (Huiskonen et al., 2010). In contrast, members of *Retroviridae*, such as human immunodeficiency virus (HIV), harbor very few glycoprotein spikes, leaving a large fraction of the membrane naked (Briggs et al., 2003). Some enveloped viruses harbor a symmetric nucleocapsid, which can have either icosahedral or helical symmetry. For example, members of *Togaviridae*, such as Semliki Forest virus (SFV), harbor an icosahedrally symmetric nucleocapsid (Fuller et al., 1995). Finally, many, but not all, enveloped viruses have a matrix protein directly under the envelope (Ke et al., 2018b; Li et al., 2016a).

In this chapter, we review recent advances in understanding enveloped virus structures by cryogenic electron microscopy (cryo-EM). We will begin by an overview of different cryo-EM data collection and processing strategies relevant to the topic and then proceed to reviewing how cryo-EM structures of enveloped virions contribute to our understanding of the molecular interactions driving assembly, the dynamic nature of viral particles, their budding and membrane fusion mechanisms in addition to virus neutralization and furthermore how these structural biology studies are informing vaccine design.

---

of *Arenaviridae* such as Lassa virus (LASV) (Li et al., 2016b), *Coronaviridae* such as severe acute respiratory syndrome-related coronavirus (SARS-CoV) (Neuman et al., 2006), and *Paramyxoviridae* such as measles virus (MeV) (Ke et al., 2018b). (I) A filamentous virion with envelope glycoprotein spikes and internal matrix layer. Examples include members of *Filoviridae*, such as Ebola virus (EBOV) (Bharat et al., 2012) and *Pneumoviridae* such as respiratory syncytial virus (RSV) (Ke et al., 2018a), in addition to filamentous forms of influenza A virus (*Orthomyxoviridae*) (Calder et al., 2010). Blue, viral structural protein; Light brown, lipid bilayer; Brown circles, nucleoprotein or other genome-associated protein; Brown line(s), viral genome segment(s); Green, matrix protein. Note that we have not attempted to accurately depict the genome type nor its organization or the symmetry and arrangement of different protein shells.

2. Cryogenic electron microscopy in membrane virus research

Cryo-EM is a well-suited method for the structural analysis of enveloped virions (Subramaniam et al., 2007). As the structures are often pleomorphic (i.e., lacking a regular shape), they are often not amenable to X-ray crystallography, another structural biology technique that relies on crystallization of the sample of interest. In fact, in only a few cases has the structure of an enveloped virion been solved by X-ray crystallography (Abrescia et al., 2004, 2008). Cryo-EM methods, however, are applicable to both regular and pleomorphic virions, as no crystals are needed and structures can be determined from a relatively small amount of purified virions. The electric potential maps (or simply "cryo-EM density maps") determined by cryo-EM have in many cases similar level of detail when compared to electron density maps determined by X-ray crystallography, although there are also subtle differences that become more significant at high resolution (Wang and Moore, 2017). Furthermore, cryo-EM allows structural investigations of more complex and rapid processes such as viral envelope–host membrane fusion and viral budding from the plasma membrane of infected cells.

As many enveloped viruses are human and animal pathogens, their production, purification and preparation for cryo-EM requires suitable containment facilities and bio-safety protocols (Sherman et al., 2013). To circumvent the need for decontaminating cryo-EM equipment, which may be in some cases impractical, purified virions can be inactivated for example by chemical fixation (Bharat et al., 2011; Halldorsson et al., 2018; Li et al., 2016b) or ultraviolet radiation (Park et al., 2011; Ye et al., 2018; Zhong et al., 2016) prior to cryo-EM sample preparation. Alternatively, virus-like particles (VLPs) that contain the relevant viral structural proteins with a lipid bilayer can be used as a model system instead of the native virion (Li et al., 2016b; Sun et al., 2013). Whether live or inactivated, virion or VLP, purified particles are prepared for cryo-EM similar to any other macromolecular complex (Thompson et al., 2016). A small aliquot of virus suspension, typically 3 μL, is pipetted on an EM sample grid, a circular metal mesh (3 mm in diameter), typically coated with a foil of holey carbon or gold. The grid, held by tweezers, is then blotted by a piece of filter paper to remove most of the sample in order to leave a very thin film of virus suspension on the grid (often not much thicker than the particle itself). The grid is then plunged in liquid ethane, which is cooled by liquid nitrogen.

The extremely cold temperature of the cryogen (around −180 °C) leads to a very rapid cooling rate, and the formation of amorphous, glass-like ice, which is compatible with the vacuum of the electron microscope column and is also transparent to the electron beam.

Typical cryo-EM data collection and processing workflows for enveloped viruses are outlined in Fig. 2. Cryo-EM data can be collected either as 2D projection images (micrographs; Fig. 2A) or as a series of tilted images (a tomographic tilt series) from which a 3D tomographic volume (or a tomogram) can be calculated (Fig. 2B). To increase the inherently low signal-to-noise ratio (SNR), signal from multiple images (or volumes) needs to be averaged in a coherent manner to reach sufficiently high resolution in the average. The target resolution depends on the types of questions addressed and ranges from better than 3 Å (required to model the polypeptide chains of proteins and to see small bound molecules) to 30 Å and even lower (sufficient for addressing the organization of envelope proteins and visualizing the lipid bilayers; Table 1). Different approaches to averaging in the context of enveloped viruses are discussed below.

2.1 Single particle averaging and localized reconstruction

Many enveloped viruses harbor an icosahedrally symmetric protein shell, which can be either external (Fig. 1A) or internal to the lipid bilayer (Fig. 1D and F). Some enveloped virions have two such shells sandwiching the lipid bilayer (Fig. 1B) and some have two internal shells (Fig. 1E). The structures of the icosahedrally symmetric protein shells in these types of virions can be determined by cryo-EM and single particle averaging (SPA). In standard SPA workflows, the virions are located in the micrographs (particle picking) and extracted in smaller images each containing one virion image in the middle (particle images). After determining the defocus value of each particle image, the exact location of the virion in the image (two coordinates) and orientation of the particle (three angles), the structure of the virion can be determined (or reconstructed) (Fig. 2A; steps 1 and 2). To increase the SNR and thus attainable resolution for the protein shells, icosahedral symmetry is normally applied at this stage. Several icosahedrally symmetric enveloped virus structures have been determined by this approach (see Section 3.1; Table 1). To obtain the highest possible resolution for large membrane viruses, it may also be crucial to take into account the thickness of the specimen in the 3D reconstruction process (Wolf et al., 2006).

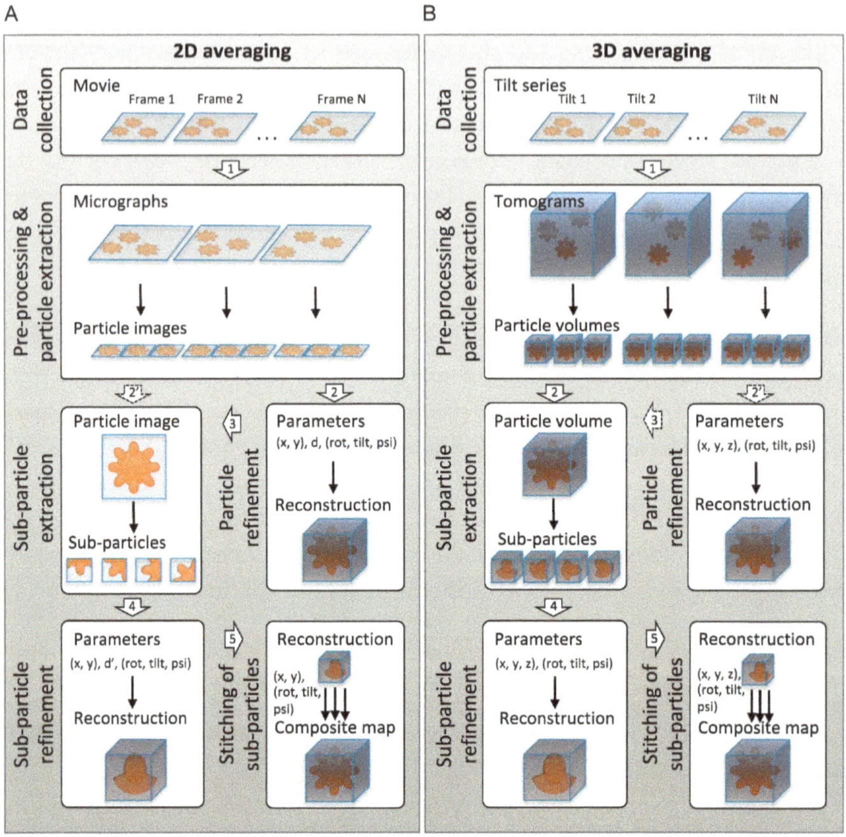

Fig. 2 Cryo-EM data processing strategies for determining structures of enveloped virions and their glycoprotein spikes. (A) Here an enveloped virion is depicted as an orange sphere with protruding blobs that depict glycoprotein spikes. Defocus of the particle is d. Defocus of the subparticle is d′. The view direction of the particle or subparticle is defined by angles rot and tilt. The in-plane rotation of the particle projection is defined by psi. (B) Note that each image in a tomography tilt series is potentially recorded first as a movie. The range between tilt 1 and tilt N is typically [−60,60°] and N is typically 41 (for 3-degree angular sampling) or 61 (for 2° angular sampling). Here and in the text we have assumed that the tomograms have been corrected for the effects of the contrast transfer function (CTF) during preprocessing but other approaches are possible. Here the three Euler angles (rot, tilt, psi) define the orientation of the particles and the subparticles. The block arrows refer to different data processing steps described in the text. The block arrows with dashed outlines depict less common data processing approaches.

Table 1 Examples of enveloped virion structures and their membrane associated components determined by different cryogenic electron microscopy data processing strategies.

Virus	Component	Method and steps[a]	Resolution (Å)	Accession codes	Reference
Simian immunodeficiency virus 1 (*Retroviridae*)	Capsomer (Env)	3D 1-2-4	28	EMD-1216 PDB:2BF1	Zanetti et al. (2006)
Zika virus (*Flaviviridae*)	Virion	2D 1-2	3.1	EMD-7543 PDB:6CO8	Sevvana et al. (2018)
Sindbis virus (*Togaviridae*)	Virion	2D 1-2-3-4-5	3.5	EMD-9693 PDB:6IMM	Chen et al. (2018)
Rift Valley fever virus (*Phenuiviridae*)	Capsomer (Gn–Gc)	2D 1-2-3-4	7.7	EMD-4201 PDB:6F9F	Halldorsson et al. (2018)
Hantaan virus (*Hantaviridae*)	Capsomer (Gn–Gc)	2D 1-2′-4	25	N/A	Battisti et al. (2011)
Tula virus (*Hantaviridae*)	Capsomer (Gn–Gc)	3D 1-2-4	16	EMD-3364	Li et al. (2016a)
Bunyamwera virus (*Orthobunyaviridae*)	Capsomer (Gn–Gc)	3D 1-2-4	30	EMD-2352	Bowden et al. (2013)
Hazara virus (*Nairoviridae*)	Capsomer (Gn–Gc)	3D 1-2-4	25	N/A	Punch et al. (2018)
Lassa virus (*Arenaviridae*)	Capsomer (GP1–GP2)	3D 1-2-4	14.0	EMD-3290	Li et al. (2016b)
Measles virus (*Paramyxoviridae*)	Virion	3D 1	N/A	N/A	Ke et al. (2018b)
Ebola virus (*Filoviridae*)	Capsomer (GP)	2D 1-2	11	EMD-8036	Beniac and Booth (2017)
Influenza virus A (*Orthomyxoviridae*)	Virion	3D 1	N/A	N/A	Calder et al. (2010)
Haloarcula hispanica SH1 virus (*Sphaerolipoviridae*)	Virion	2D 1-2-3-4-5	3.8	EMD-4633 PDB:6QT9	Colibus et al. (2019)

[a]The steps refer to different parts of the various image processing workflows in Fig. 2. For steps in 2D workflows (single particle averaging and subparticle averaging), refer to Fig. 2A. For steps in 3D workflows (tomography and subtomogram averaging), refer to Fig. 2B.

The downside of this standard SPA workflow is that those components of the virion that are not organized in a strictly symmetric fashion will get incoherently averaged and this limits the attainable resolution, or hinders their reconstruction all together (Huiskonen, 2018). For example, if the structure of a virion presented in Fig. 1A is somewhat flexible, signal both between different particles and asymmetric units within each particle will be incoherently averaged. In order to improve the resolution for flexible enveloped virions and their symmetry-mismatched components, the localized reconstruction method can be used to divide the particle in several subparticles (Fig. 2A; subparticle extraction) (Huiskonen, 2018; Ilca et al., 2015). The orientation (three angles: rot, tilt and psi) of each subparticle and location in the particle image (two coordinates: x and y) can then be calculated and used as an initial estimate to further refine subparticles around their original positions and locations (Fig. 2A; steps 1, 2, 3, and 4). Furthermore, the defocus of each subparticle is calculated to take into account the defocus gradient across the specimen (Ilca et al., 2015). This approach has allowed improving the resolution of Rift Valley fever virus (RVFV) virions that have an icosahedrally ordered, but yet highly flexible, large protein shell (diameter ~110 nm), from 13 Å to 7.7 Å (Halldorsson et al., 2018). The resolution in this study and other similar studies may be limited due to overlaps between other subparticles, the membrane and the genome. In some cases, overlapping components can be subtracted from the particle image by partial signal subtraction to improve the accuracy of subparticle alignments (Bai et al., 2015; Huiskonen et al., 2007). Also, it is worth noting that in this approach only distortions in the image plane can be handled and any movement along the beam direction (corresponding to further changes in defocus) is ignored. After aligning subparticles in the image plane and reconstructing them, a composite 3D model of the virion can be created by "stitching" the entire virion from individual subparticle reconstructions (Fig. 2A, step 5). Recently the structure of the complete Sindbis virus (SINV) virion has been reconstructed this way from three separate subparticle reconstructions at 3.5 Å resolution (Chen et al., 2018).

Some virus particles are too pleomorphic for even rough orientational alignment rendering them challenging for SPA approaches. Several studies have attempted averaging glycoprotein spikes from the edge of particle images (Fig. 2A; steps 1 and 2′). This has been done for example for HNTV (Battisti et al., 2010), DENV at acidic pH (Zhang et al., 2015), and SARS (Neuman et al., 2006). In contrast to localized reconstruction where the three angles describing the orientation of the subparticle (rot, tilt and psi) can be estimated from the orientation parameters of the entire particle, in this approach only two angles of the three angles (tilt and psi) can be estimated (tilt can be assumed to be close to 90°; psi can be estimated from the normal of the membrane projection). Due to this limitation, it is not possible to calculate a 3D reconstruction without further exhaustive alignment to determine the third angle that is unknown (rot). Another limitation of this approach is that spikes at the edge of the particle can overlap other spikes and in the absence of an approximate 3D model of the particle, these overlaps cannot be removed by partial signal subtraction (Bai et al., 2015; Huiskonen et al., 2007). Due to these reasons, most studies have classified and averaged spike side-view projections only in 2D (Fig. 2A; steps 1 and 2′). At the 2D level, such studies have been informative. For example, two distinct conformations of the SARS M protein were observed by this approach (Neuman et al., 2006). In some cases 3D averaging has also been performed (Fig. 2A; step 4). In one example, the 3D structure of the tetrameric GP spike of HNTV was resolved to 25 Å from side-view subparticles (Battisti et al., 2010) (Fig. 2A; steps 1, 2′ and 4). This structure agreed well with the structure of TULV GP spike solved by cryo-ET and subtomogram averaging (STA) (Huiskonen et al., 2010). In another example, the structure of the trimeric GP spike of Ebola virus (EBOV) was determined at 11 Å resolution from side-view projections on the virion surface (Beniac and Booth, 2017) (Fig. 2A; steps 1, 2′ and 4). Consistent with a cryo-ET and subtomogram averaging (STA) investigation of GP from EBOV VLP (Tran et al., 2014), a mucin-like domain (MLD) was located to the apex and sides of each GP1 monomer, partially shielding the receptor-binding site, while the GP1 sits atop the GP2. In these side-view averaging approaches it is only possible to include spikes from the edge of the virion projection image so the packing of the spikes on the virion surface remains unattainable.

2.2 Tomography and subtomogram averaging

Tomography is a method well suited for determining structures of flexible and truly pleomorphic virions from cryo-EM data that are challenging or unsuitable for the SPA approaches described above. When virions lack well defined shape, it is impossible to combine particles with different views extracted from 2D micrographs to reconstruct a correct 3D volume. Instead, different views must be collected for each virion as a series of tilted images (Fig. 2B). These views are then combined to calculate a tomographic 3D volume (a tomogram) of the specimen region under investigation.

Due to the slab-shaped geometry of the cryo-EM specimen holders and the cryo-EM grid itself, the specimen cannot be tilted to 90° and thus the angular range in a typical tomographic tilt series is limited from −60 to +60°. This results in incomplete sampling of information in the 3D reconstruction, which can be described as a "missing wedge" in the 3D Fourier transform of the tomogram. Because of this limitation, features in tomograms are distorted and averaging of 3D particles in different orientations is required to fully sample the information in the final 3D reconstruction (Subramaniam et al., 2007).

Similar to the 2D processing workflow described in the previous section, it is often practical to extract smaller 3D volumes, each corresponding to a single virion, from the larger tomograms (Fig. 2B; particle extraction). If these 3D particles are homogenous enough it is then possible to align, classify and average them together. For example, the first low-resolution structure of a bunyavirus (Uukuniemi virus, UUKV; *Phenuiviridae*) was determined by aligning single 3D volumes of virions and by applying icosahedral symmetry (Överby et al., 2008) (Fig. 2B; steps 1 and 2′). Once the orientation (rot, tilt, psi) of each particle is known from an initial alignment, it is possible to deal with any possible flexibility of the virion by extracting 3D subparticles (also referred to as subvolumes or subtomograms; Fig. 2B; step 3) (Castaño-Díez et al., 2017). These subparticles, corresponding for instance to envelope GP spikes can then be refined further (Fig. 2B; step 4) and finally a composite map of the entire virion can be stitched from the 3D reconstruction of the subparticles (Fig. 2B; step 5) (Huiskonen et al., 2010).

In most cases where tomography is applied, however, the 3D particles are too dissimilar to be aligned and averaged in a coherent manner. This is the case with truly pleomorphic virions. In these cases, subparticles of GP spikes are extracted from unaligned 3D volumes of enveloped virions (Fig. 2B; steps 1 and 2). The locations of the spikes first need to be determined by a 3D search, which can be restricted close to the membrane surface (Castaño-Díez et al., 2017; Huiskonen et al., 2014). Also, the direction of the spike can be estimated from the membrane surface normals. Once the 3D-subvolumes, each corresponding to a centered spike, have been extracted, only the rotation around the spike long axis remains to be determined before a 3D reconstruction of the subparticle can be calculated (Fig. 2B, step 4) (Förster et al., 2005; Zanetti et al., 2006). For example, several studies have produced low resolution reconstructions of Env from native HIV virions using STA (Liu et al., 2008; Zanetti et al., 2006; Zhu et al., 2003, 2006). As described above, these 3D subparticle reconstructions can then be plotted back onto the original particle volumes to stitch together composite models of entire virions. Here, the completeness of this stitching depends on the coverage of the 3D picking (Huiskonen et al., 2010).

2.3 Hybrid methods

As we have outlined in Fig. 2, averaging of 2D and 3D single particles follow highly analogous workflows. In some cases it may be beneficial to mix these two. In one example of such a hybrid approach, the structure of the GP spike of prototypic foamy virus (PFV) has been studied (Effantin et al., 2016). First STA was used to determine the structure of the trimeric GP at ~30 Å resolution (Fig. 2B; steps 1, 2, 3, 4). Plotting back the GP structure allowed visualizing hexagonal assemblies of six GPs on the envelope (Fig. 2B; step 5). A patch of six trimers was extracted, sixfold symmetry was imposed and this volume was used as a search model to pick and extract GP subparticles directly from untilted images (Fig. 2A; steps 1 and 2′). GP spike subparticles were then subjected to conventional SPA refinement that allowed determination of a threefold symmetrized GP density map at ~9 Å. The density map revealed a region interpreted as a coiled-coil of three α-helices, a hallmark of viral class I fusion proteins.

3. Structures of purified virions

3.1 Virions with icosahedrally symmetric protein shells

Structures of icosahedrally symmetric protein shells in enveloped virions have been a topic of several cryo-EM studies over the past two decades. The first 3D reconstructions of Semliki Forest virus (SFV; *Togaviridae*) and dengue virus (DENV; *Flaviviridae*) virions were determined at 22 Å resolution (Fuller et al., 1995) and at 24 Å resolution (Kuhn et al., 2002), respectively, by SPA with icosahedral symmetry applied (Fig. 2A; steps 1 and 2). At such limited resolution, only the rough morphology of the virions could be resolved. However, fitting of X-ray structures into cryo-EM maps has, in many cases, allowed the creation of so-called pseudo-atomic models of the glycoprotein shells. For example, fitting of the X-ray crystallographic structure of the E protein from another flavivirus, tick borne encephalitis virus (TBEV), (Rey et al., 1995) revealed the so-called herringbone arrangement of *E*-protein dimers on the DENV virion surface (Kuhn et al., 2002). The advent of direct electron detectors (DEDs) and other advances in electron microscope hardware and image processing software led to a "resolution revolution" in cryo-EM making it possible to derive atomic models from cryo-EM maps of single particles alone (Kühlbrandt, 2014). A notable exception from the pre-resolution revolution era of cryo-EM is the structure of mature DENV that was determined at 3.5 Å resolution from images collected on a charge coupled device (CCD) camera (Zhang et al., 2013a).

In the post-resolution revolution era, determining structures of icosahedrally symmetric protein shells in enveloped viruses from cryo-EM data alone has become routine. For instance, the structure of mature Zika virus (ZIKV; *Flaviviridae*) virion has been solved by cryo-EM and SPA by two research groups at ~3.8 Å resolution (Kostyuchenko et al., 2016; Sirohi et al., 2016). These studies revealed that despite the high level of structural similarity to other flaviviruses such as DENV, ZIKV particles displayed greater thermal stability at higher temperature (40 °C degrees) which could account for their survival in the semen and urine. The ZIKV protein shell also presents a unique amino acid region around the Asn154 glycosylation site which may explain the neurotropic nature of ZIKV similar to West Nile virus (WNV; *Flaviviridae*). A recent 4.3-Å cryo-EM structure of Japanese encephalitis virus (JEV; *Flaviviridae*) has allowed mapping of neurovirulence factors on the virus surface (Wang et al., 2017). The same strategy has been applied to several members of the *Togaviridae* family. A cryo-EM structure of Sindbis virus (SINV; *Togaviridae*) at 3.5 Å resolution has allowed the identification of a "pocket factor," a 20-Å long molecule, possibly a phospholipid tail, projecting from the viral lipid bilayer into a membrane-proximal hydrophobic pocket of the GP shell (Chen et al., 2018). A chikungunya virus (CHIKV; *Togaviridae*) VLP structure has been determined at 5.3-Å resolution. This study shows that togavirus VLPs reflect the structures of mature virions and revealed that CHIKV E1 and E2 glycoproteins are not associated with the E3 glycoprotein unlike other alphaviruses such as Venezuelan equine encephalomyelitis virus (VEEV) (Sun et al., 2013).

Structures of several prokaryotic viruses with icosahedral capsids and an internal membrane have also been studied by cryo-EM and SPA. Typically these viruses harbor major capsid proteins (MCPs) that consist of upright beta-barrel folds and fully cover an internal membrane following different arrangements (described by the triangulation [*T*] number). These viruses include bacteriophages PRD1 ($T=25$) (San Martín et al., 2002), Bam35 ($T=25$) (Laurinmäki et al., 2005), PM2 ($T=21d$) (Huiskonen et al., 2004) and FLiP ($T=21d$) (Laanto et al., 2017), in addition to archaeal viruses SH1 ($T=28$) (Colibus et al., 2019), STIV ($T=31d$) (Veesler et al., 2013) and STIV2 ($T=31d$) (Happonen et al., 2010). Similarity of the MCP fold and the arrangement MCPs of the virions surface have allowed grouping these viruses to a so-called "PRD1–adenovirus lineage," based on the extended similarity to the nonenveloped adenovirus. Interestingly, cryo-EM and SPA have revealed that giant eukaryotic dsDNA viruses share the same basic MCP building block with aforementioned prokaryotic viruses and adenovirus. In these giant dsDNA viruses compelling evidence exists for an internal lipid bilayer (Xiao and Rossmann, 2011). Structures of several viruses belonging to this group have been determined by cryoEM and SPA, including CIV ($T=147$) (Khayat et al., 2010), PBCV-1 ($T=169d$) (Zhang et al., 2011), PpV01 ($T=219d$) (Yan et al., 2005), CroV ($T=499$) (Xiao et al., 2017) and mimivirus ($972 \geq T \geq 1200$) (Xiao et al., 2009; Klose et al., 2010). Cryo-EM has also been used in characterizing other types of enveloped bacteriophages belonging to the *Cystoviridae* family. Members of this family have an outer lipid envelope with surface proteins enclosing two internal icosahedral protein shells (with $T=1$ and $T=13$ L architecture) enclosing a segmented dsRNA genome. Structures including Φ6 (Jäälinoja et al., 2007b; Sun et al., 2017), Φ8 (Jäälinoja et al., 2007b) and Φ12 (Wei et al., 2009) have highlighted structural similarities in their protein shells to nonenveloped reoviruses. Taken together, these studies have started to exemplify possible distant evolutionary links between enveloped and nonenveloped viruses.

3.2 Dynamic nature of enveloped virions

In addition to the high-resolution cryo-EM structures of enveloped virions and their icosahedrally symmetric protein shells, several cryo-EM studies have highlighted the dynamic nature of these shells. One realization is that enveloped virions may assemble from a fixed number of GP capsomers (such as 12 pentamers and N hexamers) on a defined icosahedral lattice but the resulting virion structure may be flexible. The first representative structure for members of *Phenuiviridae*, and for the entire order of Bunyavirales, has been studied by cryo-ET of purified UUKV virions (*Phlebovirus*, *Phenuiviridae*) (Överby et al., 2008). This study revealed that the virion has icosahedral symmetry with $T=12$ triangulation and should then in principle be amenable to SPA. Later the structure of RVFV (*Phlebovirus*, *Phenuiviridae*) virions has been studied by SPA but the resolution has been limited to 13 Å due to significant flexibility of the GP layer (Halldorsson et al., 2018). Despite this limitation, a localized reconstruction method (Ilca et al., 2015) has allowed partially dealing with flexibility to improve the resolution to 7.9 Å. This was sufficient for flexible fitting of Gn and Gc X-ray crystallographic structures, revealing how the Gn chaperone protein caps the fusion loops of the fusion protein Gc (Halldorsson et al., 2018).

Highly dynamic structural changes take place during virion maturation and entry (Hasan et al., 2018a). Cryo-EM studies have played a significant role in determining the structural changes exhibited by flaviviruses in their immature, fusogenic and mature forms. The first cryo-EM structure of a mature DENV showed that the mature infectious particles are icosahedral and ~500 Å in diameter. The surface is smooth and is comprised of 90 copies of an E protein dimer that is closely packed, suggesting that a major rearrangement is required before host cell fusion (Kuhn et al., 2002). Cryo-EM and crystallography of fusogenic virions has revealed how the parallel E protein dimers of the mature virion first rearrange into monomers and then into the E-protein fusogenic trimers with three-fold symmetry (Allison et al., 1995; Bressanelli et al., 2004; Modis et al., 2004; Zhang et al., 2015). The first cryo-EM studies of immature flavivirus particles (DENV and yellow fever virus) have shown striking differences in terms of the considerably larger diameter (600 Å) and the presence of 60 prominent trimeric spikes (Zhang et al., 2003b). More recently, the 9-Å resolution cryo-EM structure of the immature ZIKV has shown similar characteristics and spatial arrangement to DENV (Prasad et al., 2017). These studies demonstrate that flavivirus particles undergo a series of significant conformational changes during virion maturation and entry, reflecting the highly dynamic nature of the virions.

It is becoming increasingly clear that temperature may be a significant factor in the conformation of enveloped virions, yet typically samples are prepared for cryo-EM at ambient temperature or below. Two cryo-EM studies on the structure of the mature DENV virion serotype 2 (strains NGC and 16,681) have highlighted the structural changes to DENV-2 virion at elevated temperatures (Fibriansah et al., 2013; Zhang et al., 2013b). When heated to 37 °C, the envelopes of these particles changed their appearance from smooth to bumpy. This suggests that also mature DENV-2 virions involved in human infection may have bumpy structures and therefore optimal vaccines should consider also epitopes exposed on the bumpy form of the virus.

3.3 Inherently pleomorphic virions and their glycoproteins

Enveloped viruses whose structural protein shells do not exhibit icosahedral symmetry present a challenge for high resolution cryo-EM structure determination, as these virions are unsuitable to SPA processing. In these cases, cryo-ET has been used to study the overall virion ultrastructure. Examples of such studies include those on HSV-1 (Grünewald et al., 2003), vaccinia virus (Cyrklaff et al., 2005), HIV (Briggs et al., 2006), influenza virus (Calder et al., 2010; Harris et al., 2006), rabies virus (Guichard et al., 2011), EBOV (Bharat et al., 2012), MARV (Bharat et al., 2012) and baculovirus (Wang et al., 2016a). These studies have provided valuable insights into the virion morphology and high-level organization of the structural components. For example, cryo-ET of HIV has revealed that the virion is comprised of a protein core containing the viral genome surrounded by an envelope, in which the surface glycoprotein (Env) is embedded (Briggs et al., 2006). In mature virions, the HIV genome is contained within a conical capsid made up of the capsid protein CA arranged into hexamers and pentamers. Cryo-ET and STA have also been used to determine the structure of the capsid within intact mature virions (Mattei et al., 2016). Studies on influenza virions have revealed a capsular or a filamentous shape, with HA covering most of the virion surface and NA clustering in patches (Calder et al., 2010; Harris et al., 2006). A layer of matrix protein, M1, underneath the membrane and eight ribonucleoprotein (RNP) segments were also observed. Interestingly, NA and RNPs occupy opposite poles of the virion (Calder et al., 2010). Like the filamentous form of influenza, EBOV (Bharat et al., 2011) and MARV (Bharat et al., 2012) also present a strikingly filamentous morphology.

The first three-dimensional characterizations of arenavirus and their GP spikes were performed by cryo-ET and STA on University of Helsinki virus (UHV; reptarenavirus) (Hetzel et al., 2013) and on Lassa virus (LASV; mammarenavirus) (Li et al., 2016b). These studies have revealed the higher order assembly of the GP spikes, with each spike consisting of three protomers of GP1–GP2 heterodimers organized into a tripartite complex, distributed randomly over the whole virion surface. The improvement in resolution of the spike complex from 32 Å (UHV) to 14 Å (LASV) has allowed the fitting of a crystal structure of the LASV $(GP1–GP2)_3$ trimeric ectodomain (Hastie et al., 2017). The resulting model places the GP1 receptor-binding glycoproteins to the membrane-distal region of the spike complex, where they sit atop the GP2 class I fusion glycoproteins that protrude from the virion membrane. Inspection of the tomographic slices shows that the spike complex penetrates the membrane and interacts with the underlying layer of the Z matrix protein that links to the genome (Li et al., 2016b). Although association between stable signal peptide (SSP) and GP2 has been reported (Bederka et al., 2014; Shankar et al., 2016), the structure and topology of SSP remain unresolved at 14 Å resolution.

Cryo-ET and STA have also started to reveal the striking differences in GP multimerization and higher level clustering on the viral envelopes, as shown by several studies focusing on bunyaviruses (*Bunyavirales*). The organization of hantavirus GP spikes on the virion has been studied by subtomogram averaging of the TULV (*Hantaviridae*) glycoprotein spikes, followed by placing them back to their correct positions on the virion envelope (Huiskonen et al., 2010; Li et al., 2016a). These studies showed how the tetrameric spikes, consisting of four copies of Gn–Gc heterodimers, cover the envelope surface in locally ordered patches. Improved resolution in later models allowed proposing a model where the globular domains of Gn are membrane distal with a tetrameric stalk that descends into the membrane and the Gc fusion proteins occupy the space between the spikes (Li et al., 2016a). Another prototypic bunyavirus, Bunyamwera virus (BUNV; *Orthobunyaviridae*) has been studied by cryo-ET and STA (Bowden et al., 2013). This study revealed the first low-resolution structure of the trimeric glycoprotein spikes and how these spikes create locally ordered lattices on the virion surface, analogous to TULV (Huiskonen et al., 2010). Nairoviruses are another example of a pleomorphic bunyavirus displaying an ordered lattice of glycoproteins on their envelope. The first structural insights into this lattice came from STA of HAZV, revealing a tetrameric array of spikes on the virion envelope (Punch et al., 2018).

The first representation of measles virus (MeV; *Paramyxoviridae*) ultrastructure (Liljeroos et al., 2011), derived by cryo-ET of purified virions, revealed that the virus particles are highly pleomorphic, heterogeneous in size (50–510 nm) and show no obvious glycoprotein ordering, agreeing with the observations of similar structural investigations of SeV (Loney et al., 2009), NDV (Battisti et al., 2012), and HPIV3 (Gui et al., 2015) virions. Although a recent crystallographic study revealed that the fusion glycoprotein is capable of forming a pseudo-hexameric arrangement (Xu et al., 2015), this organization has yet to be visualized on native virions. Human orthopneumovirus (formerly respiratory syncytial virus [RSV]; family *Pneumoviridae*) has also been studied by cryo-ET and STA (Liljeroos et al., 2013). These virions display a wide range of morphologies, including spherical and filamentous forms of different sizes. Interestingly, a matrix layer of the M protein can only be seen sporadically in spherical particles but is common in the filamentous form. This M layer is thought to provide the membrane curvature required for budding. On the virion surface, both the fusion protein F and the attachment glycoprotein G appear to be randomly distributed. The F protein can be seen in two different conformations, pre fusion and post-fusion, as determined by classification of subvolumes. This study notes that each virion typically has only one conformation of F protein on its surface, but differences can be seen between virions. Another study goes further to suggest that the F protein is exclusively in the prefusion form on filamentous particles but in the postfusion form on spherical particles (Ke et al., 2018a). This observation suggests that the infectious form of the virus is the filamentous form and highlights the need for studying virions in their most native form, free from purification artefacts, that is by cryo-ET of budding sites on the cell surface (see Section 5).

4. Enveloped virus membrane fusion

4.1 Fusion-triggered forms of purified virions

Entry of enveloped viruses into host cells can either occur by direct fusion at the plasma membrane or by endocytic pathways (i.e., macropinocytosis, clathrin-mediated and caveolin-mediated endocytosis), leading to the formation of endocytic vesicles and eventually fusion with endosomes or lysosomes (Yamauchi and Helenius, 2013). The harsh environment of these compartments, including low pH and unusually high or low ionic concentration, often results in significant changes in the morphologies of the virions, particularly the envelope-displayed fusion proteins (Harrison, 2015).

Cryo-EM has become the principal technique to visualize the structural transitions that enveloped viruses and their fusion proteins undergo during fusion. Virions from different viral families have been imaged by cryo-EM in acidic conditions by exposing the purified virions to low pH buffer. The results appear to agree that the fusion proteins, in spite of their class (I, II, III), undergo varying degrees of structural changes. DENV virions with a bound Fab and incubated at pH 5.5 buffer have been used to reconstruct a cryo-EM map at 26 Å resolution (Zhang et al., 2015). This map revealed extended E-protein spikes with consistent shape to that of the E-protein trimer in its postfusion conformation solved by X-ray crystallography earlier (Modis et al., 2004). Averaging of side-views of the spikes at the edge of the virion allowed calculating 2D class averages of spikes (Zhang et al., 2015).

As these experiments were carried out in the absence of target membranes, these results suggested that the observed trimer is a transient prefusion trimeric state of the E fusion protein, stabilized by the bound Fab (Zhang et al., 2015). Similar side-view averaging of BUNV GP spikes from the surface of the virions, imaged at low pH, has suggested that the GP lattice loses some of its contacts at the tips of the spikes mainly formed by the class II fusion GPs (Bowden et al., 2013). Likewise, 2D averages of TULV side views has demonstrated the collapse of its ordered tetrameric lattice under low pH conditions (Rissanen et al., 2017). A recent cryo-ET and STA investigation of the LASV GP spike complex supports the sensitivity of GP1 attachment glycoprotein toward ambient pH (Cohen-Dvashi et al., 2015; Li et al., 2016b; Pryce et al., 2019). Conformational differences can be observed for GP1 structure as the pH drops from neutral (8–7) to endosomal or lysosomal pH (6.5–3.0), ultimately resulting in the shedding of the GP1 and the fusogenic rearrangement of the GP2 fusion protein (Li et al., 2016b).

Not just low pH, but other factors, either physiological or non-physiological, can also be applied to trigger the fusion protein from its prefusion condition. HRPV5 is an enveloped archeal virus with a monomeric envelope fusion protein, which does not confer to any of the existing classes I–III (El Omari et al., 2019). HRPV5 virions have been triggered to their fusogenic conformation by exposure to high temperature (55 °C). This led to a conformational change in their monomeric fusion protein, allowing it to extend far enough to conceivably reach across the host cell S-layer to bridge the virion and host cell membranes (El Omari et al., 2019). Other factors such as the concentration of potassium ions (K^+) has also been shown to be crucial for fusion events. An STA study of HAZV GP has demonstrated that the Gc fusion protein can be triggered by high K^+ concentration into an extended conformation and subsequently embedded into the target membrane of co-purified vesicles (Punch et al., 2018).

4.2 Virus–liposome complexes

Liposomes have been used extensively as mimics of cellular plasma and endosomal membranes in cryo-EM studies of virus fusion. Liposomes serve as an accessible tool as their lipid compositions are well-defined and can be easily manipulated to suit the physiological fusion environment. As compared to a whole cell, the relatively small sizes of liposomes also enable cryo-EM imaging without complicated sample preparations, such as focused ion beam (FIB) milling or cryo-sectioning. The effect of various factors such as pH, temperature, lipid composition, ion concentration, in addition to antibodies or fusion inhibitors, have been investigated using virus–liposome complexes (Calder and Rosenthal, 2016; Chlanda et al., 2016; Halldorsson et al., 2018).

Visualization of influenza virus–liposome fusion events at low pH has been achieved with cryo-ET and STA, showing a progression of sequential events from HA-liposome contact, membrane–membrane contact, full fusion, to redistribution of viral components and contents (Calder and Rosenthal, 2016). The M1 matrix layer has also been observed to undergo a conformational change at low pH prior to dissociating from the viral membrane, allowing membrane deformation and formation of a fusion pore (Fontana and Steven, 2013). In corroboration with crystal structures of HA obtained at neutral (Wilson et al., 1981) and fusion permissive pH (Bullough et al., 1994; Chen et al., 1999), cryo-EM investigations (Calder et al., 2010) have established that HA (class I fusion protein) undergoes a conformational change at low pH, exposing the fusion peptide for membrane fusion. Indeed, cryo-EM snapshots and tomographic slices of influenza virus–liposome interactions at low pH revealed an extended triple-stranded coiled coil HA structure, perpendicular to the membrane, that can be seen inserted into the target membrane. The membranes are pulled towards each other, first creating a dimple and then hemifusion of the two membranes. After the eventual formation of the fusion pore, the HA can be seen in the "foldback" conformation radiating from the membrane contact points (Calder and Rosenthal, 2016). Another cryo-ET study has proposed a hemifusion stalk-independent model of membrane fusion termed the "rupture-insertion" pathway utilized when the cholesterol content in the liposomes is low (Chlanda et al., 2016). Interestingly, the HA fusion peptide of influenza virus has been observed to insert into the viral membrane in *cis* in the absence of a target membrane (Calder et al., 2010; Calder and Rosenthal, 2016; Ruigrok et al., 1986; Skehel et al., 1982).

A recent cryo-ET and STA study of RVFV Gc (class II fusion protein) (Halldorsson et al., 2018) has shed more light on the prefusion conformation of Gc, showing how the hydrophobic fusion loop is protected prior to the fusion event, and how Gc embeds it into a target membrane upon acidification of the environment. Localized reconstructions of RVFV surface GP spikes have allowed their structures to be resolved at sufficiently high resolution (~8 Å) for flexible fitting of Gn and Gc X-ray crystallographic structures. This resulted in a model showing that the Gn glycoprotein shields the fusion loop by associating noncovalently with the Gc glycoprotein in the prefusion state at neutral pH (Halldorsson et al., 2018). Cryo-ET carried out at fusion permissive low pH and in the presence of liposomes showed that the Gn-shield shifts away to expose the fusion loop, allowing extension of the Gc molecule from a kinked, likely metastable conformation to a more straightened intermediate conformation. Extension of the Gc allows it to embed its fusion loops in the target membrane, with the aromatic side chains projected into the hydrophobic region of the lipid bilayer (Halldorsson et al., 2018). This Gn-fusion loop shielding mechanism resembles that of alphaviruses CHIKV (Sun et al., 2013) and SFV (Mancini et al., 2000), where the fusion peptide of the E1 fusion protein is shielded by the E2 receptor-binding protein but contrasts the homotypic shielding observed in E–E interactions of the flaviviruses DENV (Kuhn et al., 2002; Zhang et al., 2003a) and ZIKV (Sirohi et al., 2016).

5. Virus budding

Virus budding is a crucial step in the life cycle and propagation of enveloped viruses. Packaged with the newly synthesized genomic contents and viral proteins, virions escape from the host cell membrane prior to infecting more cells. Electron microscopy images show that budding virions can display different morphologies and often recruit viral matrix protein to the budding site. The assembly and organization of the matrix protein along with the spike GPs, NCs and in some cases also with the help of cellular ESCRT complexes induce a membrane curvature in the host cell membrane and form vesicles that are eventually pinched off to release the viral particles (Chen and Lamb, 2008). Other mechanisms that involve internal viral nonstructural proteins and lipid rafts have also been reported (Bavari et al., 2002; Hyatt et al., 1993; Leser and Lamb, 2005; Ono and Freed, 2001). Cryo-ET of infected cells has been proven to be a useful technique to study the morphologically variable process of virion budding (Bharat et al., 2011; Carlson et al., 2010; Ke et al., 2018a,b).

The matrix protein (M) of MeV has been observed, using whole-cell cryo-ET and STA, to form a well-ordered lattice that lines the inner leaflet of the plasma or virion membrane of MeV-infected cells and released virus particles, respectively (Ke et al., 2018b). The M protein has also been shown to interact with the cytoplasmic or intravirionic tails of the envelope glycoproteins and the helical ribonucleoprotein (RNP) (Ke et al., 2018b). In contrast, purified MeV virions were observed to contain ~30 nm tubular structures composed of M protein tightly coating the inner RNP, but without M protein present at the membrane (Brown et al., 1987; Liljeroos et al., 2011). Matrix oligomerization at the membrane and association on the RNP suggest that the M protein plays an essential role in MeV assembly and budding by coordinating the M–glycoprotein and M–RNP interactions and that this is likely a common assembly mechanism utilized across the *Paramyxoviridae* family. These observations suggest that the methods of virus preparation and purification may result in alterations of the virus particles and subsequently impact our interpretation and understanding of the data (Kiss et al., 2014).

The roles of matrix protein and NC in virus budding have also been studied in filoviruses such as MARV. Cryo-ET and STA investigations of intact MARV virions (Bharat et al., 2011) have revealed that the membrane-associated VP40 matrix protein interacts with other viral proteins such as VP24, VP35, and NC, and that NC displays a strong structural disparity with characteristic "pointed" and "barbed" ends by analogy with actin. The analysis of MARV-infected cells by cryo-ET has shown that the helical NCs associates laterally with the inner leaflet of the host plasma membrane and are subsequently enveloped to form filopodia-like membrane protrusions prior to excision, releasing filamentous particles. The virion VP40 lattice has been observed to undergo structural rearrangement when associated with NC as compared to VP40 in VLP without NC, indicating that the VP40-NC interaction is likely required for NC envelopment and budding. Interestingly, the bullet-shaped rhabdovirus vesicular stomatitis virus (VSV), another member of *Mononegavirales*, appears to utilize different envelopment and budding mechanisms despite the morphological homology of its NC with that of filovirus (Ge et al., 2010). Evidently, the "barbed" ends of the VSV NC bud out first, whereas, all MARV NCs are oriented with their "pointed" ends outwards. This differential budding directionality is, however, reconciled by the authors as the requirement for the first base of the genome (3′) to be synthesized to bud first, placing the absolute RNA directionality a priority over the NC directionality.

A previous study of MARV budding has shown that filamentous particles constituted infectious virions while rounded particles of lower infectivity were released during late rounds of infection (Welsch et al., 2010). This morphological distinction has also been observed in the paramyxovirus RSV using cryo-ET in the presence of a fusion inhibitor to exclude the possibility of observing fusion events (Ke et al., 2018a). The particles are observed in various stages of the budding process including initiation, elongation, and scission. Interestingly, the particles bud in the filamentous form despite spherical forms of the virus having been also reported (Liljeroos et al., 2013). The authors suggest that this provides further evidence that the infectious form of RSV is the filamentous form.

In enveloped virions that lack a matrix protein, such as in the members of *Bunyavirales* order, organization of the spike GPs may become a key contributor to viral budding. For example, hantavirus TULV tetrameric GP forms locally ordered patches as shown by cryo-ET and STA. These patches have been proposed to create membrane curvature and contribute to the budding of hantaviruses (Huiskonen et al., 2010). Interestingly, nairovirus HAZV GP has also been observed to arrange in ordered tetrameric patches (Punch et al., 2018). Analogously, locally ordered patches of GP trimers have been observed also in BUNV, an orthobunyavirus, suggesting that such local order may be a generic driver of budding in pleomorphic bunyavirus virions (Bowden et al., 2013).

For retroviruses, the Gag polyprotein mediates many essential events including membrane binding, virion assembly, and genome packaging. Cryo-ET and STA analysis of native HIV-1 budding sites revealed a consistently continuous lattice of Gag polyprotein with the same organization and structure as seen in released immature virions, indicative that the organization of these particles is determined at their intracellular assembly point (Carlson et al., 2010). Cortical actin filaments were also visualized at these sites, particularly the filopodia-assisted buds, suggesting a role of actin filaments in retrovirus assembly. A lattice lacking the NC-RNA-p6 complex was also observed in some budding sites and virions, indicating that the viral genome was absent, an observation attributed to premature proteolytic maturation and failure to recruit the ESCRT machinery in some HIV-1 infected T-cells. Interestingly, the use of whole-cell cryo-ET permitted the direct visualization of HIV-1 virions and VLPs connected to each other and to the plasma membrane by highly dynamic and filamentous proteinaceous bodies, referred as tethers (Strauss et al., 2016). The localization of tethers at budding sites supported an established restriction model for HIV-1 release and dissemination. Indeed, observation of beads-on-a-string appearance on tomographic images suggested that virion budding had occurred sequentially through a tetherin-enriched microdomain.

6. Neutralization

The increasing number of outbreaks caused by arboviruses, such as DENV, ZIKV and CHIKV, has placed tremendous pressure on scientists worldwide to generate therapeutic agents or vaccines in response to the epidemics (Weaver et al., 2018). Cryo-EM has been applied extensively to study the structures of virion–antibody complexes which have provided insights into the molecular basis of antibody-mediated neutralization mechanisms and the identification of potential therapeutic immunogens.

The cryo-EM structures of DENV in complex with monoclonal antibodies (mAb) at elevated temperatures have provided a wealth of detailed information into neutralization of different DENV serotypes. These structures show that the bumpy form of DENV2 is able to bind an anti-DENV Fab 1A1D-2 (Lok et al., 2008) and E104 (Zhang et al., 2015) only at elevated temperatures. The cryo-EM structure of Fab 1A1D–DENV2 complex demonstrates that the complex formation was temperature dependent with a higher level of Fab binding at 37 °C than at ambient temperature due to the exposure of a hidden part of the Fab binding epitope on the E proteins (Lok et al., 2008). However, it has also been shown that this temperature-dependent transition from smooth to bumpy is strain-dependent. The cryo-EM structure of the DENV4 virion at 4.1-Å resolution suggests that this serotype has higher thermal stability than other DENV strains (Kostyuchenko et al., 2014). Furthermore, unlike other DENV strains, these virions do not expose the fusion loops to allow binding to flavivirus mAbs until the temperature is increased to 40 °C (Sukupolvi-Petty et al., 2013). A DENV2-specific human mAb (2D22) complex cryo-EM structure suggests that HMAb 2D22 neutralizes DENV2 by binding E dimers and prevents the E protein rearrangement required for viral entry (Fibriansah et al., 2015a). The cryo-EM structure of Fab HMAb 5J7-DENV3 complex shows that a single Fab molecule of HMAb 5J7 simultaneously binds to three functionally significant E protein domains, each on a different E protein molecule, to prevent virus attachment (Fibriansah et al., 2015b). The cryo-EM structure of a DENV1–HMAb 1F4 complex has revealed that the HMAb 1F4 binds to DI and DI–DII hinge region on an E protein monomer to prevent different stages of viral entry (Fibriansah et al., 2014) while the DENV1 specific HM14c10 targets an adjacent surface of E protein dimers to neutralize the virus by blocking virus attachment as shown by cryo-EM of the DENV1–HMAb 14c10 complex (Teoh et al., 2012). More recently, cryo-EM has been applied to study two highly neutralizing antibodies (2C8 and 3H5) against DENV E protein and to understand the difference in their capacity to promote antibody dependent enhancement (ADE) (Renner et al., 2018; Wirawan et al., 2019). It was proposed that antibody 3H5 promoted minimal ADE by its unusual binding to the viral surface resulting in poor access to the Fc region. Another study looked at the role of anti-prM antibodies in the pathogenesis of immature DENV by the cryo-EM structures of the immature DENV in complex with a Fab fragment of HMAb 1H10, at different pH values. This suggested the mechanism by which the Fab 1H10 enhanced attachment of immature DENV to liposomes by increasing dissociation of prM from E (Wirawan et al., 2019).

Recent epidemics of ZIKV in the Americas have accelerated structural studies into the neutralization mechanism of ZIKV virion–mAb complexes (Cauchemez et al., 2016; Fauci and Morens, 2016; Mlakar et al., 2016). The first cryo-EM structure of a ZIKV–mAb complex used a cross-reactive DENV human mAb (C10) which inhibited viral fusion to acidified endosomes by locking of the E proteins (Zhang et al., 2016). This was followed by the 9-Å resolution cryo-EM complex structure of ZIKV and the ZIKV specific human mAb (Fab Z23) which demonstrated the binding of Fab Z23 to DIII domain of the E protein (Wang et al., 2016b). Another study showed that the neutralizing human mAb (ZIKV-117) cross-links monomers with E dimers as well as between neighboring E dimers to prevent the conformational changes of E dimers into fusogenic trimers in acidified endosomes (Hasan et al., 2017). More recently, a 4-Å resolution cryo-EM structure of ZIKV in complex with Fab fragments of the highly specific human mAb (ZIKV-195) suggested that Fab ZIKV-195 neutralizes ZIKV by binding to two adjacent E dimers to prevent the structural reorganization to trimers required for membrane fusion (Long et al., 2019). Cryo-EM has been applied also on other flavivirus–mAb complexes. TBEV, in complex with the neutralizing mouse mAb Fab fragment (19/1786) has been resolved to a resolution of 3.9 Å revealing that this Fab neutralizes by inhibiting virus-induced membrane fusion (Füzik et al., 2018). The 14.5 Å resolution cryo-EM structure of WNV in complex with the Fab of the strongly neutralizing antibody E16 has revealed that it binds to DIII and neutralizes by preventing the conformational change of E prior to membrane fusion (Kaufmann et al., 2006).

Many alphavirus neutralizing antibodies in complex with virions have been studied by cryo-EM. These studies have shown that most of these neutralizing mAbs target the exposed E2 protein cap (Fox et al., 2015; Hasan et al., 2018b; Long et al., 2015; Porta et al., 2016; Sun et al., 2013). One of the first studies produced four ~15-Å resolution cryo-EM structures of CHIKV VLPs complexed with the Fab fragments of neutralizing mouse MAbs (Sun et al., 2013). This study shows that the CHK-152 mAb neutralizes by stabilizing the viral surface and preventing the exposure of fusion-loop and thereby the fusion. The three mAbs CHK-9, m10 and m242 antibodies have been suggested to block the receptor-attachment site (Sun et al., 2013). More recently, two cryo-EM studies of CHIKV VLPs complexed with Fabs of highly neutralizing human mAbs have been described providing insights into the structural basis of neutralization (Long et al., 2015; Porta et al., 2016). Furthermore, it has been suggested that the Fabs 4J21 and 5M16 bind to domains B on E2 blocking virus fusion (Long et al., 2015). The Fab 8B10 inhibits attachment by covering the receptor binding site whereas 5F10 inhibits fusion by restricting B domain movement (Porta et al., 2016). Another cryo-EM study of Fab fragments of two human mAbs has revealed a mechanism for inhibiting membrane fusion and also identified E2-W64 as a key neutralizing epitope of CHIKV E protein (Jin et al., 2015). The 16-Å resolution cryo-EM structure of CHIKV in complex with CHK-265 Fab fragments has shown that the Fab binds to domain B of E2 and induces significant conformational changes in domain A and cross-links adjacent E2 spikes to block viral entry and egress steps (Fox et al., 2015). The cryo-EM study of another alphavirus, VEEV, in complex with the highly neutralizing mAb Fab fragments (F5 and 3B4C-4) at resolution of ~17.5 Å has revealed the difference in their neutralization mechanisms. While both Fab fragments neutralize by stabilizing E2 trimeric spikes and preventing the viral fusion, the F5 cross-links E2 within a trimeric spike to block the receptor binding site, whereas the 3B4C-4 Fab cross links E2 from neighboring spikes to prevent the exposure of the fusion loop by steric hindrance (Porta et al., 2014). Recent cryo-EM studies looking at the SINV-EEEV chimera in complex with five different Fabs have revealed that three Fab fragments of neutralizing mAbs (EEEV-5, >EEEV-42 and EEEV-58) bound to domain A of E2 protein, which cause intraspike cross-linking, while the other two (EEEV-3 and EEEV-69) interact only with domain B favoring interspike cross-linking (Hasan et al., 2018b).

In addition to extensive cryo-EM studies of the arboviruses belonging to *Flaviviridae* and *Togaviridae* families described above, cryo-EM has been applied to study the structures of virion–antibody complexes for several viruses from other families, notably HIV and EBOV. STA has been used to study the HIV Env trimer on native virions in complex with a broadly neutralizing antibody Fab fragment, b12, and the cell surface receptor, CD4 (Liu et al., 2008). This study shows that both the Fab and the receptor cause a large conformational change that "opens" the trimer, but this change is more dramatic with the receptor CD4. It has therefore been proposed that the b12 antibody exerts its neutralizating effect by locking Env in a confirmation that prevents a further CD4 induced conformational change that is required to continue the virus life cycle. To understand the structural basis of ZMapp (a cocktail of 3 mAbs: c24G4, c4G7 and C13C6), a promising anti-EBOV therapeutic, cryo-EM structures of c2G4, c4G7 and c13C6 IgGs bound to EBOV VLPs have been studied by STA (Tran et al., 2016). This study suggests that two of these antibodies (c2G4 and c4G7) remain bound through the endocytic pathway and exert their neutralizing effect by preventing conformational changes required for fusion.

7. Conclusions

Cryo-EM has always been the method of choice to study the structures of enveloped virions. Due to technological developments in both microscope hardware and image processing software, the rate of discoveries has been accelerating in the past 5 years. Structures of regular enveloped virions whose protein shell(s) follow icosahedral symmetry can now be determined in a matter of days by SPA. Pleomorphic virions and virions in complex with receptors or antibodies remain a challenge for structure determination. Those virions or complexes that show slight deviations from perfect icosahedral symmetry may be tractable by improved computational 2D averaging approaches that deal with structural deformations accurately. When deviations are more severe or when the virion structure is truly pleomorphic or filamentous, STA remains the method of choice and is becoming more widely utilized. In the near future, more studies are expected to utilize various hybrid data processing approaches to tackle the most challenging targets for which resolution has remained limited or which have remained entirely unattainable. The 2D and 3D averaging workflows depicted in Fig. 2 will likely merge into one framework within the same software package in the future. This should facilitate routinely resolving structural components in pleomorphic, enveloped virions to near-atomic resolution, and to further address mechanisms of virus infection and neutralization of enveloped viruses in general.

References

Abrescia, N.G.A., Cockburn, J.J.B., Grimes, J.M., Sutton, G.C., Diprose, J.M., Butcher, S.J., Fuller, S.D., San Martín, C., Burnett, R.M., Stuart, D.I., et al., 2004. Insights into assembly from structural analysis of bacteriophage PRD1. Nature 432, 68–74.

Abrescia, N.G.A., Grimes, J.M., Kivelä, H.M., Assenberg, R., Sutton, G.C., Butcher, S.J., Bamford, J.K.H., Bamford, D.H., Stuart, D.I., 2008. Insights into virus evolution and membrane biogenesis from the structure of the marine lipid-containing bacteriophage PM2. Mol. Cell 31, 749–761.

Allison, S.L., Schalich, J., Stiasny, K., Mandl, C.W., Kunz, C., Heinz, F.X., 1995. Oligomeric rearrangement of tick-borne encephalitis virus envelope proteins induced by an acidic pH. J. Virol. 69, 695–700.

Bai, X.-C., Rajendra, E., Yang, G., Shi, Y., Scheres, S.H., 2015. Sampling the conformational space of the catalytic subunit of human γ-secretase. Elife 4, e11182.

Battisti, A.J., Chu, Y.K., Chipman, P.R., Kaufmann, B., Jonsson, C.B., Rossmann, M.G., 2010. Structural studies of Hantaan virus. J. Virol. 85, 835–841.

Battisti, A.J., Chu, Y.-K., Chipman, P.R., Kaufmann, B., Jonsson, C.B., Rossmann, M.G., 2011. Structural studies of Hantaan virus. J. Virol. 85, 835–841.

Battisti, A.J., Meng, G., Winkler, D.C., McGinnes, L.W., Plevka, P., Steven, A.C., Morrison, T.G., Rossmann, M.G., 2012. Structure and assembly of a paramyxovirus matrix protein. Proc. Natl. Acad. Sci. U. S. A. 109, 13996–14000.

Bavari, S., Bosio, C.M., Wiegand, E., Ruthel, G., Will, A.B., Geisbert, T.W., Hevey, M., Schmaljohn, C., Schmaljohn, A., Aman, M.J., 2002. Lipid raft microdomains: a gateway for compartmentalized trafficking of Ebola and Marburg viruses. J. Exp. Med. 195, 593–602.

Bederka, L.H., Bonhomme, C.J., Ling, E.L., Buchmeier, M.J., 2014. Arenavirus stable signal peptide is the keystone subunit for glycoprotein complex organization. MBio 5, e02063.

Beniac, D.R., Booth, T.F., 2017. Structure of the Ebola virus glycoprotein spike within the virion envelope at 11 Å resolution. Sci. Rep. 7, 46374.

Bharat, T.A.M., Riches, J.D., Kolesnikova, L., Welsch, S., Krähling, V., Davey, N., Parsy, M.-L., Becker, S., Briggs, J.A.G., 2011. Cryo-electron tomography of Marburg virus particles and their morphogenesis within infected cells. PLoS Biol. 9, e1001196.

Bharat, T.A.M., Noda, T., Riches, J.D., Kraehling, V., Kolesnikova, L., Becker, S., Kawaoka, Y., Briggs, J.A.G., 2012. Structural dissection of Ebola virus and its assembly determinants using cryo-electron tomography. Proc. Natl. Acad. Sci. U. S. A. 109, 4275–4280.

Bowden, T.A., Bitto, D., McLees, A., Yeromonahos, C., Elliott, R.M., Huiskonen, J.T., 2013. Orthobunyavirus ultrastructure and the curious tripodal glycoprotein spike. PLoS Pathog. 9, e1003374.

Bressanelli, S., Stiasny, K., Allison, S.L., Stura, E.A., Duquerroy, S., Lescar, J., Heinz, F.X., Rey, F.A., 2004. Structure of a flavivirus envelope glycoprotein in its low-pH-induced membrane fusion conformation. EMBO J. 23, 728–738.

Briggs, J.A.G., Wilk, T., Welker, R., Kräusslich, H.-G., Fuller, S.D., 2003. Structural organization of authentic, mature HIV-1 virions and cores. EMBO J. 22, 1707–1715.

Briggs, J.A.G., Grünewald, K., Glass, B., Förster, F., Kräusslich, H.-G., Fuller, S.D., 2006. The mechanism of HIV-1 core assembly: insights from three-dimensional reconstructions of authentic virions. Structure 14, 15–20.

Brown, H.R., Goller, N., Thormar, H., Norrby, E., 1987. Fuzzy material surrounding measles virus nucleocapsids identified as matrix protein. Brief report. Arch. Virol. 94, 163–168.

Bullough, P.A., Hughson, F.M., Skehel, J.J., Wiley, D.C., 1994. Structure of influenza haemagglutinin at the pH of membrane fusion. Nature 371, 37–43.

Calder, L.J., Rosenthal, P.B., 2016. Cryomicroscopy provides structural snapshots of influenza virus membrane fusion. Nat. Struct. Mol. Biol. 23, 853–858.

Calder, L.J., Wasilewski, S., Berriman, J.A., Rosenthal, P.B., 2010. Structural organization of a filamentous influenza A virus. Proc. Natl. Acad. Sci. U. S. A. 107, 10685–10690.

Carlson, L.-A., de Marco, A., Oberwinkler, H., Habermann, A., Briggs, J.A.G., Kräusslich, H.-G., Grünewald, K., 2010. Cryo electron tomography of native HIV-1 budding sites. PLoS Pathog. 6, e1001173.

Castaño-Díez, D., Kudryashev, M., Stahlberg, H., 2017. Dynamo catalogue: geometrical tools and data management for particle picking in subtomogram averaging of cryo-electron tomograms. J. Struct. Biol. 197, 135–144.
Cauchemez, S., Besnard, M., Bompard, P., Dub, T., Guillemette-Artur, P., Eyrolle-Guignot, D., Salje, H., Van Kerkhove, M.D., Abadie, V., Garel, C., et al., 2016. Association between Zika virus and microcephaly in French Polynesia, 2013–15: a retrospective study. Lancet 387, 2125–2132.
Chen, B.J., Lamb, R.A., 2008. Mechanisms for enveloped virus budding: can some viruses do without an ESCRT? Virology 372, 221–232.
Chen, J., Skehel, J.J., Wiley, D.C., 1999. N- and C-terminal residues combine in the fusion-pH influenza hemagglutinin HA(2) subunit to form an N cap that terminates the triple-stranded coiled coil. Proc. Natl. Acad. Sci. U. S. A. 96, 8967–8972.
Chen, L., Wang, M., Zhu, D., Sun, Z., Ma, J., Wang, J., Kong, L., Wang, S., Liu, Z., Wei, L., et al., 2018. Implication for alphavirus host-cell entry and assembly indicated by a 3.5 Å resolution cryo-EM structure. Nat. Commun. 9, 5326.
Chlanda, P., Mekhedov, E., Waters, H., Schwartz, C.L., Fischer, E.R., Ryham, R.J., Cohen, F.S., Blank, P.S., Zimmerberg, J., 2016. The hemifusion structure induced by influenza virus haemagglutinin is determined by physical properties of the target membranes. Nat. Microbiol. 1, 16050.
Cohen-Dvashi, H., Cohen, N., Israeli, H., Diskin, R., 2015. Molecular mechanism for LAMP1 recognition by Lassa virus. J. Virol. 89, 7584–7592.
Colibus, L.D., Roine, E., Walter, T.S., Ilca, S.L., Wang, X., Wang, N., Roseman, A.M., Bamford, D., Huiskonen, J.T., Stuart, D.I., 2019. Assembly of complex viruses exemplified by a halophilic euryarchaeal virus. Nat. Commun. 10, 1456.
Cyrklaff, M., Risco, C., Fernández, J.J., Jiménez, M.V., Estéban, M., Baumeister, W., Carrascosa, J.L., 2005. Cryo-electron tomography of vaccinia virus. Proc. Natl. Acad. Sci. U. S. A. 102, 2772–2777.
Dryden, K.A., Wieland, S.F., Whitten-Bauer, C., Gerin, J.L., Chisari, F.V., Yeager, M., 2006. Native hepatitis B virions and capsids visualized by electron cryomicroscopy. Mol. Cell 22, 843–850.
Effantin, G., Estrozi, L.F., Aschman, N., Renesto, P., Stanke, N., Lindemann, D., Schoehn, G., Weissenhorn, W., 2016. Cryo-electron microscopy structure of the native prototype foamy virus glycoprotein and virus architecture. PLoS Pathog. 12, e1005721.
El Omari, K., Li, S., Kotecha, A., Walter, T.S., Bignon, E.A., Harlos, K., Somerharju, P., De Haas, F., Clare, D.K., Molin, M., et al., 2019. The structure of a prokaryotic viral envelope protein expands the landscape of membrane fusion proteins. Nat. Commun. 10, 846.
Fauci, A.S., Morens, D.M., 2016. Zika virus in the Americas—yet another arbovirus threat. N. Engl. J. Med. 374, 601–604.
Fibriansah, G., Ng, T.-S., Kostyuchenko, V.A., Lee, J., Lee, S., Wang, J., Lok, S.-M., 2013. Structural changes in dengue virus when exposed to a temperature of 37 °C. J. Virol. 87, 7585–7592.
Fibriansah, G., Tan, J.L., Smith, S.A., de Alwis, A.R., Ng, T.-S., Kostyuchenko, V.A., Ibarra, K.D., Wang, J., Harris, E., de Silva, A., et al., 2014. A potent anti-dengue human antibody preferentially recognizes the conformation of E protein monomers assembled on the virus surface. EMBO Mol. Med. 6, 358–371.
Fibriansah, G., Ibarra, K.D., Ng, T.-S., Smith, S.A., Tan, J.L., Lim, X.-N., Ooi, J.S.G., Kostyuchenko, V.A., Wang, J., de Silva, A.M., et al., 2015a. Dengue virus. Cryo-EM structure of an antibody that neutralizes dengue virus type 2 by locking E protein dimers. Science 349, 88–91.

Fibriansah, G., Tan, J.L., Smith, S.A., de Alwis, R., Ng, T.-S., Kostyuchenko, V.A., Jadi, R.S., Kukkaro, P., de Silva, A.M., Crowe, J.E., et al., 2015b. A highly potent human antibody neutralizes dengue virus serotype 3 by binding across three surface proteins. Nat. Commun. 6, 6341.

Fontana, J., Steven, A.C., 2013. At low pH, influenza virus matrix protein M1 undergoes a conformational change prior to dissociating from the membrane. J. Virol. 87, 5621–5628.

Förster, F., Medalia, O., Zauberman, N., Baumeister, W., Fass, D., 2005. Retrovirus envelope protein complex structure in situ studied by cryo-electron tomography. Proc. Natl. Acad. Sci. U. S. A. 102, 4729–4734.

Fox, J.M., Long, F., Edeling, M.A., Lin, H., van Duijl-Richter, M.K.S., Fong, R.H., Kahle, K.M., Smit, J.M., Jin, J., Simmons, G., et al., 2015. Broadly neutralizing alphavirus antibodies bind an epitope on E2 and inhibit entry and egress. Cell 163, 1095–1107.

Fuller, S.D., Berriman, J.A., Butcher, S.J., Gowen, B.E., 1995. Low pH induces swiveling of the glycoprotein heterodimers in the Semliki forest virus spike complex. Cell 81, 715–725.

Füzik, T., Formanová, P., Růžek, D., Yoshii, K., Niedrig, M., Plevka, P., 2018. Structure of tick-borne encephalitis virus and its neutralization by a monoclonal antibody. Nat. Commun. 9, 436.

Ge, P., Tsao, J., Schein, S., Green, T.J., Luo, M., Zhou, Z.H., 2010. Cryo-EM model of the bullet-shaped vesicular stomatitis virus. Science 327, 689–693.

Grünewald, K., Desai, P., Winkler, D., Heymann, J., Belnap, D., Baumeister, W., Steven, A., 2003. Three-dimensional structure of herpes simplex virus from cryo-electron tomography. Science 302, 1396–1398.

Gui, L., Jurgens, E.M., Ebner, J.L., Porotto, M., Moscona, A., Lee, K.K., 2015. Electron tomography imaging of surface glycoproteins on human parainfluenza virus 3: association of receptor binding and fusion proteins before receptor engagement. MBio 6, e02393-14.

Guichard, P., Krell, T., Chevalier, M., Vaysse, C., Adam, O., Ronzon, F., Marco, S., 2011. Three dimensional morphology of rabies virus studied by cryo-electron tomography. J. Struct. Biol. 176, 32–40.

Halldorsson, S., Li, S., Li, M., Harlos, K., Bowden, T.A., Huiskonen, J.T., 2018. Shielding and activation of a viral membrane fusion protein. Nat. Commun. 9, 349.

Happonen, L.J., Redder, P., Peng, X., Reigstad, L.J., Prangishvili, D., Butcher, S.J., 2010. Familial relationships in hyperthermo- and acidophilic archaeal viruses. J. Virol. 84, 4747–4754.

Harris, A., Cardone, G., Winkler, D.C., Heymann, J.B., Brecher, M., White, J.M., Steven, A.C., 2006. Influenza virus pleiomorphy characterized by cryoelectron tomography. Proc. Natl. Acad. Sci. U. S. A. 103, 19123–19127.

Harrison, S.C., 2015. Viral membrane fusion. Virology 479–480, 498–507.

Hasan, S.S., Saif Hasan, S., Miller, A., Sapparapu, G., Fernandez, E., Klose, T., Long, F., Fokine, A., Porta, J.C., Jiang, W., et al., 2017. A human antibody against Zika virus crosslinks the E protein to prevent infection. Nat. Commun. 8, 14722.

Hasan, S.S., Sevvana, M., Kuhn, R.J., Rossmann, M.G., 2018a. Structural biology of Zika virus and other flaviviruses. Nat. Struct. Mol. Biol. 25, 13–20.

Hasan, S.S., Sun, C., Kim, A.S., Watanabe, Y., Chen, C.-L., Klose, T., Buda, G., Crispin, M., Diamond, M.S., Klimstra, W.B., et al., 2018b. Cryo-EM structures of eastern equine encephalitis virus reveal mechanisms of virus disassembly and antibody neutralization. Cell Rep. 25, 3136–3147.e5.

Hastie, K.M., Zandonatti, M.A., Kleinfelter, L.M., Heinrich, M.L., Rowland, M.M., Chandran, K., Branco, L.M., Robinson, J.E., Garry, R.F., Saphire, E.O., 2017. Structural basis for antibody-mediated neutralization of Lassa virus. Science 356, 923–928.

Hetzel, U., Sironen, T., Laurinmäki, P., Liljeroos, L., Patjas, A., Henttonen, H., Vaheri, A., Artelt, A., Kipar, A., Butcher, S.J., et al., 2013. Isolation, identification, and characterization of novel arenaviruses, the etiological agents of boid inclusion body disease. J. Virol. 87, 10918–10935.

Huiskonen, J.T., 2018. Image processing for cryogenic transmission electron microscopy of symmetry-mismatched complexes. Biosci. Rep. 38, BSR20170203.

Huiskonen, J.T., Kivelä, H.M., Bamford, D.H., Butcher, S.J., 2004. The PM2 virion has a novel organization with an internal membrane and pentameric receptor binding spikes. Nat. Struct. Mol. Biol. 11, 850–856.

Huiskonen, J.T., Jäälinoja, H.T., Briggs, J.A.G., Fuller, S.D., Butcher, S.J., 2007. Structure of a hexameric RNA packaging motor in a viral polymerase complex. J. Struct. Biol. 158, 156–164.

Huiskonen, J.T., Overby, A.K., Weber, F., Grünewald, K., 2009. Electron cryo-microscopy and single-particle averaging of Rift Valley fever virus: evidence for GN-GC glycoprotein heterodimers. J. Virol. 83, 3762–3769.

Huiskonen, J.T., Hepojoki, J., Laurinmäki, P., Vaheri, A., Lankinen, H., Butcher, S.J., Grünewald, K., 2010. Electron cryotomography of Tula hantavirus suggests a unique assembly paradigm for enveloped viruses. J. Virol. 84, 4889–4897.

Huiskonen, J.T., Parsy, M.-L., Li, S., Bitto, D., Renner, M., Bowden, T.A., 2014. Averaging of viral envelope glycoprotein spikes from electron cryotomography reconstructions using Jsubtomo. J. Vis. Exp. 92, e51714.

Hyatt, A.D., Zhao, Y., Roy, P., 1993. Release of bluetongue virus-like particles from insect cells is mediated by BTV nonstructural protein NS3/NS3A. Virology 193, 592–603.

Ilca, S.L., Kotecha, A., Sun, X., Poranen, M.M., Stuart, D.I., Huiskonen, J.T., 2015. Localized reconstruction of subunits from electron cryomicroscopy images of macromolecular complexes. Nat. Commun. 6, 8843.

Jäälinoja, H.T., Huiskonen, J.T., Butcher, S.J., 2007a. Electron cryomicroscopy comparison of the architectures of the enveloped bacteriophages phi6 and phi8. Structure 15, 157–167.

Jäälinoja, H.T., Huiskonen, J.T., Butcher, S.J., 2007b. Electron cryomicroscopy comparison of the architectures of the enveloped bacteriophages ϕ6 and ϕ8. Structure 15, 157–167.

Jin, J., Liss, N.M., Chen, D.-H., Liao, M., Fox, J.M., Shimak, R.M., Fong, R.H., Chafets, D., Bakkour, S., Keating, S., et al., 2015. Neutralizing monoclonal antibodies block Chikungunya virus entry and release by targeting an epitope critical to viral pathogenesis. Cell Rep. 13, 2553–2564.

Kaufmann, B., Nybakken, G.E., Chipman, P.R., Zhang, W., Diamond, M.S., Fremont, D.H., Kuhn, R.J., Rossmann, M.G., 2006. West Nile virus in complex with the Fab fragment of a neutralizing monoclonal antibody. Proc. Natl. Acad. Sci. U. S. A. 103, 12400–12404.

Ke, Z., Strauss, J.D., Hampton, C.M., Brindley, M.A., Dillard, R.S., Leon, F., Lamb, K.M., Plemper, R.K., Wright, E.R., 2018a. Promotion of virus assembly and organization by the measles virus matrix protein. Nat. Commun. 9, 1736.

Ke, Z., Dillard, R.S., Chirkova, T., Leon, F., Stobart, C.C., Hampton, C.M., Strauss, J.D., Rajan, D., Rostad, C.A., Taylor, J.V., et al., 2018b. The morphology and assembly of respiratory syncytial virus revealed by cryo-electron tomography. Viruses 10, 446.

Khayat, R., Fu, C.-Y., Ortmann, A.C., Young, M.J., Johnson, J.E., 2010. The architecture and chemical stability of the archaeal Sulfolobus turreted icosahedral virus. J. Virol. 84, 9575–9583.

Kielian, M., 2014. Mechanisms of virus membrane fusion proteins. Annu. Rev. Virol. 1, 171–189.

Kiss, G., Chen, X., Brindley, M.A., Campbell, P., Afonso, C.L., Ke, Z., Holl, J.M., Guerrero-Ferreira, R.C., Byrd-Leotis, L.A., Steel, J., et al., 2014. Capturing enveloped viruses on affinity grids for downstream cryo-electron microscopy applications. Microsc. Microanal. 20, 164–174.

Klose, T., Kuznetsov, Y.G., Xiao, C., Sun, S., McPherson, A., Rossmann, M.G., 2010. The three-dimensional structure of Mimivirus. Intervirology 53, 268–273.
Kostyuchenko, V.A., Chew, P.L., Ng, T.-S., Lok, S.-M., 2014. Near-atomic resolution cryo-electron microscopic structure of dengue serotype 4 virus. J. Virol. 88, 477–482.
Kostyuchenko, V.A., Lim, E.X.Y., Zhang, S., Fibriansah, G., Ng, T.-S., Ooi, J.S.G., Shi, J., Lok, S.-M., 2016. Structure of the thermally stable Zika virus. Nature 533, 425–428.
Kühlbrandt, W., 2014. Biochemistry. The resolution revolution. Science 343, 1443–1444.
Kuhn, R.J., Zhang, W., Rossmann, M.G., Pletnev, S.V., Corver, J., Lenches, E., Jones, C.T., Mukhopadhyay, S., Chipman, P.R., Strauss, E.G., et al., 2002. Structure of dengue virus: implications for flavivirus organization, maturation, and fusion. Cell 108, 717–725.
Laanto, E., Mäntynen, S., De Colibus, L., Marjakangas, J., Gillum, A., Stuart, D.I., Ravantti, J.J., Huiskonen, J.T., Sundberg, L.-R., 2017. Virus found in a boreal lake links ssDNA and dsDNA viruses. Proc. Natl. Acad. Sci. U. S. A. 114, 8378–8383.
Laurinmäki, P.A., Huiskonen, J.T., Bamford, D.H., Butcher, S.J., 2005. Membrane proteins modulate the bilayer curvature in the bacterial virus Bam35. Structure 13, 1819–1828.
Leser, G.P., Lamb, R.A., 2005. Influenza virus assembly and budding in raft-derived microdomains: a quantitative analysis of the surface distribution of HA, NA and M2 proteins. Virology 342, 215–227.
Li, S., Sun, Z., Pryce, R., Parsy, M.-L., Fehling, S.K., Schlie, K., Siebert, C.A., Garten, W., Bowden, T.A., Strecker, T., et al., 2016a. Acidic pH-Induced conformations and LAMP1 binding of the Lassa virus glycoprotein spike. PLoS Pathog. 12, e1005418.
Li, S., Rissanen, I., Zeltina, A., Hepojoki, J., Raghwani, J., Harlos, K., Pybus, O.G., Huiskonen, J.T., Bowden, T.A., 2016b. A molecular-level account of the antigenic hantaviral surface. Cell Rep. 15, 959–967.
Liljeroos, L., Huiskonen, J.T., Ora, A., Susi, P., Butcher, S.J., 2011. Electron cryotomography of measles virus reveals how matrix protein coats the ribonucleocapsid within intact virions. Proc. Natl. Acad. Sci. U. S. A. 108, 18085–18090.
Liljeroos, L., Krzyzaniak, M.A., Helenius, A., Butcher, S.J., 2013. Architecture of respiratory syncytial virus revealed by electron cryotomography. Proc. Natl. Acad. Sci. U. S. A. 110, 11133–11138.
Liu, J., Bartesaghi, A., Borgnia, M.J., Sapiro, G., Subramaniam, S., 2008. Molecular architecture of native HIV-1 gp120 trimers. Nature 455, 109–113.
Lok, S.-M., Kostyuchenko, V., Nybakken, G.E., Holdaway, H.A., Battisti, A.J., Sukupolvi-Petty, S., Sedlak, D., Fremont, D.H., Chipman, P.R., Roehrig, J.T., et al., 2008. Binding of a neutralizing antibody to dengue virus alters the arrangement of surface glycoproteins. Nat. Struct. Mol. Biol. 15, 312–317.
Loney, C., Mottet-Osman, G., Roux, L., Bhella, D., 2009. Paramyxovirus ultrastructure and genome packaging: cryo-electron tomography of Sendai virus. J. Virol. 83, 8191–8197.
Long, F., Fong, R.H., Austin, S.K., Chen, Z., Klose, T., Fokine, A., Liu, Y., Porta, J., Sapparapu, G., Akahata, W., et al., 2015. Cryo-EM structures elucidate neutralizing mechanisms of anti-chikungunya human monoclonal antibodies with therapeutic activity. Proc. Natl. Acad. Sci. U. S. A. 112, 13898–13903.
Long, F., Doyle, M., Fernandez, E., Miller, A.S., Klose, T., Sevvana, M., Bryan, A., Davidson, E., Doranz, B.J., Kuhn, R.J., et al., 2019. Structural basis of a potent human monoclonal antibody against Zika virus targeting a quaternary epitope. Proc. Natl. Acad. Sci. U. S. A. 116, 1591–1596.
Mancini, E.J., Clarke, M., Gowen, B.E., Rutten, T., Fuller, S.D., 2000. Cryo-electron microscopy reveals the functional organization of an enveloped virus, Semliki Forest virus. Mol. Cell 5, 255–266.

Mattei, S., Glass, B., Hagen, W.J.H., Kräusslich, H.-G., Briggs, J.A.G., 2016. The structure and flexibility of conical HIV-1 capsids determined within intact virions. Science 354, 1434–1437.
Mlakar, J., Korva, M., Tul, N., Popović, M., Poljšak-Prijatelj, M., Mraz, J., Kolenc, M., Resman Rus, K., Vesnaver Vipotnik, T., Fabjan Vodušek, V., et al., 2016. Zika virus associated with microcephaly. N. Engl. J. Med. 374, 951–958.
Modis, Y., Ogata, S., Clements, D., Harrison, S.C., 2004. Structure of the Dengue virus envelope protein after membrane fusion. Nature 427, 313–319.
Neuman, B.W., Adair, B.D., Yoshioka, C., Quispe, J.D., Orca, G., Kuhn, P., Milligan, R.A., Yeager, M., Buchmeier, M.J., 2006. Supramolecular architecture of severe acute respiratory syndrome coronavirus revealed by electron cryomicroscopy. J. Virol. 80, 7918–7928.
Ono, A., Freed, E.O., 2001. Plasma membrane rafts play a critical role in HIV-1 assembly and release. Proc. Natl. Acad. Sci. U. S. A. 98, 13925–13930.
Överby, A.K., Pettersson, R.F., Grünewald, K., Huiskonen, J.T., 2008. Insights into bunyavirus architecture from electron cryotomography of Uukuniemi virus. Proc. Natl. Acad. Sci. U. S. A. 105, 2375–2379.
Park, G.W., Linden, K.G., Sobsey, M.D., 2011. Inactivation of murine norovirus, feline calicivirus and echovirus 12 as surrogates for human norovirus (NoV) and coliphage (F) MS2 by ultraviolet light (254 nm) and the effect of cell association on UV inactivation. Lett. Appl. Microbiol. *52*, 162–167.
Porta, J., Jose, J., Roehrig, J.T., Blair, C.D., Kuhn, R.J., Rossmann, M.G., 2014. Locking and blocking the viral landscape of an alphavirus with neutralizing antibodies. J. Virol. 88, 9616–9623.
Porta, J., Mangala Prasad, V., Wang, C.-I., Akahata, W., Ng, L.F.P., Rossmann, M.G., 2016. Structural studies of Chikungunya virus-like particles complexed with human antibodies: neutralization and cell-to-cell transmission. J. Virol. 90, 1169–1177.
Prasad, V.M., Miller, A.S., Klose, T., Sirohi, D., Buda, G., Jiang, W., Kuhn, R.J., Rossmann, M.G., 2017. Structure of the immature Zika virus at 9 Å resolution. Nat. Struct. Mol. Biol. 24, 184–186.
Pryce, R., Ng, W.M., Zeltina, A., Watanabe, Y., El Omari, K., Wagner, A., Bowden, T.A., 2019. Structure-based classification defines the discrete conformational classes adopted by the arenaviral GP1. J. Virol. 93, e01048-18.
Punch, E.K., Hover, S., Blest, H.T.W., Fuller, J., Hewson, R., Fontana, J., Mankouri, J., Barr, J.N., 2018. Potassium is a trigger for conformational change in the fusion spike of an enveloped RNA virus. J. Biol. Chem. 293, 9937–9944.
Renner, M., Flanagan, A., Dejnirattisai, W., Puttikhunt, C., Kasinrerk, W., Supasa, P., Wongwiwat, W., Chawansuntati, K., Duangchinda, T., Cowper, A., et al., 2018. Characterization of a potent and highly unusual minimally enhancing antibody directed against dengue virus. Nat. Immunol. 19, 1248–1256.
Rey, F.A., Heinz, F.X., Mandl, C., Kunz, C., Harrison, S.C., 1995. The envelope glycoprotein from tick-borne encephalitis virus at 2 A resolution. Nature 375, 291–298.
Rissanen, I., Stass, R., Zeltina, A., Li, S., Hepojoki, J., Harlos, K., Gilbert, R.J.C., Huiskonen, J.T., Bowden, T.A., 2017. Structural transitions of the conserved and metastable hantaviral glycoprotein envelope. J. Virol. 91, e00378-17.
Ruigrok, R.W., Wrigley, N.G., Calder, L.J., Cusack, S., Wharton, S.A., Brown, E.B., Skehel, J.J., 1986. Electron microscopy of the low pH structure of influenza virus haemagglutinin. EMBO J. 5, 41–49.
San Martín, C., Huiskonen, J.T., Bamford, J.K.H., Butcher, S.J., Fuller, S.D., Bamford, D.H., Burnett, R.M., 2002. Minor proteins, mobile arms and membrane-capsid interactions in the bacteriophage PRD1 capsid. Nat. Struct. Biol. 9, 756–763.
Sevvana, M., Long, F., Miller, A.S., Klose, T., Buda, G., Sun, L., Kuhn, R.J., Rossmann, M.G., 2018. Refinement and analysis of the mature Zika virus Cryo-EM structure at 3.1 Å resolution. Structure 26, 1169–1177.e3.

Shankar, S., Whitby, L.R., Casquilho-Gray, H.E., York, J., Boger, D.L., Nunberg, J.H., 2016. Small-molecule fusion inhibitors bind the pH-sensing stable signal peptide-GP2 subunit interface of the Lassa virus envelope glycoprotein. J. Virol. 90, 6799–6807.

Sherman, M.B., Trujillo, J., Leahy, I., Razmus, D., Dehate, R., Lorcheim, P., Czarneski, M.A., Zimmerman, D., Newton, J.T.M., Haddow, A.D., et al., 2013. Construction and organization of a BSL-3 cryo-electron microscopy laboratory at UTMB. J. Struct. Biol. 181, 223–233.

Sirohi, D., Chen, Z., Sun, L., Klose, T., Pierson, T.C., Rossmann, M.G., Kuhn, R.J., 2016. The 3.8 Å resolution cryo-EM structure of Zika virus. Science 352, 467–470.

Skehel, J.J., Bayley, P.M., Brown, E.B., Martin, S.R., Waterfield, M.D., White, J.M., Wilson, I.A., Wiley, D.C., 1982. Changes in the conformation of influenza virus hemagglutinin at the pH optimum of virus-mediated membrane fusion. Proc. Natl. Acad. Sci. U. S. A. 79, 968–972.

Strauss, J.D., Hammonds, J.E., Yi, H., Ding, L., Spearman, P., Wright, E.R., 2016. Three-dimensional structural characterization of HIV-1 tethered to human cells. J. Virol. 90, 1507–1521.

Subramaniam, S., Bartesaghi, A., Liu, J., Bennett, A.E., Sougrat, R., 2007. Electron tomography of viruses. Curr. Opin. Struct. Biol. 17, 596–602.

Sukupolvi-Petty, S., Brien, J.D., Austin, S.K., Shrestha, B., Swayne, S., Kahle, K., Doranz, B.J., Johnson, S., Pierson, T.C., Fremont, D.H., et al., 2013. Functional analysis of antibodies against dengue virus type 4 reveals strain-dependent epitope exposure that impacts neutralization and protection. J. Virol. 87, 8826–8842.

Sun, S., Xiang, Y., Akahata, W., Holdaway, H., Pal, P., Zhang, X., Diamond, M.S., Nabel, G.J., Rossmann, M.G., 2013. Structural analyses at pseudo atomic resolution of Chikungunya virus and antibodies show mechanisms of neutralization. Elife 2 e00435.

Sun, Z., El Omari, K., Sun, X., Ilca, S.L., Kotecha, A., Stuart, D.I., Poranen, M.M., Huiskonen, J.T., 2017. Double-stranded RNA virus outer shell assembly by bona fide domain-swapping. Nat. Commun. 8, 14814.

Teoh, E.P., Kukkaro, P., Teo, E.W., Lim, A.P.C., Tan, T.T., Yip, A., Schul, W., Aung, M., Kostyuchenko, V.A., Leo, Y.S., et al., 2012. The structural basis for serotype-specific neutralization of dengue virus by a human antibody. Sci. Transl. Med. 4, 139ra83.

Thompson, R.F., Walker, M., Siebert, C.A., Muench, S.P., Ranson, N.A., 2016. An introduction to sample preparation and imaging by cryo-electron microscopy for structural biology. Methods 100, 3–15.

Tran, E.E.H., Simmons, J.A., Bartesaghi, A., Shoemaker, C.J., Nelson, E., White, J.M., Subramaniam, S., 2014. Spatial localization of the Ebola virus glycoprotein mucin-like domain determined by cryo-electron tomography. J. Virol. 88, 10958–10962.

Tran, E.E.H., Nelson, E.A., Bonagiri, P., Simmons, J.A., Shoemaker, C.J., Schmaljohn, C.S., Kobinger, G.P., Zeitlin, L., Subramaniam, S., White, J.M., 2016. Mapping of ebolavirus neutralization by monoclonal antibodies in the ZMapp cocktail using cryo-electron tomography and studies of cellular entry. J. Virol. 90, 7618–7627.

Veesler, D., Ng, T.-S., Sendamarai, A.K., Eilers, B.J., Lawrence, C.M., Lok, S.-M., Young, M.J., Johnson, J.E., Fu, C.-Y., 2013. Atomic structure of the 75 MDa extremophile Sulfolobus turreted icosahedral virus determined by CryoEM and X-ray crystallography. Proc. Natl. Acad. Sci. U. S. A. 110, 5504–5509.

Wang, J., Moore, P.B., 2017. On the interpretation of electron microscopic maps of biological macromolecules. Protein Sci. 26, 122–129.

Wang, Q., Bosch, B.-J., Vlak, J.M., van Oers, M.M., Rottier, P.J., van Lent, J.W.M., 2016a. Budded baculovirus particle structure revisited. J. Invertebr. Pathol. 134, 15–22.

Wang, Q., Yang, H., Liu, X., Dai, L., Ma, T., Qi, J., Wong, G., Peng, R., Liu, S., Li, J., et al., 2016b. Molecular determinants of human neutralizing antibodies isolated from a patient infected with Zika virus. Sci. Transl. Med. 8, 369ra179.

Wang, X., Li, S.-H., Zhu, L., Nian, Q.-G., Yuan, S., Gao, Q., Hu, Z., Ye, Q., Li, X.-F., Xie, D.-Y., et al., 2017. Near-atomic structure of Japanese encephalitis virus reveals critical determinants of virulence and stability. Nat. Commun. 8, 14.
Weaver, S.C., Charlier, C., Vasilakis, N., Lecuit, M., 2018. Zika, Chikungunya, and other emerging vector-borne viral diseases. Annu. Rev. Med. 69, 395–408.
Wei, H., Holland Cheng, R., Berriman, J., Rice, W.J., Stokes, D.L., Katz, A., Morgan, D.G., Gottlieb, P., 2009. Three-dimensional structure of the enveloped bacteriophage Φ12: an incomplete T = 13 lattice is superposed on an enclosed T = 1 shell. PLoS One 4, e6850.
Welsch, S., Kolesnikova, L., Krähling, V., Riches, J.D., Becker, S., Briggs, J.A.G., 2010. Electron tomography reveals the steps in filovirus budding. PLoS Pathog. 6 e1000875.
Wilson, I.A., Skehel, J.J., Wiley, D.C., 1981. Structure of the haemagglutinin membrane glycoprotein of influenza virus at 3 A resolution. Nature 289, 366–373.
Wirawan, M., Fibriansah, G., Marzinek, J.K., Lim, X.X., Ng, T.-S., Sim, A.Y.L., Zhang, Q., Kostyuchenko, V.A., Shi, J., Smith, S.A., Verma, C.S., Anand, G., Crowe, J.E., Bond, P.J., Lok, S.-M., 2019. Mechanism of enhanced immature dengue virus attachment to endosomal membrane induced by prM antibody. Structure 27, 253–267. e258.
Wolf, M., DeRosier, D.J., Grigorieff, N., 2006. Ewald sphere correction for single-particle electron microscopy. Ultramicroscopy 106, 376–382.
Xiao, C., Rossmann, M.G., 2011. Structures of giant icosahedral eukaryotic dsDNA viruses. Curr. Opin. Virol. 1, 101–109.
Xiao, C., Kuznetsov, Y.G., Sun, S., Hafenstein, S.L., Kostyuchenko, V.A., Chipman, P.R., Suzan-Monti, M., Raoult, D., McPherson, A., Rossmann, M.G., 2009. Structural studies of the giant mimivirus. PLoS Biol. 7, e92.
Xiao, C., Fischer, M.G., Bolotaulo, D.M., Ulloa-Rondeau, N., Avila, G.A., Suttle, C.A., 2017. Cryo-EM reconstruction of the Cafeteria roenbergensis virus capsid suggests novel assembly pathway for giant viruses. Sci. Rep. 7, 5484.
Xu, K., Chan, Y.-P., Bradel-Tretheway, B., Akyol-Ataman, Z., Zhu, Y., Dutta, S., Yan, L., Feng, Y., Wang, L.-F., Skiniotis, G., et al., 2015. Crystal structure of the pre-fusion Nipah virus fusion glycoprotein reveals a novel hexamer-of-trimers assembly. PLoS Pathog. 11, e1005322.
Yamauchi, Y., Helenius, A., 2013. Virus entry at a glance. J. Cell Sci. 126, 1289–1295.
Yan, X., Chipman, P.R., Castberg, T., Bratbak, G., Baker, T.S., 2005. The marine algal virus PpV01 has an icosahedral capsid with T=219 quasisymmetry. J. Virol. 79, 9236–9243.
Ye, Y., Chang, P.H., Hartert, J., Wigginton, K.R., 2018. Reactivity of enveloped virus genome, proteins, and lipids with free chlorine and UV254. Environ. Sci. Technol. 52, 7698–7708.
Zanetti, G., Briggs, J.A.G., Grünewald, K., Sattentau, Q.J., Fuller, S.D., 2006. Cryo-electron tomographic structure of an immunodeficiency virus envelope complex in situ. PLoS Pathog. 2, e83.
Zhang, W., Chipman, P.R., Corver, J., Johnson, P.R., Zhang, Y., Mukhopadhyay, S., Baker, T.S., Strauss, J.H., Rossmann, M.G., Kuhn, R.J., 2003a. Visualization of membrane protein domains by cryo-electron microscopy of dengue virus. Nat. Struct. Biol. 10, 907–912.
Zhang, Y., Corver, J., Chipman, P.R., Zhang, W., Pletnev, S.V., Sedlak, D., Baker, T.S., Strauss, J.H., Kuhn, R.J., Rossmann, M.G., 2003b. Structures of immature flavivirus particles. EMBO J. 22, 2604–2613.
Zhang, X., Xiang, Y., Dunigan, D.D., Klose, T., Chipman, P.R., Van Etten, J.L., Rossmann, M.G., 2011. Three-dimensional structure and function of the *Paramecium bursaria* chlorella virus capsid. Proc. Natl. Acad. Sci. U. S. A. 108, 14837–14842.

Zhang, X., Ge, P., Yu, X., Brannan, J.M., Bi, G., Zhang, Q., Schein, S., Zhou, Z.H., 2013a. Cryo-EM structure of the mature dengue virus at 3.5-Å resolution. Nat. Struct. Mol. Biol. 20, 105–110.

Zhang, X., Sheng, J., Plevka, P., Kuhn, R.J., Diamond, M.S., Rossmann, M.G., 2013b. Dengue structure differs at the temperatures of its human and mosquito hosts. Proc. Natl. Acad. Sci. U. S. A. 110, 6795–6799.

Zhang, X., Sheng, J., Austin, S.K., Hoornweg, T.E., Smit, J.M., Kuhn, R.J., Diamond, M.S., Rossmann, M.G., 2015. Structure of acidic pH dengue virus showing the fusogenic glycoprotein trimers. J. Virol. 89, 743–750.

Zhang, S., Kostyuchenko, V.A., Ng, T.-S., Lim, X.-N., Ooi, J.S.G., Lambert, S., Tan, T.Y., Widman, D.G., Shi, J., Baric, R.S., et al., 2016. Neutralization mechanism of a highly potent antibody against Zika virus. Nat. Commun. 7 13679.

Zhong, Q., Carratalà, A., Nazarov, S., Guerrero-Ferreira, R.C., Piccinini, L., Bachmann, V., Leiman, P.G., Kohn, T., 2016. Genetic, structural, and phenotypic properties of MS2 coliphage with resistance to ClO disinfection. Environ. Sci. Technol. 50, 13520–13528.

Zhu, P., Chertova, E., Bess Jr., J., Lifson, J.D., Arthur, L.O., Liu, J., Taylor, K.A., Roux, K.H., 2003. Electron tomography analysis of envelope glycoprotein trimers on HIV and simian immunodeficiency virus virions. Proc. Natl. Acad. Sci. U. S. A. 100, 15812–15817.

Zhu, P., Liu, J., Bess Jr., J., Chertova, E., Lifson, J.D., Grisé, H., Ofek, G.A., Taylor, K.A., Roux, K.H., 2006. Distribution and three-dimensional structure of AIDS virus envelope spikes. Nature 441, 847–852.

CHAPTER THREE

CryoEM reconstruction approaches to resolve asymmetric features

Daniel J. Goetschius[a,b], Hyunwook Lee[a,c], Susan Hafenstein[a,b,c,*]
[a]Department of Biochemistry and Molecular Biology, The Pennsylvania State University, University Park, PA, United States
[b]Department of Medicine, College of Medicine, The Pennsylvania State University, Hershey, PA, United States
[c]Huck Institutes of the Life Sciences, The Pennsylvania State University, University Park, PA, United States
*Corresponding author: e-mail address: shafenstein@psu.edu

Contents

Abstract

Although icosahedral viruses have highly symmetrical capsid features, asymmetric structural elements are also present since the genome and minor structural proteins are usually incorporated without adhering to icosahedral symmetry. Besides this inherent asymmetry, interactions with the host during the virus life cycle are also asymmetric. However, until recently it was impossible to resolve high resolution asymmetric features during single-particle cryoEM image processing. This review summarizes the current approaches that can be used to visualize asymmetric structural features. We have included examples of advanced structural strategies developed to reveal unique features and asymmetry in icosahedral viruses.

Advances in Virus Research, Volume 105
ISSN 0065-3527
https://doi.org/10.1016/bs.aivir.2019.07.007

Abbreviations

AP protein	maturation or A-protein (MS2 phage)
CAR	coxsackie- and adenovirus receptor
CryoEM	cryogenic electron microscopy
CVB3	coxsackievirus B3
Fab	antigen-binding fragment
HSV	herpes simplex virus
MS2	bacteriophage MS2
MurA	UDP-*N*-acetylglucosamine 1-carboxyvinyltransferase (*E. coli*)
Murine CMV	murine cytomegalovirus
Protein A2	maturation protein 2 (Q beta phage)
RdRp	RNA-dependent RNA polymerase
SH1	*Haloarcula hispanica* SH1 virus
T number	triangulation number
TBEV	tick-borne encephalitis virus
VP4	viral protein 4 (rotovirus)

1. Icosahedral symmetry

Throughout the history of cryogenic electron microscopy (cryoEM) 3D image reconstruction, icosahedral viruses have long played a prominent role due to the symmetry of the repeating protein subunits that make up the capsid. Early work by Caspar and Klug established the "quasi-equivalence" principle, which is the foundation for modern structural virology (Caspar and Klug, 1962). The most common form for small spherical viruses, the icosahedron, has five-, three- and twofold rotational symmetry axes, and the icosahedral lattice presents the most efficient organization of repeating subunits. The simplest architecture is comprised of only 12 pentamers, but larger icosahedral shells are constructed with additional hexamers in arrangements that depend on quasi-equivalent interactions due to variable environments for the chemically identical structural proteins. Most icosahedral viruses follow the quasi-equivalence principle of Caspar and Klug, defined by a triangulation number, or T number, which describes how many subunits make up the capsid. Assembling an icosahedral capsid from many of the same building blocks, allows the virus to package a small genome that encodes the main structural proteins, which are expressed in multiple copies.

The structural biologist has benefited by using icosahedral structural redundancy to gain resolution by averaging all the many subunits together. However, imposing icosahedral symmetry averaging results in the loss of any unique feature, which is overwhelmed by signal of the predominant features. Thus, even in the case of structures known to have unique features, strict imposition of icosahedral averaging can still result in an atomic resolution map (Kostyuchenko et al., 2016; Lee et al., 2016; Sirohi et al., 2016; Therkelsen et al., 2018), albeit with the asymmetry obliterated.

2. Asymmetry in icosahedral viruses

Upon packaging, parts of the genome may become icosahedrally-ordered and thus partially visualized during a traditional reconstruction process, but for the most part the virus genome folds into an energetically favorable conformation allowed by interactions with the internal capsid surface. Thus, the genome with or without its packaging proteins does not completely follow the entire symmetry of the capsid and represents an inherent asymmetry that has seldom been resolved. Besides the asymmetry sometimes required for genome packaging, there are "minor" structural proteins, which are present in smaller numbers per capsid than the proteins that assemble the main capsid shell. Many minor structural proteins are incorporated asymmetrically into the mature virus particle.

Many steps during the virus life cycle involve asymmetric changes. Viruses recognize host molecules to facilitate attachment and entry. Genome release may only be successful if the changes leading up to release are asymmetric. Conversely, host immune molecules are triggered that recognize the virus. In a physiological setting these interactions are almost certainly asymmetric or stoichiometric, but have rarely been visualized at atomic resolution using a structural approach. However, with recent advances in cryoEM, asymmetric features can be resolved that were previously missed. This is clearly an exciting advance in structural virology since all viruses contain or develop asymmetric features that play important functions. Furthermore, since asymmetry has been so difficult to study using a structural approach, most of these features are generally poorly understood. Those structures often perform essential functions for viral infection, replication, assembly, and transmission, so these recent advances will provide us with crucial information for future efforts to understand virus infections and to develop new methods for their control.

3. Reconstruction approaches

Standard icosahedral refinement obliterates any underlying asymmetric features due to the averaging process. In the best case, heterogeneous features may leave partial density or poor local resolution as a clue to their existence. Once one has been clued into the potential existence of structural heterogeneity, the next step is to attempt one of the various asymmetric reconstruction approaches. So far in the field, many different terms have been used to describe techniques solving symmetry mismatch in icosahedral viruses. Building on efforts to categorize this diversity of terms (Huiskonen, 2018), the basic asymmetric approaches can be classified into the following categories:

(a) naive asymmetric refinement.
(b) symmetry relaxation.
(c) symmetry expansion and focused classification/refinement.
(d) subparticle classification/refinement.

Within these categories, additional subroutines can be implemented to visualize asymmetric features. A good example is partial image subtraction. Density may be subtracted from the particle images via a 2D projection of a masked reconstruction to enhance the relative strength of the asymmetric viral features. For instance, the capsid density may be subtracted as a first step to solve the fundamentally asymmetric structure of the viral genome (Liu and Cheng, 2015; Li et al., 2016; Zhang et al., 2015).

Reconstruction terms

Focused = masked
Classification = no further changes in angles or offsets
Localized = subvolume extraction
Blockbased = subvolume extraction
Refinement = includes changes in angles and offsets
Local refinement = changes in angles and offsets applied to masked or extracted subvolumes
Symmetry expansion = reorientation of individual particles
Partial image subtraction = removes densities through use of a masked map

3.1 Standard icosahedral refinement

Even when asymmetry is suspected, an icosahedral refinement is often conducted as a starting point. Importantly, icosahedrally averaging an asymmetric particle can result in atomic resolution (Lee et al., 2016) of a map that has incorporated clues to asymmetry. Details from the icosahedral map may provide additional information on the asymmetry to fuel further decisions about the reconstruction process. Depending on the context, weak density or poor local resolution may suggest which of several potential problems need to be addressed, such as subunit heterogeneity, partial occupancy of a binding partner, and local or global capsid flexibility. Additional clues to a heterogeneous dataset include a high b-factor or poorer than expected resolution given specific parameters of the dataset, especially particle number and pixel size. Such an example would be having a million particles with a fine pixel size, but when reconstructing the map the resolution stalls out at a modest value. As a matter of practical workflow, the particle orientations solved in the icosahedral refinement serve as an initial starting point for the more advanced asymmetric techniques. Thus, in an "Icosahedral plus" approach, once an initial icosahedral reconstruction has been completed to align the data according to the global capsid symmetry, subsequent operations identify local deviations from that symmetry. These sampling methods currently take the form of single-particle (masked, i.e., focused classification) or subparticle techniques that rely on the extraction of smaller subvolumes (localized classification and block-based reconstruction).

3.2 Naïve asymmetric refinement

In an icosahedral reconstruction, each particle contributes to averaging in 60 different orientations, described by a set of Euler angles defined as rot, tilt, and psi (Scheres, 2012). The simplest approach to resolve asymmetry in otherwise icosahedral viruses is omitting the imposition of symmetry averaging during data processing, which has sometimes been termed a C1 reconstruction approach. This process is computationally expensive, as it requires sampling the entire globe of possible orientations. The naïve asymmetric refinement also sacrifices resolution, because each icosahedral particle contributes to the final reconstruction only once rather than multiple times (i.e., once for every symmetry-related orientation). Additionally, there is a risk of becoming trapped in a local minimum, because each particle contributes only one orientation. However, this orientation may not be the one that emphasizes the asymmetric feature, which allows the signal from local asymmetric features to be overwhelmed by the global icosahedral architecture of the virus capsid during sampling. Thus the algorithm may prematurely settle on a "good" solution while precluding the opportunity to find the ideal solution that emphasizes the underlying asymmetric feature.

The naïve asymmetric approach is especially error prone for deeply heterogeneous datasets, such as an undersaturated complex of an icosahedral virus that only a few binding sites occupied by a host protein molecule. However, if the naïve asymmetric reconstruction is able to reveal unique features, the risks and trade-offs are clearly warranted. Such an approach was used to reconstruct MS2 phage to 8.7 Å resolution, revealing the location and structure of the asymmetrically incorporated AP protein and genome, whereas an icosahedral reconstruction of the same data reached 4.1 Å while obscuring these asymmetric features (Koning et al., 2016). Likewise, asymmetric reconstructions of immature and mature Kunjin virus (20 Å and 35 Å, respectively) showed an icosahedral configuration of the glycoprotein shell with variable positioning of the nucleocapsid core that was eccentrically positioned in the immature virus, but which became centered in the mature virus (Therkelsen et al., 2018). Additionally, an otherwise naïve asymmetric refinement may be purposefully biased by introducing an extra density to the reference model at the site of the expected asymmetry, as was done in the case of the RNA polymerase within cytoplasmic polyhedrosis virus (Zhang et al., 2015).

3.3 Symmetry relaxation

Symmetry relaxation begins with an icosahedral refinement in order to assign orientations, which allows subsequent iterative sampling of the 60 symmetry-related orientations per particle. This approach effectively "relaxes" the higher-order symmetry and allows for the emergence of underlying asymmetric features. In addition, the process of symmetry relaxation circumvents the high computational costs associated with naive asymmetric refinements because it removes the need to sample the full globe of conformation space.

Liu and Cheng used a symmetry-relaxation approach to identify and reconstruct the 10 RdRp molecules of cypovirus to 4 Å resolution, along with their associated segmented RNA genome (Li et al., 2017; Liu and Cheng, 2015) (Fig. 1). This approach was performed for both non-transcribing and transcribing virus, providing the first in situ structural details of genome replication in cypovirus. In this case, the genome and RdRps were found to exhibit pseudo-D3 symmetry, compared to the overall icosahedral symmetry of the capsid itself. A separate paper (Li et al., 2016) details the software implementation of their reconstruction protocol.

Lee et al. used a similar symmetry-relaxation approach to solve the 7.8 Å asymmetric structure of coxsackievirus B3 (CVB3) interacting with the host receptor, coxsackie- and adenovirus receptor (CAR) inserted in nanodiscs. That structure identified a local reorganization of the viral genome upon receptor engagement (Lee et al., 2016) and formation of a unique pore, likely for genome release. Thus, the 7.8 Å symmetry-mismatch reconstruction revealed details that were abolished by symmetry averaging in the 3.7 Å icosahedral map.

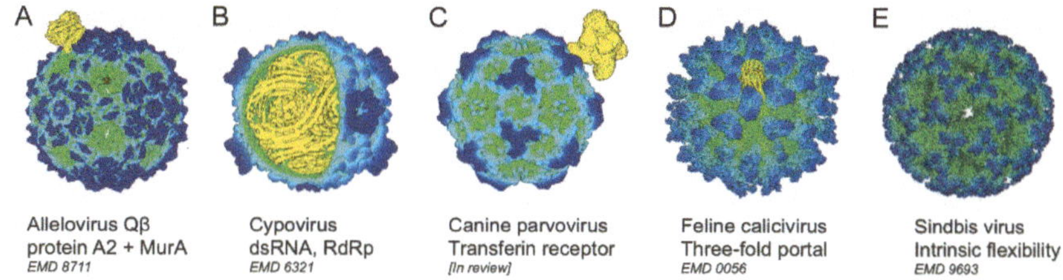

Fig. 1 Recent examples of asymmetric features resolved in icosahedral viruses: (A) Alleovirus Qβ, featuring asymmetrical incorporated maturation protein A2 and MurA (Cui et al., 2017), was reconstructed using a focused (masked) approach. Previous studies resolved protein A2 via naïve asymmetric refinement (Gorzelnik et al., 2016). (B) Cypovirus (reoviridae), with asymmetries including the segmented dsRNA genome and RdRp, responded well to a symmetry-relaxation reconstruction approach (Liu and Cheng, 2015). (C) Canine parvovirus interacting asymmetrically with transferrin-bound transferrin receptor, solved using focused classification (in review). (D) Feline calicivirus, with a unique threefold portal vertex, presumably for genome release, was likewise solved using a masked classification approach (Conley et al., 2019). (E) Sindbis virus (alphaviridae), a large 700 Å diameter virus with intrinsic flexibility that was refined to high resolution using a subparticle approach (Chen et al., 2018).

3.4 Symmetry expansion and focused classification/refinement

3.4.1 Symmetry expansion

Enforcement of icosahedral symmetry during refinement generally limits the computational search range for particle orientations to that of a single asymmetric unit (RELION) (Scheres, 2012), though some programs omit this step (cryoSPARC) (Punjani et al., 2017). Fundamentally, this means that every asymmetric unit contributes equally to the idealized reconstruction, irrespective of unique features that a given portion of the capsid might possess.

To allow interrogation of unique features each particle is assigned every possible symmetry-related orientation, called symmetry expansion. This process aligns each asymmetric unit within the dataset, allowing for subsequent operations to investigate deviation from the idealized icosahedral symmetry. Completing symmetry expansion is an essential first step to adequately perform functions such as focused classification, localized reconstruction, or local refinement.

3.4.2 Discrete vs. continuous heterogeneity

The type of heterogeneity in the target particle will dictate whether to do classification or refinement. Structural heterogeneities may be either discrete or continuous. Discrete heterogeneities include the presence of a unique vertex (HSV) (McElwee et al., 2018), incorporation of minor capsid proteins (A-protein in MS2) (Koning et al., 2016), or partial occupancy of a binding partner (TBEV) (Füzik et al., 2018). Such heterogeneities can be readily addressed with focused classification, without the need for local refinement beyond the initial icosahedral refinement and symmetry expansion operations.

Continuous heterogeneities may be more difficult to deal with. For instance, a large viral particle may not exhibit perfect structural rigidity, either as a result of the high forces experienced during purification (centrifugation), or as an intrinsic characteristic. Thus any given viral particle may exhibit anisotropy, with deviations in diameter depending on the axis that is chosen. In the case of such squishy particles, local deviations from rigid icosahedral symmetry (x-, y-offsets, orientation angles) exist on a continuous spectrum, rather than as discrete states. Thus a perfect focused classification would result in infinitely many classes, with each possessing a vanishingly small number of particles. In such cases, local refinement of angles and offsets is a more sensible approach. Successful application of local searches during the reconstruction process resulted in high resolution maps for HSV (Dai and Zhou, 2018), sindbis virus (Chen et al., 2018), murine CMV (Liu et al., 2019), and SH1 (De Colibus et al., 2019).

3.4.3 Focused classification

During masked or focused classification, the data underlying the icosahedral reconstruction are expanded to generate all symmetry-related orientations of individual particles (Bai et al., 2015). A mask is then used to define a smaller subvolume within the refined structure, for example, an individual capsomer, unique feature, or capsid region attached to its ligand. These masked volumes can then be subjected to 3D classification as if they were individual particles, without further refinement of angles or x-, y-offsets. This approach allows for identification of distinct classes based solely on the characteristics of the masked volumes and within the context of the otherwise icosahedral structure of the virus.

Although used initially to identify subunit variation in multimeric proteins such as γ-secretase (Bai et al., 2015) as well as particles with higher-order symmetry including GroEL (seven-fold symmetry) (Roh et al., 2017), focused classification has also been used recently to identify deviations from symmetry within icosahedral viruses. This approach was used to reconstruct herpes simplex virus 1 (HSV-1) to 8 Å resolution, identifying a unique five-fold vertex corresponding to the portal as well as the portal-vertex-associated tegument proteins (McElwee et al., 2018). Since the portal itself contains fivefold symmetry, the final reconstruction could also be accomplished with fivefold symmetry imposed (C5) rather than as a C1 reconstruction, retaining the resolution-boosting benefits of higher-order symmetry. This same approach was used to reveal the threefold vertex portal of calicivirus to high resolution (Conley et al., 2019). Recently, a similar approach was used to reconstruct the structure of the canine parvovirus capsid bound to TfR, which revealed distinct variations in the orientation of the bound receptor with respect to the viral capsid (Lee et al., 2019). In this case, the virus-receptor complex possesses no higher symmetry, limiting the final resolution of the map.

3.4.4 Focused refinement

Once an initial icosahedral refinement followed by symmetry expansion has been completed, local refinement may be used to permit subregions of the viral capsid to make small deviations from strict icosahedral symmetry. These deviations take the form of changes to the x-, y-offset and orientations (angles). To define the subvolume of interest, a soft-edged mask must be generated by a process that currently requires user intervention and optimization. Sjors Scheres (MRC, Cambridge, UK) provides a helpful rule of thumb for successfully implementing masked refinement, namely, that the masked region must contain at least as much mass as would be required to refine an isolated complex. Thus a mass of about 150 kDa (Scheres, 2016) or greater is typically required. It is likely that this size constraint will continue to decrease with further improvements in hardware and software.

3.4.5 Multibody refinement

Multibody refinement (Nakane et al., 2018) is a further development of focused refinement, though application to many asymmetric questions with icosahedral viruses is currently problematic. The technique allows the user to define multiple distinct volumes within the reconstruction via masking. Each masked region is locally refined independently, with the other regions subtracted at the end of each iteration. This allows the image subtraction to become more precise with each subsequent iteration, improving the quality of the local refinement. Practically speaking, the approach can be used to investigate interactions between capsid and a binding partner, by defining a few masked volumes for refinement. Currently; however, it is impractical to use multibody refinement to allow independent, simultaneous iterative local refinement of 60 symmetry-related subvolumes, due to computational limits.

3.4.6 *Software*

Symmetry expansion, focused classification, focused refinement, and multibody refinement are currently implemented in RELION 3 (Zivanov et al., 2018). Symmetry expansion is trivial to accomplish in RELION using the command relion_particle_symmetry_expand. At an algorithmic level, each particle orientation (Psi, Rot, Tilt in RELION) is transformed by the selected icosahedral rotation matrix, and a new expanded star file is written with 60 sets of rlnAnglePsi, rlnAngleRot, and rlnAngleTilt values for every original particle. For custom purposes, the same results can be achieved using the freely available python module pyquaternion (http://kieranwynn.github.io/pyquaternion), or the scipy.spatial.transform sub-module (https://docs.scipy.org/doc/scipy/reference/spatial.transform.html).

Focused refinement is also implemented as a beta feature in cryoSPARC v2 (Punjani et al., 2017), where it is termed local refinement. However, symmetry expansion is not currently implemented as a native feature in cryoSPARC, making it challenging to achieve focused refinement of icosahedral viruses with this program. Custom scripts have been developed to implement symmetry expansion in cryoSPARC, however, this process is not currently user-friendly (http://php.scripts.psu.edu/dept/hafenstein/software). A unique feature of cryoSPARC is in allowing the user to define a point within the reconstruction volume as a customized fulcrum, effectively assigning a new center point to the focused region. This advantage allows for much more precise adjustments to orientation angles, as the rotations can be performed about the new fulcrum, rather than the center point of the viral capsid. Liu and Cheng's symmetry-mismatch program is provided as a supplement to their paper describing the technique (Li et al., 2016). Implementation has also been achieved with a modified version of RELION, symbreak (Lee et al., 2016) (http://php.scripts.psu.edu/dept/hafenstein/software).

3.5 Subparticle classification/refinement

Subparticle approaches are fundamentally similar to focused classification and refinement, with a few additional benefits: the ability to adjust for the local defocus of the subparticle of interest (Caspar and Klug, 1962), lower computational burden due to reextract at a smaller box size (Lee et al., 2016), and finer angular adjustment due to reassignment of the center point from bulk capsid to the subparticle of interest (Therkelsen et al., 2018). Computational benefits of a smaller box size are partially offset by the increased storage requirements from extraction of up to 60 subparticles for every 1 particle in the original dataset. Subparticle approaches are frequently coupled with partial image subtraction.

3.5.1 Localized reconstruction

Although focused approaches use a mask to narrow the region of investigation, localized reconstruction targets the volume via extraction of smaller subparticles from the larger virus particle (Ilca et al., 2015). The extracted subvolumes are centered and reextracted from the original micrographs, with or without partial image subtraction. Recentering and reextracting allows for finer adjustments to be made to the angles and offsets during localized refinement, compared to a focused refinement (c. Section 3.4.4). However, it is possible merely to classify the subvolumes without making further refinement to angles and offsets. This localized classification of the extracted subvolumes makes for faster computation and is preferred when the subvolume is too small to allow for further refinement. The final result is a small output subvolume map, which is out of context from the entire map.

Ilca et al. used localized reconstruction to identify the symmetry-mismatched P2 polymerase that is located within the capsid of bacteriophage phi6 under the icosahedral threefold vertex. In the same work, this approach was also used to analyze the partial occupancy of the VP4 spike in rotavirus (Ilca et al., 2015). Localized reconstruction approaches were also used to identify and characterize bound antibody fragments (Fabs) in a partially occupied complex that had about 11 Fab bound per virus. Dai and Zhou also used llca's approach with local refinements to account for local capsid flexibility and refine the HSV-1 capsid to 3.5 Å (Dai and Zhou, 2018). Phi6, rotovirus, and the virus-Fab projects all benefited from removing the density of the bulk particle (partial image subtraction) to aid the process of local angle searches, thus improving the resolution of the subvolume of interest.

3.5.2 Block-based reconstruction

To achieve atomic resolution reconstructions for large icosahedral virus capsids (greater than 50 nm diameter) there are two significant bottlenecks inherent to the molecular mass. First, the capsid shell is a multiple-protein complex, composed of hundreds to thousands of copies of protein subunits. The large size and multimeric nature causes intrinsic structural flexibility that contributes to overall particle deformation, limiting the attainable resolution. Second, and more importantly, the large capsid generates a defocus gradient across the particle because the thickness of the specimen exceeds the microscope depth of field. Thus, the grounding assumption for 3D reconstruction, the projection theorem, does not hold true at high resolution for large viruses because the Fourier transform of the 2D projection cannot truly represent the central section of the Fourier transform of the 3D object (DeRosier, 2000).

This physical phenomenon can be described by the Ewald sphere effect in X-ray crystallography (Ewald, 1921). Normally, particles smaller than 30 nm are not affected by the Ewald sphere effect (Zhu et al., 2018).

Both of these two problems can be resolved by using a "block-based" reconstruction method, in which the object is divided into multiple blocks and each block is reconstructed separately with its own local mean defocus value (Zhu et al., 2018). The separation of the structural components resolves the structural flexibility and can solve the Ewald problem (Russo and Henderson, 2018).

The process for the block-based reconstruction is fundamentally similar to the localized reconstruction approach. After conventional icosahedral refinement, each asymmetric unit is split into multiple "blocks," or sub-volumes, by determining the block positions and centers. The 2D images for all blocks are clipped, or reextracted, from the raw cryoEM images by using the calculated 2D centers. The local defocus of individual blocks is calculated accordingly. The clipped 2D images are used for local refining and reconstructing of the blocks. The culmination of this process is when the reconstructions of all block are combined to generate a full capsid structure, which is unique to the block-based reconstruction method.

The blockbased method was successfully applied to the of HSV-2 B- and C-capsids, producing 3.1 Å and 3.75 Å resolution maps, respectively, as well as atomic models (Wang et al., 2018; Yuan et al., 2018). The HSV capsid shell is 1250 Å in diameter and composed of about 3000 proteins, forming extensive intermolecular interactions. Conventional icosahedral reconstruction limits the resolution to ~4 Å for these large diameter capsids. Subsequently, the same technique was used to solve the structure of Sindbis virus to 3.5 Å (alphaviridae, ~700 Å diameter) (Chen et al., 2018). Future advances in image processing algorithms might improve the achievable resolution even further for the large icosahedral capsids.

3.5.3 Software

A software package, localized reconstruction (Ilca et al., 2015), was written to be used in conjunction with scipion (de la Rosa-Trevín et al., 2016) and RELION (Scheres, 2012). It is available for download at github (https://github.com/OPIC-Oxford/localrec). Block-based reconstruction software is likewise available at github (https://github.com/homurachan/Block-based-recontruction), with the local refinement and reconstruction being accomplished using EMAN (Ludtke et al., 1999), JSPR (Guo and Jiang, 2014), or RELION.

4. Complementary methods

4.1 Molecular dynamics

Another area that shows promise for the analysis of imperfect icosahedral symmetry is molecular dynamics simulations. Some large icosahedral viruses such as hepatitis B virus exhibit deviations from symmetry, resulting in distorted capsids that respond poorly to symmetry averaging, limiting the attainable resolution (Schlicksup et al., 2018). All-atom molecular dynamics simulations have helped define the extent of capsid flexibility and distortion, providing constraints on the resolution achievable by strict symmetry averaging of cryoEM specimens (Hadden et al., 2018). This analysis further supports the utility of subparticle-based local refinement techniques, which compensate for global flexibility. It should also be possible for local subparticle refinement to inform molecular dynamics simulations by comparing parameters from individual locally refined subparticles to their original icosahedral consensus.

4.2 Correlation of local classification

Local classification provides an opportunity to identify structurally distinct regions that are otherwise geometrically identical. This could take the form of an asymmetrically incorporated minor capsid protein (Gorzelnik et al., 2016; Koning et al., 2016), polymerase (Li et al., 2017; Liu and Cheng, 2015; Zhang et al., 2015), or an interacting host protein such as a receptor or antibody fragment (Füzik et al., 2018). Although subvolume classification can be used for purely 3D structural reasons, namely, to reduce heterogeneity and achieve high resolution structures for the contents of each classified volume, the underlying data retains potentially valuable information regarding the original setting of the subvolumes in the context of the individual particles. By returning the distinct subvolume structures to their original whole-particle environments, relationships between components of the entire map can be revealed.

4.2.1 Population histograms

At the most basic level, one can identify the number of occurrences of subvolume class A vs. class B at both a population and a per-particle level. For example, a polymerase might be reasonably expected to be present once-and-only-once in any given capsid. By querying the underlying 3D classification files for the chosen subvolumes (e.g., the relion .star file) one at a time for both class A and class B, one can generate histograms for the number of times each particle identifier is present for a given class. For the case of a polymerase, if class A corresponds to P+ (polymerase present) and class B corresponds to P− (polymerase absent), this should result in a strong 1:59 ratio in the histograms. Some complexity will necessarily arise if overlapping subvolumes are chosen for the analysis. For instance, in the case of a polymerase under the threefold symmetry axis, three distinct P+ classes will likely be present, one for each 120° rotation about the local symmetry axis, thus resulting in a 1:20 polymerase present to absent ratio. While this seemingly adds complexity, it is nevertheless trivial to compensate for in the analytical pipeline.

4.2.2 Analysis of cooperativity

More advanced analytical techniques can assess interactions between subvolumes on individual capsids, i.e., the relative spatial distribution of subvolume structures in class A vs. class B. We have developed this approach to assess the spatial distribution of antibody fragments (Fab) in undersaturated canine parvovirus + Fab datasets (unpublished). This relies on the *localized reconstruction* package (Ilca et al., 2015) to first extract and recenter all symmetry-related subvolumes. Subsequently, localized 3D classification is performed to distinguish Fab occupied vs. Fab unoccupied sites of the capsid on a per-particle basis. Each individual viral particle is then assigned a nondegenerate fingerprint consisting of an integer string detailing the 3D configuration of Fab fragments. Crucially, this integer string contains the inter-subvolume distances between each occupied site, such that a particle with n occurrences of class A is represented by a non-degenerate string of length $\{n(n-1)/2\}$. These fingerprints are then pooled over the entire dataset and normalized against a fingerprint representing all valid symmetry-related relationships. This process reveals deviations in the distribution of the chosen subvolume from random-chance behavior, specifically, whether certain configurations are favored, under-represented, or forbidden in the dataset. At a biochemical level, this analysis reveals information about underlying positive or negative cooperativity of ligand binding within the sample. As currently implemented, the correlative classification approach relies on binary discrimination between two distinct states, but could be adapted to more complex situations.

5. Summary

Although we have long known that there are many important functions controlled by asymmetric structures in icosahedral viruses, they have been difficult to analyze at a structural level. The development of key new technologies has made it possible to gain high enough resolution to provide a detailed analysis (Fig. 2; Table 1). Undoubtedly, there will be

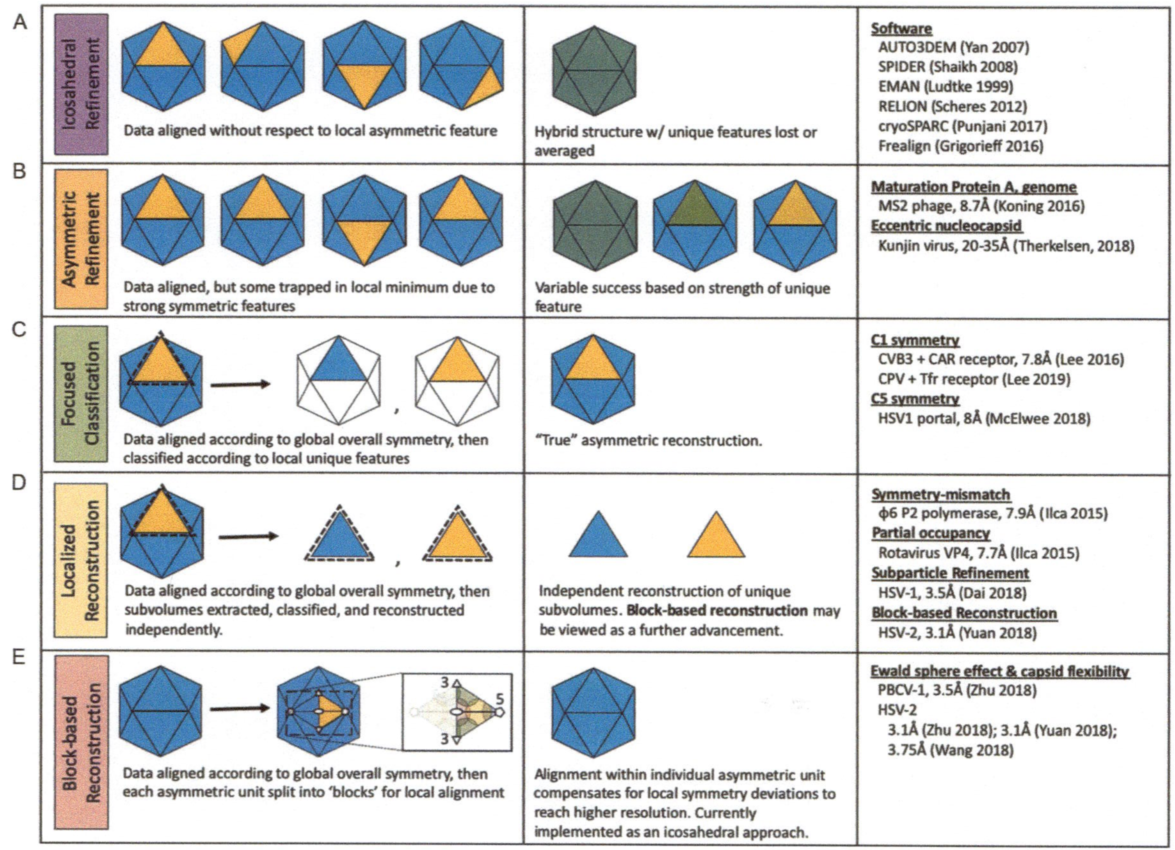

Fig. 2 See legend on next page.

Table 1 Approaches to asymmetry in icosahedral viruses with examples from the literature.

Asymmetric refinement			
• Cytoplasmic polyhedrosis virus	dsRNA genome, RdRp	PMID 26503045	2015
• MS2	ssRNA genome, A-protein	PMID 27561669	2016
• Q-beta	ssRNA genome, A_2	PMID 27671640	2016
• Kunjin virus (flaviviridae)	Eccentric core position	PMID 30348794	2018
Symmetry mismatch			
• Cypovirus (reoviridae)	dsRNA genome, RdRp	PMID 26383954	2015
• Coxsackievirus	CAR receptor	PMID 27574701	2016
• Cypovirus (reoviridae)	dsRNA genome, RdRp	PMID 27914893	2017

Fig. 2 Reconstruction strategies with examples from the literature. (A) In traditional *icosahedral refinement* (I) the higher symmetry is useful for reaching higher resolution, but any asymmetric features are lost. A smaller range of angles is sampled, aiding computation. Multiple software suites have been developed for cryoEM (Grigorieff, 2016; Ludtke et al., 1999; Punjani et al., 2017; Scheres, 2012; Shaikh et al., 2008; Yan et al., 2007) (B) In traditional *asymmetric refinement* (C1) (Koning et al., 2016; Therkelsen et al., 2018), the full range of angles (rot, tilt, psi) must be sampled during refinement. Failure to impose symmetry effectively reduces particle number 60-fold, reducing overall resolution. Refinement may or may not resolve asymmetric features, as strong features present in the otherwise icosahedral structure may trap the algorithm in a local (false) minimum. Success depends on the relative strength of the asymmetric feature. (C) *Focused classification* begins with a traditional icosahedral refinement, and then subsequently implements a mask to sample and 3D-classify symmetry-related locations within the map. Following classification, a full reconstruction can be completed, revealing the asymmetric feature within the context of the overall map (Lee et al., 2016; McElwee et al., 2018). (D) *Localized reconstruction* is similar to focused classification, but subvolumes of interest are centered and extracted before 3D-classification. The process is compatible with partial image subtraction and local refinement to boost the resolution of the unique feature of interest (Dai and Zhou, 2018; Ilca et al., 2015; Yuan et al., 2018). (E) After initial icosahedral refinement, *block-based reconstruction* instead divides the asymmetric unit into smaller blocks for local refinement. Following local refinement, blocks are rebuilt into an icosahedral map, thus compensating for local imperfections of the icosahedral symmetric due to intrinsic capsid flexibility of large viruses (Wang et al., 2018; Yuan et al., 2018; Zhu et al., 2018).

Table 1 Approaches to asymmetry in icosahedral viruses with examples from the literature.—cont'd

Focused/masked approaches			
• HIV-1	Intasome	PMID 28059769	2017
• Q-beta	ssRNA genome, A_2	PMID 29078304	2017
• Herpes Simplex Virus	Portal-vertex-assoc. tegument	PMID 29924793	2018
• Calicivirus	Portal	PMID 30626974	2019
Symmetry expansion and classification			
• Bacteriophage IKe <not Icos>	ssDNA genome	PMID 30819888	2019
Subparticle approaches			
• Bacteriophage phi6	RdRp	PMID 26534841	2015
• Adenovirus D26	Fiber	PMID 28508067	2017
• Tick-borne encephalitis virus	Fab, partial occupancy	PMID 29382836	2018
• Rift Valley fever virus	Capsomers	PMID 29367607	2018
• HSV-1	Capsid "blocks"	PMID 29622628	2018
• HSV-2	Capsid "blocks"	PMID 29674632	2018
• Sindbis virus (alphaviridae)	Capsid "blocks"	PMID 30552337	2018
• Murine CMV	Capsid "blocks"	PMID 30779794	2019
• SH1 virus	Capsid "blocks"	PMID 30926810	2019
• HCIV-1, HHIV-2	Spikes	PMID 30862777	2019

additional refinements on the methods outlined here. These approaches provide new opportunities to gain a more complete understanding of asymmetric structure-function relationships. The many dynamic, partially occupied, and asymmetric features of viruses will also certainly provide important new targets for antiviral interventions.

Acknowledgments

This work was funded in part by the Pennsylvania Department of Health CURE funds, Huck Institutes at The Pennsylvania State University, and by the Office of the Director, National Institutes of Health, under Award Numbers 1 R01 AI107121-01 (SH). The content is solely the responsibility of the authors and does not necessarily represent the official views of the National Institutes of Health.

References

Bai, X., Rajendra, E., Yang, G., Shi, Y., Scheres, S.H.W., 2015. Sampling the conformational space of the catalytic subunit of human γ-secretase. eLife 4, e11182. https://doi.org/10.7554/eLife.11182.

Caspar, D.L., Klug, A., 1962. Physical principles in the construction of regular viruses. Cold Spring Harb. Symp. Quant. Biol. 27, 1–24.

Chen, L., et al., 2018. Implication for alphavirus host-cell entry and assembly indicated by a 3.5Å resolution cryo-EM structure. Nat. Commun. 9, 5326.

Conley, M.J., et al., 2019. Calicivirus VP2 forms a portal-like assembly following receptor engagement. Nature 565, 377–381.

Cui, Z., et al., 2017. Structures of Qβ virions, virus-like particles, and the Qβ-MurA complex reveal internal coat proteins and the mechanism of host lysis. Proc. Natl. Acad. Sci. U. S. A. 114, 11697–11702.

Dai, X., Zhou, Z.H., 2018. Structure of the herpes simplex virus 1 capsid with associated tegument protein complexes. Science 360 eaao7298. https://doi.org/10.1126/science.aao7298.

De Colibus, L., et al., 2019. Assembly of complex viruses exemplified by a halophilic euryarchaeal virus. Nat. Commun. 10, 1456.

de la Rosa-Trevín, J.M., et al., 2016. Scipion: a software framework toward integration, reproducibility and validation in 3D electron microscopy. J. Struct. Biol. 195, 93–99.

DeRosier, D.J., 2000. Correction of high-resolution data for curvature of the Ewald sphere. Ultramicroscopy 81, 83–98.

Ewald, P., 1921. Evaluation of optical and electrostatic lattice potentials. Ann. Phys. 64, 253–287.

Füzik, T., et al., 2018. Structure of tick-borne encephalitis virus and its neutralization by a monoclonal antibody. Nat. Commun. 9, 436.

Gorzelnik, K.V., et al., 2016. Asymmetric cryo-EM structure of the canonical Allolevivirus Qβ reveals a single maturation protein and the genomic ssRNA in situ. Proc. Natl. Acad. Sci. U. S. A. 113, 11519–11524.

Grigorieff, N., 2016. Frealign: an exploratory tool for single-particle cryo-EM. Methods Enzymol. 579, 191–226.

Guo, F., Jiang, W., 2014. Single particle cryo-electron microscopy and 3-D reconstruction of viruses. Methods Mol. Biol. 1117, 401–443.

Hadden, J.A., et al., 2018. All-atom molecular dynamics of the HBV capsid reveals insights into biological function and cryo-EM resolution limits. eLife 7, e32478. https://doi.org/10.7554/eLife.32478.

Huiskonen, J.T., 2018. Image processing for cryogenic transmission electron microscopy of symmetry-mismatched complexes. Biosci. Rep. 38, BSR20170203. https://doi.org/10.1042/BSR20170203.

Ilca, S.L., et al., 2015. Localized reconstruction of subunits from electron cryomicroscopy images of macromolecular complexes. Nat. Commun. 6, 8843.

Koning, R.I., et al., 2016. Asymmetric cryo-EM reconstruction of phage MS2 reveals genome structure in situ. Nat. Commun. 7, 12524.

Kostyuchenko, V.A., et al., 2016. Structure of the thermally stable Zika virus. Nature 533, 425–428.

Lee, H., et al., 2016. The novel asymmetric entry intermediate of a picornavirus captured with nanodiscs. Sci. Adv. 2, e1501929.

Lee, H., et al., 2019. Transferrin receptor binds virus capsid with dynamic motion. Proc. Natl. Acad. Sci. (in press; accepted August 29, 2019).

Li, X., Liu, H., Cheng, L., 2016. Symmetry-mismatch reconstruction of genomes and associated proteins within icosahedral viruses using cryo-EM. Biophys. Rep. 2, 25–32.

Li, X., et al., 2017. Near-atomic resolution structure determination of a cypovirus capsid and polymerase complex using cryo-EM at 200 kV. J. Mol. Biol. 429, 79–87.
Liu, H., Cheng, L., 2015. Cryo-EM shows the polymerase structures and a nonspooled genome within a dsRNA virus. Science 349, 1347–1350.
Liu, W., et al., 2019. Atomic structures and deletion mutant reveal different capsid-binding patterns and functional significance of tegument protein pp150 in murine and human cytomegaloviruses with implications for therapeutic development. PLoS Pathog. 15, e1007615.
Ludtke, S.J., Baldwin, P.R., Chiu, W., 1999. EMAN: semiautomated software for high-resolution single-particle reconstructions. J. Struct. Biol. 128, 82–97.
McElwee, M., Vijayakrishnan, S., Rixon, F., Bhella, D., 2018. Structure of the herpes simplex virus portal-vertex. PLoS Biol. 16, e2006191.
Nakane, T., Kimanius, D., Lindahl, E., Scheres, S.H., 2018. Characterisation of molecular motions in cryo-EM single-particle data by multi-body refinement in RELION. eLife 7, e36861. https://doi.org/10.7554/eLife.36861.
Punjani, A., Rubinstein, J.L., Fleet, D.J., Brubaker, M.A., 2017. cryoSPARC: algorithms for rapid unsupervised cryo-EM structure determination. Nat. Methods 14, 290–296.
Roh, S.-H., et al., 2017. Subunit conformational variation within individual GroEL oligomers resolved by cryo-EM. Proc. Natl. Acad. Sci. U. S. A. 114, 8259–8264.
Russo, C.J., Henderson, R., 2018. Ewald sphere correction using a single side-band image processing algorithm. Ultramicroscopy 187, 26–33.
Scheres, S.H.W., 2012. RELION: implementation of a Bayesian approach to cryo-EM structure determination. J. Struct. Biol. 180, 519–530.
Scheres, S.H.W., 2016. Processing of structurally heterogeneous cryo-EM data in RELION. Methods Enzymol. 579, 125–157.
Schlicksup, C.J., et al., 2018. Hepatitis B virus core protein allosteric modulators can distort and disrupt intact capsids. eLife 7, e31473. https://doi.org/10.7554/eLife.31473.
Shaikh, T.R., et al., 2008. SPIDER image processing for single-particle reconstruction of biological macromolecules from electron micrographs. Nat. Protoc. 3, 1941–1974.
Sirohi, D., et al., 2016. The 3.8 Å resolution cryo-EM structure of Zika virus. Science 352, 467–470.
Therkelsen, M.D., et al., 2018. Flaviviruses have imperfect icosahedral symmetry. Proc. Natl. Acad. Sci. U. S. A. 115, 11608–11612.
Wang, J., et al., 2018. Structure of the herpes simplex virus type 2 C-capsid with capsid-vertex-specific component. Nat. Commun. 9, 3668.
Yan, X., Sinkovits, R.S., Baker, T.S., 2007. AUTO3DEM—an automated and high throughput program for image reconstruction of icosahedral particles. J. Struct. Biol. 157, 73–82.
Yuan, S., et al., 2018. Cryo-EM structure of a herpesvirus capsid at 3.1 Å. Science 360, eaao7283. https://doi.org/10.1126/science.aao7283.
Zhang, X., et al., 2015. In situ structures of the segmented genome and RNA polymerase complex inside a dsRNA virus. Nature 527, 531–534.
Zhu, D., et al., 2018. Pushing the resolution limit by correcting the Ewald sphere effect in single-particle cryo-EM reconstructions. Nat. Commun. 9, 1552.
Zivanov, J., et al., 2018. New tools for automated high-resolution cryo-EM structure determination in RELION-3. eLife 7, e42166. https://doi.org/10.7554/eLife.42166.

CHAPTER FOUR

Structural insights into coronavirus entry

M. Alejandra Tortorici[a,b,c], David Veesler[a,*]
[a]Department of Biochemistry, University of Washington, Seattle, WA, United States
[b]Institut Pasteur, Unité de Virologie Structurale, Paris, France
[c]CNRS UMR 3569, Unité de Virologie Structurale, Paris, France
*Corresponding author: e-mail address: dveesler@uw.edu

Contents

Abstract

Coronaviruses (CoVs) have caused outbreaks of deadly pneumonia in humans since the beginning of the 21st century. The severe acute respiratory syndrome coronavirus (SARS-CoV) emerged in 2002 and was responsible for an epidemic that spread to five continents with a fatality rate of 10% before being contained in 2003 (with additional cases reported in 2004). The Middle-East respiratory syndrome coronavirus (MERS-CoV) emerged in the Arabian Peninsula in 2012 and has caused recurrent outbreaks in humans with a fatality rate of 35%. SARS-CoV and MERS-CoV are zoonotic viruses that crossed the species barrier using bats/palm civets and dromedary camels, respectively. No specific treatments or vaccines have been approved against any of the six human coronaviruses, highlighting the need to investigate the principles governing viral entry and cross-species transmission as well as to prepare for zoonotic outbreaks which are likely to occur due to the large reservoir of CoVs found in mammals and birds. Here, we review our understanding of the infection mechanism used by coronaviruses derived from recent structural and biochemical studies.

Advances in Virus Research, Volume 105
ISSN 0065-3527
https://doi.org/10.1016/bs.aivir.2019.08.002

Abbreviations

3D	three dimensional
9-O-Ac-Sia	5-*N*-acetyl-9-*O*-acetyl-sialosides
ACE2	angiotensin-converting enzyme 2
APN	amino-peptidase N
CoV	Coronavirus
cryoEM	cryo-electron microscopy
DPP4	dipeptidyl peptidase 4
Fab	antigen-binding fragment of an antibody
MERS-CoV	Middle-East respiratory syndrome coronavirus
MHV	mouse hepatitis virus
PDCoV	porcine δ-CoV
PEDV	porcine epidemic diarrhea virus
SARS-CoV	severe acute respiratory syndrome coronavirus
TGEV	transmissible gastroenteritis virus

1. Introduction

Coronaviruses (CoVs) are enveloped viruses with a positive sense RNA genome, that belong to the subfamily *Coronavirinae* within the family *Coronaviridae*, which is part of the Nidovirales order. They are classified in four genera (α, β, γ, and δ) and four lineages are recognized whithin the β-CoV genus (A, B, C and D). CoVs cause a variety of respiratory and enteric diseases in mammalian and avian species. Until recently, CoVs were considered to be pathogens with a largely veterinary relevance but with limited impact on human health. However, outbreaks of severe acute respiratory syndrome (SARS) in 2002–2004 and of Middle-East respiratory syndrome (MERS) starting in 2012, with fatality rates of 10% and 35%, respectively, led CoVs to be recognized as zoonotic threats with pandemic potential. Four other CoVs are endemic in the human population and cause up to 30% of mild respiratory tract infections as well as occasional severe disease in young children, the elderly or immunocompromised individuals (Isaacs et al., 1983; Su et al., 2016). These viruses are HCoV-NL63 and HCoV-229E (α-CoVs) as well as HCoV-OC43 and HCoV-HKU1 (β-CoVs). Numerous SARS-CoV and MERS-CoV-like viruses currently circulate in bats and dromedaries making outbreaks of highly pathogenic human CoVs a global health threat (Ge et al., 2013; Haagmans et al., 2014; Hu et al., 2017; Menachery et al., 2015, 2016; Sabir et al., 2016).

The CoV virion contains at least four structural proteins: spike (S), envelope (E), membrane (M) and nucleocapsid (N). In contrast to other β-CoV lineages, lineage A CoVs also encode a hemagglutinin–esterase which serves as receptor-destroying enzyme to facilitate release of viral progeny from infected cells and escape from attachment to non-permissive host cells or decoys (Bakkers et al., 2017, 2016). S is a class 1 viral fusion protein that promotes host attachment and fusion of the viral and cellular membranes during entry (Bosch et al., 2003). As a consequence, S determines host range and cell tropism. S is also the main target of neutralizing antibodies elicited during infection and the focus of vaccine design. S is trimeric and each protomer is synthesized as a single polypeptide chain of 1100–1600 residues, depending on the CoV species. For many CoVs, S is processed by host proteases to generate two functional subunits, designated S_1 and S_2, which remain non-covalently bound in the prefusion conformation (Bosch et al., 2003). The S_1 subunit comprises the apex of the S trimer, including the receptor-binding domains, and stabilizes the prefusion state of the S_2 fusion machinery, which is anchored in the viral membrane. For all CoVs, S is further cleaved by host proteases at the so-called S_2' site located immediately upstream of the fusion peptide. This cleavage has been proposed to activate the protein for membrane fusion *via* large-scale, irreversible conformational changes (Heald-Sargent and Gallagher, 2012; Millet and Whittaker, 2015).

Although several class 1 viral fusion proteins have been extensively studied, CoV S proteins have proven reluctant to structural characterization until recently. Structural studies were largely limited to X-ray crystallographic analysis of isolated receptor-binding domains in complex with viral receptor ectodomains or neutralizing antibodies (Li et al., 2005a; Lu et al., 2013; Peng et al., 2011; Prabakaran et al., 2006; Reguera et al., 2012; Wang et al., 2013; Wong et al., 2017; Wu et al., 2009; Yu et al., 2015) and of the S_2 postfusion core (Duquerroy et al., 2005; Gao et al., 2013; Supekar et al., 2004; Xu et al., 2004a, b; Zheng et al., 2006) with the exception of two low-resolution electron microscopy reports (Beniac et al., 2006, 2007). In the past few years, however, technical advances in single-particle cryo-electron microscopy (cryoEM) (Bai et al., 2013; Brilot et al., 2012; Campbell et al., 2012, 2015; Li et al., 2013; Punjani et al., 2017; Scheres, 2012) together with the implementation of strategies for the stabilization of CoV S proteins in prefusion conformation (Pallesen et al., 2017; Walls et al., 2017a) led to a surge of structural data for multiple S ectodomain trimers. We review here our current understanding of the mechanism used by CoVs to infect host cells based on recent structural and biochemical studies of S glycoprotein ectodomains in prefusion and postfusion states as well as complexes with known receptors or neutralizing antibodies.

2. Prefusion S architecture

CryoEM studies of the S glycoproteins of mouse hepatitis virus (MHV) and HKU1 led to the first structures at high-enough resolution to obtain an atomic model of the prefusion state (Kirchdoerfer et al., 2016; Walls et al., 2016a). These structures revealed that prefusion S ectodomains are ~160 Å-long trimers with a triangular cross-section (Fig. 1A and B).

The S_1 subunit adopts a "V" shaped architecture for β and γ CoVs (Gui et al., 2017; Kirchdoerfer et al., 2016; Shang et al., 2018a; Tortorici et al., 2019; Walls et al., 2016a; Yuan et al., 2017) (Fig. 1C), or a square-shaped organization for α- and δ-CoVs (Shang et al., 2018b; Walls et al., 2016b; Xiong et al., 2018). The S_1 subunit folds as β-rich domains designated A, B, C, D. Several α-CoVs harbor a likely duplication of their domain A at the N-terminus of the S glycoprotein (Hulswit et al., 2016; Walls et al., 2016b). This additional domain, designated domain 0, was visualized in the NL63 S structure and hypothesized to interact with heparan sulfate present at the host cell surface during viral entry (Milewska et al., 2014; Walls et al., 2016b). Domain A and domain 0 adopt a galectin-like β-sandwich fold conserved across all CoV genera (Kirchdoerfer et al., 2016; Peng et al., 2011, 2012; Walls et al., 2016a) (Fig. 1D). Domain B, which shows the highest sequence variability within CoV S_1 subunits, has a markedly different architecture between α-, β-, γ- and δ-CoVs. B domains of β-CoVs contain a β-sheet core subdomain decorated with a highly variable external subdomain mediating receptor engagement (Chen et al., 2013; Kirchdoerfer et al., 2016; Li et al., 2005b; Lu et al., 2013; Tortorici et al., 2019; Walls et al., 2016a; Wang et al., 2013) (Fig. 1E). B domains of α-, γ- and δ-CoV form a β-sandwich decorated with loops mediating receptor attachment (Reguera et al., 2012; Shang et al., 2018a, b; Walls et al., 2016b; Wong et al., 2017; Xiong et al., 2018). In the context of the S trimer, β/γ CoV B domains interact with the A and B domains of another protomer, whereas they pack against the A domain of the same protomer in α/δ-CoVs (Shang et al., 2018a; Walls et al., 2016a, b; Xiong et al., 2018).

The S_2 subunit, which is more conserved than S_1, comprises the fusion machinery and connects to the viral membrane. It is assembled from a large

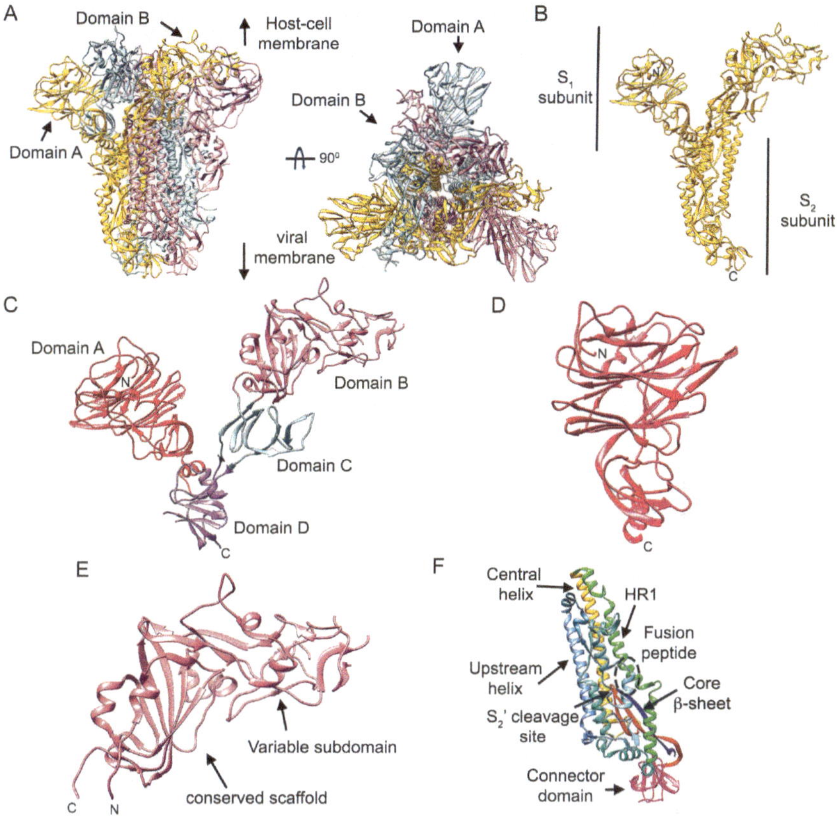

Fig. 1 CryoEM structure of the apo-HCoV-OC43 S glycoprotein. (A) Ribbon diagrams of the apo HCoV-OC43 S ectodomain trimer (PDB: 6OHW) in two orthogonal orientations, from the side (left) and from the top, looking towards the viral membrane (right). (B) Side view of one S protomer. (C) Ribbon diagram of the HCoV-OC43 S_1 subunit. (D–E) Close-up view of HCoV-OC43 domain A (D) and domain B (E). (F) Ribbon diagram of the HCoV-OC43 S_2 subunit in the prefusion conformation. The N- and C-termini are labeled in panels (B–E).

number of α-helices, an antiparallel core β-sheet, a β-rich connector domain and a stem helix leading to the heptad-repeat 2 (HR2) and the transmembrane region (Fig. 1F) (Gui et al., 2017; Kirchdoerfer et al., 2016, 2018; Pallesen et al., 2017; Shang et al., 2018a, b; Tortorici et al., 2019; Walls et al., 2016a, b, 2017b; Xiong et al., 2018; Yuan et al., 2017). Key S_2 features facilitating virus-cell fusion include the fusion peptide, two heptad repeat regions (named HR1 and HR2) and the transmembrane domain. In the prefusion S conformation, a central helix stretches along the threefold axis, perpendicular to the viral membrane, and is located downstream the HR1 motif, which folds as four consecutive α-helices (Kirchdoerfer et al., 2016; Walls et al., 2016a). Moreover, an upstream helix runs parallel to and is zipped against the central helix *via* hydrophobic contacts (Fig. 1F). The CoV S_2 subunit shares similarity with the pneumovirus/paramyxovirus F proteins—including a comparable 3D organization of the core β-sheet, the upstream helix and the central helix—suggesting an evolutionary relatedness between the viral fusion proteins of these different viruses and a conservation of their fusion mechanism (McLellan et al., 2013; Walls et al., 2016a, 2017b; Wong et al., 2016; Xu et al., 2015; Yin et al., 2006).

A conserved tryptophan-rich segment (Y(V/I)KWPW(Y/W)VWL) directly preceding the CoV S transmembrane region is crucial for proper trimerization. This segment is also required functionally for formation of a fusion pore (Schroth-Diez et al., 2000). Furthermore, transmembrane domain interactions within and possibly between S trimers have been proposed to be essential to complete the membrane fusion process (Schroth-Diez et al., 2000). The transmembrane domain is followed by an intraviral/cytoplasmic tail of variable length (36–46 residues) depending on the coronavirus species, which contains a palmitoylated cysteine-rich region (of about 18–24 residues with 7–10 cysteines) and a variable C-terminal end (Thorp et al., 2006). The cytoplasmic tail is involved in assembly, intracellular transport, cell-surface expression and cell-cell fusion (Bos et al., 1995; Bosch et al., 2005; Chang et al., 2000; Lontok et al., 2004; Petit et al., 2005; Ye et al., 2004; Youn et al., 2005). Currently, no structural information is available for any CoV full-length S, hindering our understanding of the influence of the transmembrane and cytoplasmic domains on the conformation of exposed antigenic sites, as previously studied for HIV-1 envelope (Chen et al., 2015;Dev et al., 2016).

3. Diversity of CoV receptors and entry mechanisms

CoV entry into susceptible cells is a complex process that requires the concerted action of receptor-binding and proteolytic processing of the S protein to promote virus-cell fusion (Heald-Sargent and Gallagher, 2012; Millet and Whittaker, 2015). Domain 0, domain A and/or domain B can act as receptor-binding domains and both attachment and entry receptors have been described, depending on the CoV species.

Lineage A β-CoVs attach *via* their S domain A to 5-*N*-acetyl-9-*O*-acetyl-sialosides (9-*O*-Ac-Sia) found on glycoproteins and glycolipids at the

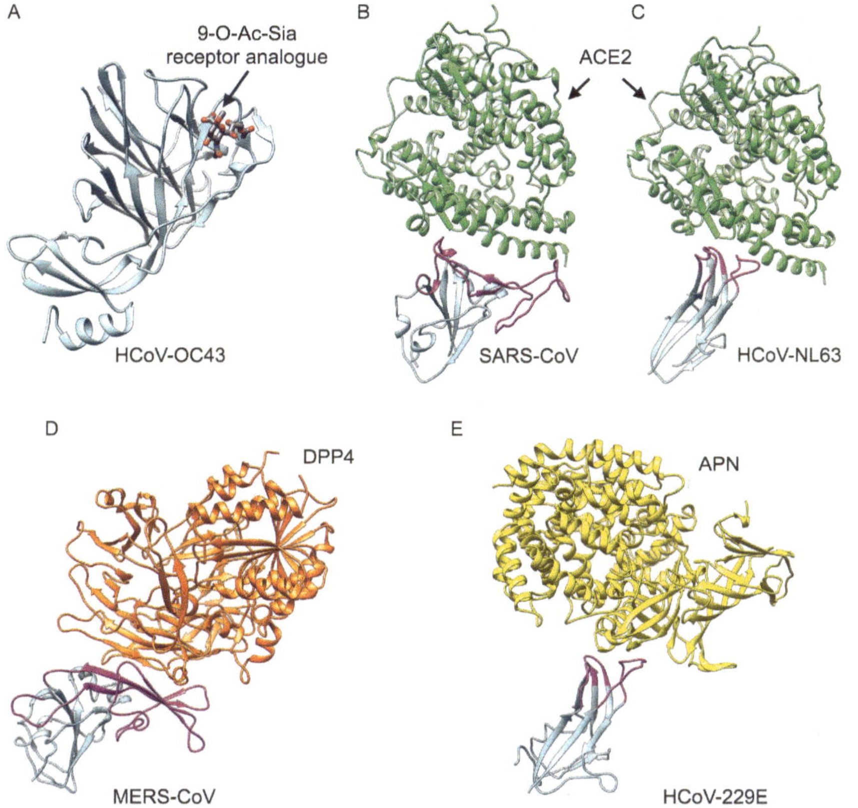

Fig. 2 Structural studies of human CoV attachment to host receptors. (A–E), Ribbon diagrams of the complex between domain A of HCoV-OC43 S with a 9-*O*-Ac-Sia receptor analogue ((A) PDB: 6NZK), or the domain B of SARS-CoV S with ACE2 ((B) PDB: 2AJF), HCoV-NL63 S with ACE2 ((C) PDB: 3KBH), MERS-CoV S with DPP4 ((D) PDB: 4L72) and HCoV-229E S with APN ((E) PDB: 6ATK). In panels (B–E), each domain B is rendered in light blue and the receptor binding-motifs are colored purple.

host cell surface to promote entry into susceptible cells (Vlasak et al., 1988). These include human CoVs OC43 and HKU1, bovine CoV (BCoV) and porcine hemagglutinating encephalomyelitis virus. We recently identified and visualized by cryoEM the HCoV-OC43 S sialoside-binding site, which is located in a groove at the surface of domain A (Fig. 2A) (Hulswit et al., 2019; Tortorici et al., 2019). This site is conserved in all other CoVs known to attach to 9-O-Ac-Sia (β-CoVs, lineage A) and shares architectural similarity with the ligand-binding pockets of CoV hemagglutinin-esterases and influenza virus C/D hemagglutinin-esterase-fusion glycoproteins, highlighting common structural principles of recognition (Bakkers et al., 2017, 2016;

Hulswit et al., 2019; Rosenthal et al., 1998; Tortorici et al., 2019). The current consensus in the field is that HCoV-OC43 only utilizes 9-*O*-Ac-sialosides as host receptors. In line with this statement, ligand-interacting residues were shown to be essential for S-mediated viral entry (Hulswit et al., 2019; Tortorici et al., 2019) and 9-*O*-Ac-Sia depletion from target cells resulted in severe decrease in virus infectivity (Krempl et al., 1995; Vlasak et al., 1988). Free 9-*O*-Ac-Sia, however, did not trigger S conformational changes associated with membrane fusion (Tortorici et al., 2019). This observation contrasts with data for SARS-CoV S, for which addition of the human angiotensin-converting enzyme 2 (ACE2) ectodomain (the proteinaceous receptor) promoted S refolding to the postfusion state (Song et al., 2018; Walls et al., 2019). These findings suggested that either 9-*O*-Ac-Sia-containing receptors differ from proteinaceous receptors in their mode of action, or that an interaction with a yet unidentified proteinaceous receptor is required before or after virus internalization for HCoV-OC43 entry into target cells.

The sialoside-binding site identified in HCoV-OC43 S is not conserved among CoVs which are also known to interact with sialoglycans to initiate host cell infection but are outside of the lineage A of β-CoVs, such as MERS-CoV (β-CoV, lineage C) or infectious bronchitis virus (IBV, δ-CoV) (Li et al., 2017; Wickramasinghe et al., 2011). Some α-CoVs such as transmissible gastroenteritis virus (TGEV) and porcine epidemic diarrhea virus (PEDV) use domain 0 to attach to sialoglycans, presumably to increase virus concentration at the cell surface and enhance subsequent attachment to proteinaceous receptors (Liu et al., 2015; Schwegmann-Wessels et al., 2003). Carbohydrate binding *via* this domain has been proposed to be a determinant of the TGEV enteric tropism since loss of domain 0 appears to correlate with a loss of enteric tropism for porcine respiratory coronavirus (PRCoV), the latter virus being a naturally occurring TGEV variant (Krempl et al., 1997).

CoVs exploit a limited variety of proteinaceous receptors compared with the large number and diversity of viral species. All CoVs known to engage proteinaceous receptors do so using domain B with the exception of MHV, which binds CEACAM1a using domain A (Dveksler et al., 1991; Peng et al., 2011; Williams et al., 1991). Remarkably, viruses from different genera, such as HCoV-NL63 (α-CoV) and SARS-CoV (β-CoV), can recognize the same region of ACE2 (entry receptor) using structurally distinct B domains (Hofmann et al., 2005; Li et al., 2005a, 2003; Wu et al., 2009). Many α-CoVs, including HCoV-229E, TGEV and PRCV, as well as porcine δ-CoV (PDCoV) utilize aminopeptidase N (APN) as entry receptor (Delmas et al., 1992; Delmas et al., 1993; Li et al., 2018; Reguera et al., 2012;

Wong et al., 2017; Yeager et al., 1992) whereas MERS-CoV uses dipeptidyl peptidase 4 (DPP4) (Lu et al., 2013; Raj et al., 2013; Wang et al., 2013). Crystal structures of SARS-CoV, HCoV-NL63, MERS-CoV, HCoV-229E B domains in complex with their cognate receptors provided atomic details of the interacting-interface and identified key residues for cross-species transmission and infection (Fig. 2) (Li et al., 2005a; Lu et al., 2013; Wang et al., 2013; Wong et al., 2017; Wu et al., 2009). This information will be useful to guide the development of therapeutics and vaccines against human CoVs.

Recent cryoEM studies revealed that MERS-CoV S and SARS-CoV S can adopt open and closed conformations in which the receptor binding site of domain B is exposed and occluded, respectively (Gui et al., 2017; Kirchdoerfer et al., 2018; Pallesen et al., 2017; Song et al., 2018; Walls et al., 2019; Yuan et al., 2017). In contrast, the MHV, HCoV-NL63, HCoV-HKU1, PDCoV, IBV and HCoV-OC43 S glycoproteins appear to only adopt a closed conformation (Kirchdoerfer et al., 2016; Shang et al., 2018a, b; Tortorici et al., 2019; Walls et al., 2016a, b; Xiong et al., 2018) and unknown trigger(s), besides proteolytic activation, might be necessary for these viruses to expose their receptor-binding motifs for recognition to occur. These findings suggest that CoVs have evolved a fine-tuned mechanism to balance masking of the receptor-binding motifs, putatively to avoid neutralization by the host humoral immune response, and their necessary exposure to enable receptor recognition and infection of host cells (Walls et al., 2016b, 2019; Wong et al., 2017).

Upon host recognition, CoVs are internalized *via* receptor-mediated clathrin-dependent, caveolin-dependent or other uptake pathways (Burkard et al., 2014; Eifart et al., 2007; Inoue et al., 2007; Nomura et al., 2004). For instance, both clathrin-dependent and clathrin/caveolae-independent entry pathways have been reported for SARS-CoV (Inoue et al., 2007; Wang et al., 2008). Feline infectious peritonitis virus was suggested to enter host cells *via* a clathrin/caveolin-independent internalization route (Regan et al., 2008; Van Hamme et al., 2008) whereas a caveolin-dependent endocytic uptake has been suggested for HCoV-229E and HCoV-OC43 (Nomura et al., 2004; Owczarek et al., 2018).

4. S proteolytic cleavage

Several reports have demonstrated the key role of proteolytic processing of CoV S for cell-cell fusion activity and/or virus entry into host cells using experiments of inhibition of intracellular proteases

(Burkard et al., 2014; Frana et al., 1985; Simmons et al., 2005; Yamada and Liu, 2009) and/or substitutions of residues at the S_1/S_2 or S_2' cleavage sites (Belouzard et al., 2009; Li et al., 2015; Millet and Whittaker, 2014; Wicht et al., 2014; Yang et al., 2015).

Prior to and/or after uptake of the virion by a host cell, the S protein is proteolytically processed by host proteases at one or two cleavage sites and both receptor-binding and proteolytic processing act in synergy to induce large-scale S conformational changes promoting CoV entry. One of the cleavage sites is located at the boundary between the S_1 and S_2 subunits (S_1/S_2 cleavage site), whereas the other is located immediately upstream of the fusion peptide (S_2' cleavage site), reviewed in (Millet and Whittaker, 2015). Cleavage at the S_1/S_2 site can occur upon viral egress, such as for MHV (Frana et al., 1985), or upon encounter with a target cell, such as for SARS-CoV (Belouzard et al., 2009; Bosch et al., 2008; Shulla et al., 2011), to yield two non-covalently associated subunits. This first cleavage event, along with binding to the host receptor, promotes further cleavage at the S_2' site for SARS-CoV S (Belouzard et al., 2009) and MERS-CoV S (Millet and Whittaker, 2014;Park et al., 2016). Proteolysis at the conserved S_2' site is essential for fusion activation of all characterized CoV S proteins, and it can occur at the host membrane or in internal cellular compartments of the target cell (Belouzard et al., 2009; Burkard et al., 2014; Millet and Whittaker, 2015; Park et al., 2016).

Cleavage at the MERS-CoV S_1/S_2 site by furin during viral egress enables subsequent exposure of the S_2' site upon binding to the host receptor and a second cleavage step by serine proteases anchored in the membrane of the target cells, eventually leading to fusion at the cytoplasmic membrane (early entry) (Park et al., 2016). Conversely, MERS-CoV budding with uncleaved S glycoproteins traffic to the endosomes of target cells where cathepsin L or other proteases promote membrane fusion (late entry) (Park et al., 2016). The former mechanism has been proposed to be the route of MERS-CoV entry into cell types relevant to lung infection, and therefore a significant determinant of MERS-CoV virulence (Park et al., 2016). Moreover, tetraspanin CD9 has been implicated in clustering DPP4 and transmembrane serine proteases to promote early entry of MERS-CoV (Earnest et al., 2017, 2015). PEDV, which replicates in the epithelial cells of the small intestine, undergoes S proteolytic activation by trypsin, which is highly abundant in the intestinal lumen (Wicht et al., 2014). The critical importance of cleavage at the S_1/S_2 site was also exemplified in studies with the MERS-CoV-related bat coronavirus HKU4. Although HKU4

S recognizes human DPP4, *in vitro* infectivity assays revealed that entry into human cells required addition of exogenous trypsin, suggesting proteolytic activation of this bat virus did not occur in human cells (Wang et al., 2014). In line with these findings, various DPP4 mammalian orthologues, with variable binding affinities for the MERS-CoV S receptor-binding domain, were shown to support virus or pseudovirus entry into target cells in the presence of an activating protease (Barlan et al., 2014). These results collectively illustrate how specific S proteolytic cleavage participates in determining the intracellular site of fusion and also viral tropism and pathogenesis of CoVs. Therefore, the zoonotic potential of CoVs is not only determined by receptor engagement, but also by proteolytic processing of the S protein required for fusion activation.

5. Mechanism of fusion activation

We showed that *in vitro* trypsin cleavage of MHV, SARS-CoV and MERS-CoV S, under limited proteolysis conditions, recapitulated fusion activation by inducing the pre-to postfusion transition (Walls et al., 2017b). The cryoEM structure of the MHV S_2 subunit ectodomain trimer revealed that membrane fusion involves large-scale S conformational changes that are reminiscent of the ones described for other class 1 fusion proteins, including the pneumovirus/paramyxovirus F glycoproteins (Fig. 3) (McLellan et al., 2011; Swanson et al., 2010, 2011; Walls et al., 2016b; Yin et al., 2005). These experiments also demonstrated that (i) the S_1 subunits stabilize the S_2 fusion machinery in the spring-loaded, metastable prefusion state before initiation of infection; and (ii) postfusion S is the ground state of the fusion reaction. Similarly to the organization of influenza virus hemagglutinins (Xiong et al., 2013), domain B interacts with the HR1-central helix hairpin in prefusion closed S structures likely to stabilize S_2 in the spring-loaded prefusion state. This interaction appears to coordinate receptor engagement with fusion. Upon receptor binding and proteolytic cleavage at the S_1/S_2 and S_2' sites, the S_1 crown is likely shed (as observed for MERS-CoV S by Yuan et al., 2017) to facilitate a conformational change of S_2, which involves projection of the fusion peptide to a distance of ~100 Å and its insertion into the target membrane (Fig. 3) (Walls et al., 2016a, 2017b). The free energy released upon S_2 refolding from the prefusion to the postfusion state is believed to bring the viral and host membranes in close proximity and promote membrane merger (Harrison, 2008).

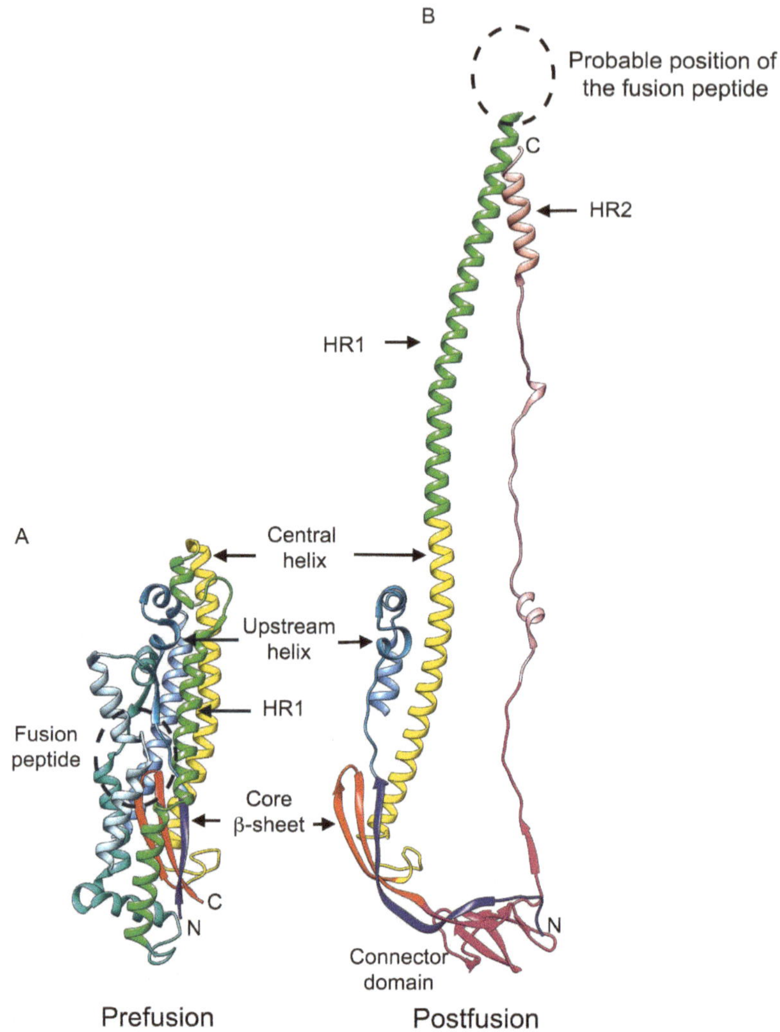

Fig. 3 CoV S conformational changes driving the fusion reaction. (A), Ribbon diagram of the MHV S_2 subunit in the prefusion conformation, PDB: 3JCL. (B), Ribbon diagram of the MHV S_2 subunit in the postfusion conformation, PDB: 6B3O. The prefusion to postfusion transition involves a "jack-knife" refolding of the HR1 helices and intervening regions into a single continuous helix appended to the central helix. The connector domain and HR2 in the prefusion structure and the fusion peptide in the postfusion structure of MHV were not resolved and are therefore not shown.

Recent structural work comparing recombinant S proteins from SARS-CoV and MERS-CoV in isolation and in complex with their cognate receptors or neutralizing antibodies suggested an activation mechanism for coronavirus fusion (Gui et al., 2017; Kirchdoerfer et al., 2018;

Pallesen et al., 2017; Song et al., 2018; Walls et al., 2019; Yuan et al., 2017). Specifically, SARS-CoV and MERS-CoV S structures in complex with neutralizing antibodies isolated from survivors showed both antibodies competitively blocked receptor interaction, in agreement with previous surface plasmon resonance data (Corti et al., 2015; Rockx et al., 2008; Traggiai et al., 2004; Walls et al., 2019). The anti-SARS-CoV S230 antibody, however, functionally mimicked the receptor by promoting S fusogenic conformational rearrangements through a molecular ratcheting mechanism (Walls et al., 2019) (Fig. 4). These observations suggested that upon receptor recognition, bound B domains are locked in the open state, thereby releasing the constraints imposed on the HR1-central helix hairpin, allowing refolding of the S_2 fusion machinery and membrane fusion to occur (Pallesen et al., 2017; Song et al., 2018; Walls et al., 2019; Yuan et al., 2017) (Fig. 4). Proteolytic activation is likely required to ensure that S glycoproteins will work in synergy, with proper spatial and temporal coordination, to drive fusion of the viral and host membranes.

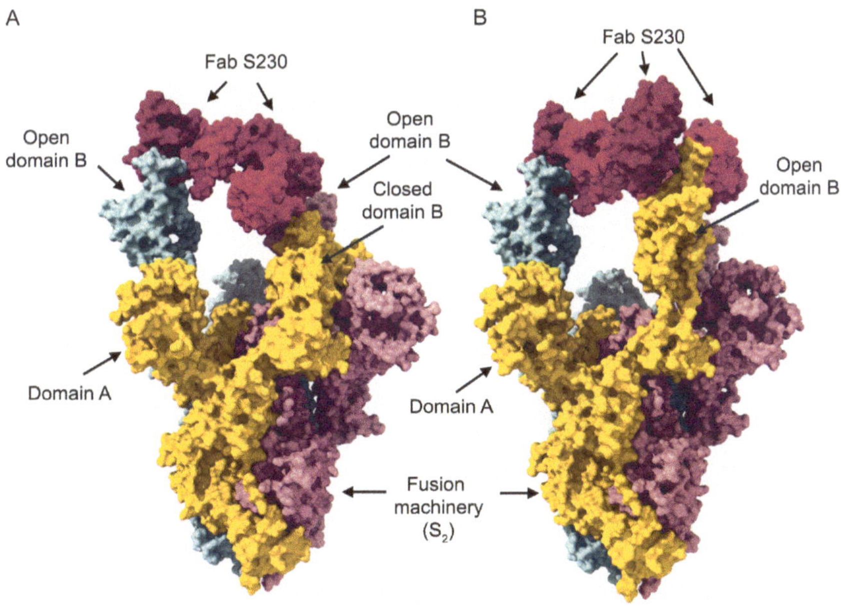

Fig. 4 CryoEM structures of the SARS-CoV S glycoprotein in complex with the S230 neutralizing antibody. (A–B), Molecular surface representation of a complex with one open, one partially open, and one closed B domain, PDB: 6NB6 (left) and with three open B domains that do not follow threefold symmetry, PDB: 6NB7 (right). The structures are rendered with different colors for each S protomer (light blue, plum and gold) and the S230 Fab heavy (dark magenta) and light (magenta) chains (only the variable domains are shown).

6. Epitope masking and glycan shielding

A deep knowledge of the organization and chemical composition of carbohydrates obstructing the surface of CoV S glycoproteins is key for understanding accessibility to neutralizing antibodies and for guiding the rational development of subunit vaccines and therapeutics. S glycoproteins feature ~20–35 predicted N-linked oligosaccharides per protomer. A cryoEM structure of the HCoV-NL63 S ectodomain allowed to visualize for the first time the extensive N-linked glycans covering the surface of a CoV S trimer (Walls et al., 2016b) (Fig. 5). A subsequent study revealed that numerous glycosylation sites are strictly or topologically conserved between PDCoV S and HCoV-NL63 S although the two glycoproteins share only

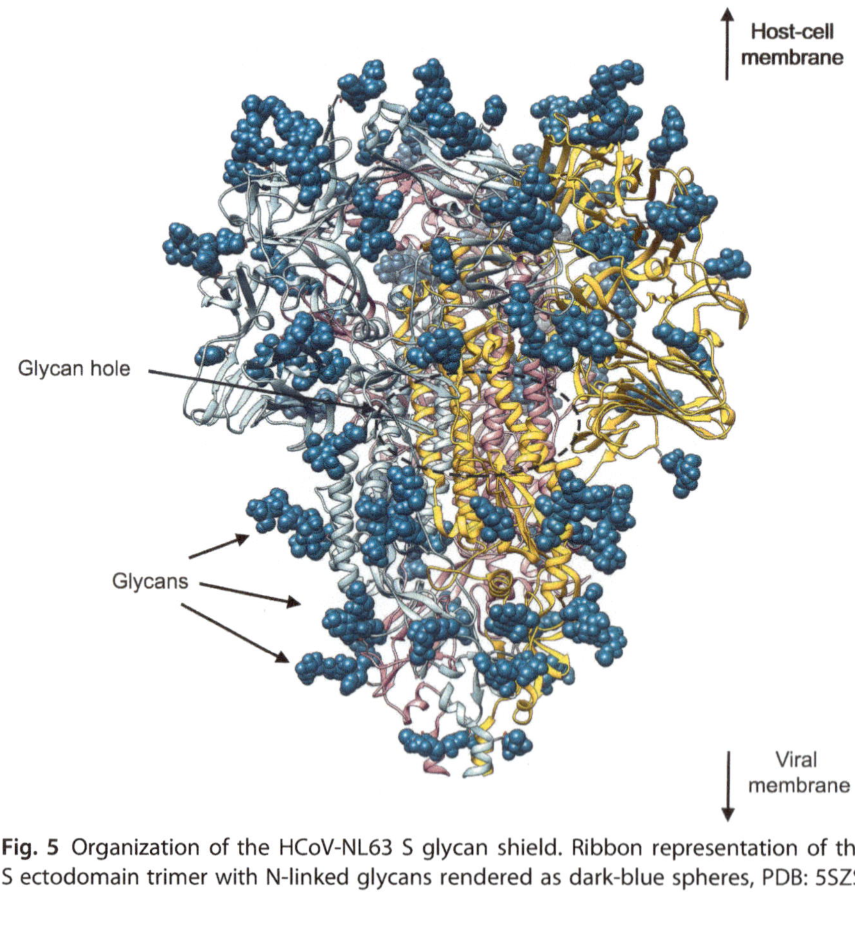

Fig. 5 Organization of the HCoV-NL63 S glycan shield. Ribbon representation of the S ectodomain trimer with N-linked glycans rendered as dark-blue spheres, PDB: 5SZS.

43% amino acid sequence identity and the two viruses belong to different genera infecting different hosts (Xiong et al., 2018). This observation suggested that all CoVs face similar immune pressure in their respective hosts, and that the areas that are masked by the conserved glycans might be key to the function of S. Based on the information gained from the HCoV-NL63 S structure, in which a glycan participates to masking the receptor-binding loops, it was proposed that the S glycan shield is involved in immune evasion, similarly to the well-characterized HIV-1 envelope trimer (Walls et al., 2016b).

Comparison of the N-linked oligosaccharides of full-length MERS-CoV S derived from virions produced in African green monkey VeroE6 cells, or of a purified MERS-CoV S ectodomain recombinantly produced in HEK293F cells, revealed an extensive overlap of glycan composition, including the presence of hybrid and complex glycans (Walls et al., 2019). Processed oligosaccharides were also observed decorating S trimers at the surface of authentic SARS-CoV virions (Krokhin et al., 2003; Ritchie et al., 2010). These data indicated that at least a fraction of the MERS-CoV and SARS-CoV virions produced in a cell are exposed to the glycan-processing enzymes residing in the Golgi apparatus during assembly and budding, in contrast with previous models of CoV budding (Ng et al., 2003; Stertz et al., 2007).

A common feature observed in the glycosylation patterns of S glycoproteins is the presence of less densely glycosylated regions surrounding the S_1/S_2 cleavage site and the conserved fusion peptide, near the S_2' cleavage site, probably to allow access to activating host proteases and for membrane fusion to take place (Walls et al., 2016b; Walls et al., 2019) (Fig. 5). These "glycan holes" could be targeted for epitope-focused immunogen design or new therapeutic development against CoV, as supported by the identification of a neutralization epitope within a comparable breach of the HIV-1 envelope glycan shield (McCoy et al., 2016).

7. Concluding remarks

Recent structural and functional characterization of CoV S glycoproteins provided insights into the mechanism used by these viruses to infect host cells and suggested possible strategies for rational design of vaccines and therapeutics. Introducing stabilizing mutations, which prevent the prefusion to postfusion S transition, led to the elicitation of improved neutralization titers in mice and will be a key tool for the design of subunit vaccines against CoVs (Kirchdoerfer et al., 2018; Pallesen et al., 2017). Furthermore, the exposure of the fusion peptide at the surface of prefusion S trimers (Walls et al., 2016a) and its conservation among CoVs indicate it might be an attractive target for broad inhibition of CoV entry. Major antigenic determinants of MHV and SARS-CoV S overlap with the fusion peptide region (Daniel et al., 1993; Zhang et al., 2004) and binding of neutralizing antibodies to this site could putatively prevent fusogenic conformational changes, as proposed for influenza virus hemagglutinin or HIV envelope (Corti et al., 2011; Kong et al., 2016; Lang et al., 2017). Finally, masking strain-specific antigenic regions *via* engineering of additional N-linked glycosylation sites, as implemented for the MERS-CoV domain B (Du et al., 2016), bears the promise of focusing the immune response on highly conserved epitopes and eliciting broadly neutralizing antibodies against CoVs.

Acknowledgments

We acknowledge support from the National Institute of General Medical Sciences (R01GM120553, D.V.), the National Institute of Allergy and Infectious Diseases (HHSN272201700059C, DV), a Pew Biomedical Scholars Award (D.V.), an Investigators in the Pathogenesis of Infectious Disease Award from the Burroughs Wellcome Fund (D.V.) and the Pasteur Institute (M.A.T.).

References

Bai, X.C., Fernandez, I.S., McMullan, G., Scheres, S.H., 2013. Ribosome structures to near-atomic resolution from thirty thousand cryo-EM particles. elife 2, e00461.

Bakkers, M.J., Lang, Y., Feitsma, L.J., Hulswit, R.J., de Poot, S.A., van Vliet, A.L., Margine, I., de Groot-Mijnes, J.D., van Kuppeveld, F.J., Langereis, M.A., Huizinga, E.G., de Groot, R.J., 2017. Betacoronavirus adaptation to humans involved progressive loss of hemagglutinin-esterase lectin activity. Cell Host Microbe 21, 356.

Bakkers, M.J., Zeng, Q., Feitsma, L.J., Hulswit, R.J., Li, Z., Westerbeke, A., van Kuppeveld, F.J., Boons, G.J., Langereis, M.A., Huizinga, E.G., de Groot, R.J., 2016. Coronavirus receptor switch explained from the stereochemistry of protein-carbohydrate interactions and a single mutation. Proc. Natl. Acad. Sci. U. S. A. 113, E3111.

Barlan, A., Zhao, J., Sarkar, M.K., Li, K., McCray, P.B., Perlman, S., Gallagher, T., 2014. Receptor variation and susceptibility to Middle East respiratory syndrome coronavirus infection. J. Virol. 88, 4953.

Belouzard, S., Chu, V.C., Whittaker, G.R., 2009. Activation of the SARS coronavirus spike protein via sequential proteolytic cleavage at two distinct sites. Proc. Natl. Acad. Sci. U. S. A. 106, 5871.

Beniac, D.R., Andonov, A., Grudeski, E., Booth, T.F., 2006. Architecture of the SARS coronavirus prefusion spike. Nat. Struct. Mol. Biol. 13, 751.

Beniac, D.R., deVarennes, S.L., Andonov, A., He, R., Booth, T.F., 2007. Conformational reorganization of the SARS coronavirus spike following receptor binding: implications for membrane fusion. PLoS One 2, e1082.

Bos, E.C., Heijnen, L., Spaan, W.J., 1995. Site directed mutagenesis of the murine coronavirus spike protein. Effects on fusion. Adv. Exp. Med. Biol. 380, 283.

Bosch, B.J., Bartelink, W., Rottier, P.J., 2008. Cathepsin L functionally cleaves the severe acute respiratory syndrome coronavirus class I fusion protein upstream of rather than adjacent to the fusion peptide. J. Virol. 82, 8887.

Bosch, B.J., de Haan, C.A., Smits, S.L., Rottier, P.J., 2005. Spike protein assembly into the coronavirion: exploring the limits of its sequence requirements. Virology 334, 306.

Bosch, B.J., van der Zee, R., de Haan, C.A., Rottier, P.J., 2003. The coronavirus spike protein is a class I virus fusion protein: structural and functional characterization of the fusion core complex. J. Virol. 77, 8801.

Brilot, A.F., Chen, J.Z., Cheng, A., Pan, J., Harrison, S.C., Potter, C.S., Carragher, B., Henderson, R., Grigorieff, N., 2012. Beam-induced motion of vitrified specimen on holey carbon film. J. Struct. Biol. 177, 630.

Burkard, C., Verheije, M.H., Wicht, O., van Kasteren, S.I., van Kuppeveld, F.J., Haagmans, B.L., Pelkmans, L., Rottier, P.J., Bosch, B.J., de Haan, C.A., 2014. Coronavirus cell entry occurs through the endo-/lysosomal pathway in a proteolysis-dependent manner. PLoS Pathog. 10, e1004502.

Campbell, M.G., Cheng, A., Brilot, A.F., Moeller, A., Lyumkis, D., Veesler, D., Pan, J., Harrison, S.C., Potter, C.S., Carragher, B., Grigorieff, N., 2012. Movies of ice-embedded particles enhance resolution in electron cryo-microscopy. Structure 20, 1823.

Campbell, M.G., Veesler, D., Cheng, A., Potter, C.S., Carragher, B., 2015. 2.8 Å resolution reconstruction of the Thermoplasma acidophilum 20S proteasome using cryo-electron microscopy. Elife 4. https://doi.org/10.7554/eLife.06380.

Chang, K.W., Sheng, Y., Gombold, J.L., 2000. Coronavirus-induced membrane fusion requires the cysteine-rich domain in the spike protein. Virology 269, 212.

Chen, J., Kovacs, J.M., Peng, H., Rits-Volloch, S., Lu, J., Park, D., Zablowsky, E., Seaman, M.S., Chen, B., 2015. HIV-1 envelope. Effect of the cytoplasmic domain on antigenic characteristics of HIV-1 envelope glycoprotein. Science 349, 191.

Chen, Y., Rajashankar, K.R., Yang, Y., Agnihothram, S.S., Liu, C., Lin, Y.L., Baric, R.S., Li, F., 2013. Crystal structure of the receptor-binding domain from newly emerged Middle East respiratory syndrome coronavirus. J. Virol. 87, 10777.

Corti, D., Voss, J., Gamblin, S.J., Codoni, G., Macagno, A., Jarrossay, D., Vachieri, S.G., Pinna, D., Minola, A., Vanzetta, F., Silacci, C., Fernandez-Rodriguez, B.M., Agatic, G., Bianchi, S., Giacchetto-Sasselli, I., Calder, L., Sallusto, F., Collins, P., Haire, L.F., Temperton, N., Langedijk, J.P., Skehel, J.J., Lanzavecchia, A., 2011. A neutralizing antibody selected from plasma cells that binds to group 1 and group 2 influenza A hemagglutinins. Science 333, 850.

Corti, D., Zhao, J., Pedotti, M., Simonelli, L., Agnihothram, S., Fett, C., Fernandez-Rodriguez, B., Foglierini, M., Agatic, G., Vanzetta, F., Gopal, R., Langrish, C.J., Barrett, N.A., Sallusto, F., Baric, R.S., Varani, L., Zambon, M., Perlman, S., Lanzavecchia, A., 2015. Prophylactic and postexposure efficacy of a potent human monoclonal antibody against MERS coronavirus. Proc. Natl. Acad. Sci. U. S. A. 112, 10473.

Daniel, C., Anderson, R., Buchmeier, M.J., Fleming, J.O., Spaan, W.J., Wege, H., Talbot, P.J., 1993. Identification of an immunodominant linear neutralization domain on the S2 portion of the murine coronavirus spike glycoprotein and evidence that it forms part of complex tridimensional structure. J. Virol. 67, 1185.

Delmas, B., Gelfi, J., L'Haridon, R., Vogel, L.K., Sjöström, H., Norén, O., Laude, H., 1992. Aminopeptidase N is a major receptor for the entero-pathogenic coronavirus TGEV. Nature 357, 417.

Delmas, B., Gelfi, J., Sjöström, H., Noren, O., Laude, H., 1993. Further characterization of aminopeptidase-N as a receptor for coronaviruses. Adv. Exp. Med. Biol. 342, 293.

Dev, J., Park, D., Fu, Q., Chen, J., Ha, H.J., Ghantous, F., Herrmann, T., Chang, W., Liu, Z., Frey, G., Seaman, M.S., Chen, B., Chou, J.J., 2016. Structural basis for membrane anchoring of HIV-1 envelope spike. Science 353, 172.

Du, L., Tai, W., Yang, Y., Zhao, G., Zhu, Q., Sun, S., Liu, C., Tao, X., Tseng, C.K., Perlman, S., Jiang, S., Zhou, Y., Li, F., 2016. Introduction of neutralizing immunogenicity index to the rational design of MERS coronavirus subunit vaccines. Nat. Commun. 7, 13473.

Duquerroy, S., Vigouroux, A., Rottier, P.J., Rey, F.A., Bosch, B.J., 2005. Central ions and lateral asparagine/glutamine zippers stabilize the post-fusion hairpin conformation of the SARS coronavirus spike glycoprotein. Virology 335, 276.

Dveksler, G.S., Pensiero, M.N., Cardellichio, C.B., Williams, R.K., Jiang, G.S., Holmes, K.V., Dieffenbach, C.W., 1991. Cloning of the mouse hepatitis virus (MHV) receptor: expression in human and hamster cell lines confers susceptibility to MHV. J. Virol. 65, 6881.

Earnest, J.T., Hantak, M.P., Li, K., McCray, P.B., Perlman, S., Gallagher, T., 2017. The tetraspanin CD9 facilitates MERS-coronavirus entry by scaffolding host cell receptors and proteases. PLoS Pathog. 13, e1006546.

Earnest, J.T., Hantak, M.P., Park, J.E., Gallagher, T., 2015. Coronavirus and influenza virus proteolytic priming takes place in tetraspanin-enriched membrane microdomains. J. Virol. 89, 6093.

Eifart, P., Ludwig, K., Bottcher, C., de Haan, C.A., Rottier, P.J., Korte, T., Herrmann, A., 2007. Role of endocytosis and low pH in murine hepatitis virus strain A59 cell entry. J. Virol. 81, 10758.

Frana, M.F., Behnke, J.N., Sturman, L.S., Holmes, K.V., 1985. Proteolytic cleavage of the E2 glycoprotein of murine coronavirus: host-dependent differences in proteolytic cleavage and cell fusion. J. Virol. 56, 912.

Gao, J., Lu, G., Qi, J., Li, Y., Wu, Y., Deng, Y., Geng, H., Li, H., Wang, Q., Xiao, H., Tan, W., Yan, J., Gao, G.F., 2013. Structure of the fusion core and inhibition of fusion by a heptad repeat peptide derived from the S protein of Middle East respiratory syndrome coronavirus. J. Virol. 87, 13134.

Ge, X.Y., Li, J.L., Yang, X.L., Chmura, A.A., Zhu, G., Epstein, J.H., Mazet, J.K., Hu, B., Zhang, W., Peng, C., Zhang, Y.J., Luo, C.M., Tan, B., Wang, N., Zhu, Y., Crameri, G., Zhang, S.Y., Wang, L.F., Daszak, P., Shi, Z.L., 2013. Isolation and characterization of a bat SARS-like coronavirus that uses the ACE2 receptor. Nature 503, 535.

Gui, M., Song, W., Zhou, H., Xu, J., Chen, S., Xiang, Y., Wang, X., 2017. Cryo-electron microscopy structures of the SARS-CoV spike glycoprotein reveal a prerequisite conformational state for receptor binding. Cell Res. 27, 119.

Haagmans, B.L., Al Dhahiry, S.H., Reusken, C.B., Raj, V.S., Galiano, M., Myers, R., Godeke, G.J., Jonges, M., Farag, E., Diab, A., Ghobashy, H., Alhajri, F., Al-Thani, M., Al-Marri, S.A., Al Romaihi, H.E., Al Khal, A., Bermingham, A., Osterhaus, A.D., AlHajri, M.M., Koopmans, M.P., 2014. Middle East respiratory syndrome coronavirus in dromedary camels: an outbreak investigation. Lancet Infect. Dis. 14, 140.

Harrison, S.C., 2008. Viral membrane fusion. Nat. Struct. Mol. Biol. 15, 690.

Heald-Sargent, T., Gallagher, T., 2012. Ready, set, fuse! The coronavirus spike protein and acquisition of fusion competence. Viruses 4, 557.

Hofmann, H., Pyrc, K., van der Hoek, L., Geier, M., Berkhout, B., Pohlmann, S., 2005. Human coronavirus NL63 employs the severe acute respiratory syndrome coronavirus receptor for cellular entry. Proc. Natl. Acad. Sci. U. S. A. 102, 7988.

Hu, B., Zeng, L.P., Yang, X.L., Ge, X.Y., Zhang, W., Li, B., Xie, J.Z., Shen, X.R., Zhang, Y.Z., Wang, N., Luo, D.S., Zheng, X.S., Wang, M.N., Daszak, P., Wang, L.F., Cui, J., Shi, Z.L., 2017. Discovery of a rich gene pool of bat SARS-related

coronaviruses provides new insights into the origin of SARS coronavirus. PLoS Pathog. 13, e1006698.
Hulswit, R.J., de Haan, C.A., Bosch, B.J., 2016. Coronavirus spike protein and tropism changes. Adv. Virus Res. 96, 29.
Hulswit, R.J.G., Lang, Y., Bakkers, M.J.G., Li, W., Li, Z., Schouten, A., Ophorst, B., van Kuppeveld, F.J.M., Boons, G.J., Bosch, B.J., Huizinga, E.G., de Groot, R.J., 2019. Human coronaviruses OC43 and HKU1 bind to 9-O-acetylated sialic acids via a conserved receptor-binding site in spike protein domain A. Proc. Natl. Acad. Sci. U. S. A. 116, 2681.
Inoue, Y., Tanaka, N., Tanaka, Y., Inoue, S., Morita, K., Zhuang, M., Hattori, T., Sugamura, K., 2007. Clathrin-dependent entry of severe acute respiratory syndrome coronavirus into target cells expressing ACE2 with the cytoplasmic tail deleted. J. Virol. 81, 8722.
Isaacs, D., Flowers, D., Clarke, J.R., Valman, H.B., MacNaughton, M.R., 1983. Epidemiology of coronavirus respiratory infections. Arch. Dis. Child. 58, 500.
Kirchdoerfer, R.N., Cottrell, C.A., Wang, N., Pallesen, J., Yassine, H.M., Turner, H.L., Corbett, K.S., Graham, B.S., McLellan, J.S., Ward, A.B., 2016. Pre-fusion structure of a human coronavirus spike protein. Nature 531, 118.
Kirchdoerfer, R.N., Wang, N., Pallesen, J., Wrapp, D., Turner, H.L., Cottrell, C.A., Corbett, K.S., Graham, B.S., McLellan, J.S., Ward, A.B., 2018. Stabilized coronavirus spikes are resistant to conformational changes induced by receptor recognition or proteolysis. Sci. Rep. 8, 15701.
Kong, R., Xu, K., Zhou, T., Acharya, P., Lemmin, T., Liu, K., Ozorowski, G., Soto, C., Taft, J.D., Bailer, R.T., Cale, E.M., Chen, L., Choi, C.W., Chuang, G.Y., Doria-Rose, N.A., Druz, A., Georgiev, I.S., Gorman, J., Huang, J., Joyce, M.G., Louder, M.K., Ma, X., McKee, K., O'Dell, S., Pancera, M., Yang, Y., Blanchard, S.C., Mothes, W., Burton, D.R., Koff, W.C., Connors, M., Ward, A.B., Kwong, P.D., Mascola, J.R., 2016. Fusion peptide of HIV-1 as a site of vulnerability to neutralizing antibody. Science 352, 828.
Krempl, C., Schultze, B., Herrler, G., 1995. Analysis of cellular receptors for human coronavirus OC43. Adv. Exp. Med. Biol. 380, 371.
Krempl, C., Schultze, B., Laude, H., Herrler, G., 1997. Point mutations in the S protein connect the sialic acid binding activity with the enteropathogenicity of transmissible gastroenteritis coronavirus. J. Virol. 71, 3285.
Krokhin, O., Li, Y., Andonov, A., Feldmann, H., Flick, R., Jones, S., Stroeher, U., Bastien, N., Dasuri, K.V., Cheng, K., Simonsen, J.N., Perreault, H., Wilkins, J., Ens, W., Plummer, F., Standing, K.G., 2003. Mass spectrometric characterization of proteins from the SARS virus: a preliminary report. Mol. Cell. Proteomics 2, 346.
Lang, S., Xie, J., Zhu, X., Wu, N.C., Lerner, R.A., Wilson, I.A., 2017. Antibody 27F3 broadly targets influenza A group 1 and 2 hemagglutinins through a further variation in VH1-69 antibody orientation on the HA stem. Cell Rep. 20, 2935.
Li, F., Li, W., Farzan, M., Harrison, S.C., 2005a. Structure of SARS coronavirus spike receptor-binding domain complexed with receptor. Science 309, 1864.
Li, W., Hulswit, R.J.G., Kenney, S.P., Widjaja, I., Jung, K., Alhamo, M.A., van Dieren, B., van Kuppeveld, F.J.M., Saif, L.J., Bosch, B.J., 2018. Broad receptor engagement of an emerging global coronavirus may potentiate its diverse cross-species transmissibility. Proc. Natl. Acad. Sci. U. S. A. 115, E5135.
Li, W., Hulswit, R.J.G., Widjaja, I., Raj, V.S., McBride, R., Peng, W., Widagdo, W., Tortorici, M.A., van Dieren, B., Lang, Y., van Lent, J.W.M., Paulson, J.C., de Haan, C.A.M., de Groot, R.J., van Kuppeveld, F.J.M., Haagmans, B.L., Bosch, B.J., 2017. Identification of sialic acid-binding function for the Middle East respiratory syndrome coronavirus spike glycoprotein. Proc. Natl. Acad. Sci. U. S. A. 114, E8508.

Li, W., Moore, M.J., Vasilieva, N., Sui, J., Wong, S.K., Berne, M.A., Somasundaran, M., Sullivan, J.L., Luzuriaga, K., Greenough, T.C., Choe, H., Farzan, M., 2003. Angiotensin-converting enzyme 2 is a functional receptor for the SARS coronavirus. Nature 426, 450.

Li, W., Wicht, O., van Kuppeveld, F.J., He, Q., Rottier, P.J., Bosch, B.J., 2015. A single point mutation creating a furin cleavage site in the spike protein renders porcine epidemic diarrhea coronavirus trypsin independent for cell entry and fusion. J. Virol. 89, 8077.

Li, W., Zhang, C., Sui, J., Kuhn, J.H., Moore, M.J., Luo, S., Wong, S.K., Huang, I.C., Xu, K., Vasilieva, N., Murakami, A., He, Y., Marasco, W.A., Guan, Y., Choe, H., Farzan, M., 2005b. Receptor and viral determinants of SARS-coronavirus adaptation to human ACE2. EMBO J. 24, 1634.

Li, X., Mooney, P., Zheng, S., Booth, C.R., Braunfeld, M.B., Gubbens, S., Agard, D.A., Cheng, Y., 2013. Electron counting and beam-induced motion correction enable near-atomic-resolution single-particle cryo-EM. Nat. Methods 10, 584.

Liu, C., Tang, J., Ma, Y., Liang, X., Yang, Y., Peng, G., Qi, Q., Jiang, S., Li, J., Du, L., Li, F., 2015. Receptor usage and cell entry of porcine epidemic diarrhea coronavirus. J. Virol. 89, 6121.

Lontok, E., Corse, E., Machamer, C.E., 2004. Intracellular targeting signals contribute to localization of coronavirus spike proteins near the virus assembly site. J. Virol. 78, 5913.

Lu, G., Hu, Y., Wang, Q., Qi, J., Gao, F., Li, Y., Zhang, Y., Zhang, W., Yuan, Y., Bao, J., Zhang, B., Shi, Y., Yan, J., Gao, G.F., 2013. Molecular basis of binding between novel human coronavirus MERS-CoV and its receptor CD26. Nature 500, 227.

McCoy, L.E., van Gils, M.J., Ozorowski, G., Messmer, T., Briney, B., Voss, J.E., Kulp, D.W., Macauley, M.S., Sok, D., Pauthner, M., Menis, S., Cottrell, C.A., Torres, J.L., Hsueh, J., Schief, W.R., Wilson, I.A., Ward, A.B., Sanders, R.W., Burton, D.R., 2016. Holes in the glycan shield of the native HIV envelope are a target of trimer-elicited neutralizing antibodies. Cell Rep. 16, 2327.

McLellan, J.S., Chen, M., Leung, S., Graepel, K.W., Du, X., Yang, Y., Zhou, T., Baxa, U., Yasuda, E., Beaumont, T., Kumar, A., Modjarrad, K., Zheng, Z., Zhao, M., Xia, N., Kwong, P.D., Graham, B.S., 2013. Structure of RSV fusion glycoprotein trimer bound to a prefusion-specific neutralizing antibody. Science 340, 1113.

McLellan, J.S., Yang, Y., Graham, B.S., Kwong, P.D., 2011. Structure of respiratory syncytial virus fusion glycoprotein in the postfusion conformation reveals preservation of neutralizing epitopes. J. Virol. 85, 7788.

Menachery, V.D., Yount, B.L., Debbink, K., Agnihothram, S., Gralinski, L.E., Plante, J.A., Graham, R.L., Scobey, T., Ge, X.Y., Donaldson, E.F., Randell, S.H., Lanzavecchia, A., Marasco, W.A., Shi, Z.L., Baric, R.S., 2015. A SARS-like cluster of circulating bat coronaviruses shows potential for human emergence. Nat. Med. 21, 1508.

Menachery, V.D., Yount, B.L., Sims, A.C., Debbink, K., Agnihothram, S.S., Gralinski, L.E., Graham, R.L., Scobey, T., Plante, J.A., Royal, S.R., Swanstrom, J., Sheahan, T.P., Pickles, R.J., Corti, D., Randell, S.H., Lanzavecchia, A., Marasco, W.A., Baric, R.S., 2016. SARS-like WIV1-CoV poised for human emergence. Proc. Natl. Acad. Sci. U. S. A. 113, 3048.

Milewska, A., Zarebski, M., Nowak, P., Stozek, K., Potempa, J., Pyrc, K., 2014. Human coronavirus NL63 utilizes heparan sulfate proteoglycans for attachment to target cells. J. Virol. 88, 13221.

Millet, J.K., Whittaker, G.R., 2014. Host cell entry of Middle East respiratory syndrome coronavirus after two-step, furin-mediated activation of the spike protein. Proc. Natl. Acad. Sci. U. S. A. 111, 15214.

Millet, J.K., Whittaker, G.R., 2015. Host cell proteases: critical determinants of coronavirus tropism and pathogenesis. Virus Res. 202, 120.

Ng, M.L., Tan, S.H., See, E.E., Ooi, E.E., Ling, A.E., 2003. Proliferative growth of SARS coronavirus in Vero E6 cells. J. Gen. Virol. 84, 3291.
Nomura, R., Kiyota, A., Suzaki, E., Kataoka, K., Ohe, Y., Miyamoto, K., Senda, T., Fujimoto, T., 2004. Human coronavirus 229E binds to CD13 in rafts and enters the cell through caveolae. J. Virol. 78, 8701.
Owczarek, K., Szczepanski, A., Milewska, A., Baster, Z., Rajfur, Z., Sarna, M., Pyrc, K., 2018. Early events during human coronavirus OC43 entry to the cell. Sci. Rep. 8, 7124.
Pallesen, J., Wang, N., Corbett, K.S., Wrapp, D., Kirchdoerfer, R.N., Turner, H.L., Cottrell, C.A., Becker, M.M., Wang, L., Shi, W., Kong, W.P., Andres, E.L., Kettenbach, A.N., Denison, M.R., Chappell, J.D., Graham, B.S., Ward, A.B., McLellan, J.S., 2017. Immunogenicity and structures of a rationally designed prefusion MERS-CoV spike antigen. Proc. Natl. Acad. Sci. U. S. A. 114, E7348.
Park, J.E., Li, K., Barlan, A., Fehr, A.R., Perlman, S., McCray, P.B., Gallagher, T., 2016. Proteolytic processing of Middle East respiratory syndrome coronavirus spikes expands virus tropism. Proc. Natl. Acad. Sci. U. S. A. 113, 12262.
Peng, G., Sun, D., Rajashankar, K.R., Qian, Z., Holmes, K.V., Li, F., 2011. Crystal structure of mouse coronavirus receptor-binding domain complexed with its murine receptor. Proc. Natl. Acad. Sci. U. S. A. 108, 10696.
Peng, G., Xu, L., Lin, Y.L., Chen, L., Pasquarella, J.R., Holmes, K.V., Li, F., 2012. Crystal structure of bovine coronavirus spike protein lectin domain. J. Biol. Chem. 287, 41931.
Petit, C.M., Melancon, J.M., Chouljenko, V.N., Colgrove, R., Farzan, M., Knipe, D.M., Kousoulas, K.G., 2005. Genetic analysis of the SARS-coronavirus spike glycoprotein functional domains involved in cell-surface expression and cell-to-cell fusion. Virology 341, 215.
Prabakaran, P., Gan, J., Feng, Y., Zhu, Z., Choudhry, V., Xiao, X., Ji, X., Dimitrov, D.S., 2006. Structure of severe acute respiratory syndrome coronavirus receptor-binding domain complexed with neutralizing antibody. J. Biol. Chem. 281, 15829.
Punjani, A., Rubinstein, J.L., Fleet, D.J., Brubaker, M.A., 2017. cryoSPARC: algorithms for rapid unsupervised cryo-EM structure determination. Nat. Methods 14, 290.
Raj, V.S., Mou, H., Smits, S.L., Dekkers, D.H., Muller, M.A., Dijkman, R., Muth, D., Demmers, J.A., Zaki, A., Fouchier, R.A., Thiel, V., Drosten, C., Rottier, P.J., Osterhaus, A.D., Bosch, B.J., Haagmans, B.L., 2013. Dipeptidyl peptidase 4 is a functional receptor for the emerging human coronavirus-EMC. Nature 495, 251.
Regan, A.D., Shraybman, R., Cohen, R.D., Whittaker, G.R., 2008. Differential role for low pH and cathepsin-mediated cleavage of the viral spike protein during entry of serotype II feline coronaviruses. Vet. Microbiol. 132, 235.
Reguera, J., Santiago, C., Mudgal, G., Ordoño, D., Enjuanes, L., Casasnovas, J.M., 2012. Structural bases of coronavirus attachment to host aminopeptidase N and its inhibition by neutralizing antibodies. PLoS Pathog. 8, e1002859.
Ritchie, G., Harvey, D.J., Feldmann, F., Stroeher, U., Feldmann, H., Royle, L., Dwek, R.A., Rudd, P.M., 2010. Identification of N-linked carbohydrates from severe acute respiratory syndrome (SARS) spike glycoprotein. Virology 399, 257.
Rockx, B., Corti, D., Donaldson, E., Sheahan, T., Stadler, K., Lanzavecchia, A., Baric, R., 2008. Structural basis for potent cross-neutralizing human monoclonal antibody protection against lethal human and zoonotic severe acute respiratory syndrome coronavirus challenge. J. Virol. 82, 3220.
Rosenthal, P.B., Zhang, X., Formanowski, F., Fitz, W., Wong, C.H., Meier-Ewert, H., Skehel, J.J., Wiley, D.C., 1998. Structure of the haemagglutinin-esterase-fusion glycoprotein of influenza C virus. Nature 396, 92.
Sabir, J.S., Lam, T.T., Ahmed, M.M., Li, L., Shen, Y., Abo-Aba, S.E., Qureshi, M.I., Abu-Zeid, M., Zhang, Y., Khiyami, M.A., Alharbi, N.S., Hajrah, N.H., Sabir, M.J., Mutwakil, M.H., Kabli, S.A., Alsulaimany, F.A., Obaid, A.Y., Zhou, B.,

Smith, D.K., Holmes, E.C., Zhu, H., Guan, Y., 2016. Co-circulation of three camel coronavirus species and recombination of MERS-CoVs in Saudi Arabia. Science 351, 81.

Scheres, S.H., 2012. RELION: implementation of a Bayesian approach to cryo-EM structure determination. J. Struct. Biol. 180, 519.

Schroth-Diez, B., Ludwig, K., Baljinnyam, B., Kozerski, C., Huang, Q., Herrmann, A., 2000. The role of the transmembrane and of the intraviral domain of glycoproteins in membrane fusion of enveloped viruses. Biosci. Rep. 20, 571.

Schwegmann-Wessels, C., Zimmer, G., Schröder, B., Breves, G., Herrler, G., 2003. Binding of transmissible gastroenteritis coronavirus to brush border membrane sialoglycoproteins. J. Virol. 77, 11846.

Shang, J., Zheng, Y., Yang, Y., Liu, C., Geng, Q., Luo, C., Zhang, W., Li, F., 2018a. Cryo-EM structure of infectious bronchitis coronavirus spike protein reveals structural and functional evolution of coronavirus spike proteins. PLoS Pathog. 14, e1007009.

Shang, J., Zheng, Y., Yang, Y., Liu, C., Geng, Q., Tai, W., Du, L., Zhou, Y., Zhang, W., Li, F., 2018b. Cryo-electron microscopy structure of porcine deltacoronavirus spike protein in the prefusion state. J. Virol. 92 (4), pii: e01556-17. https://doi.org/10.1128/JVI.01556-17.

Shulla, A., Heald-Sargent, T., Subramanya, G., Zhao, J., Perlman, S., Gallagher, T., 2011. A transmembrane serine protease is linked to the severe acute respiratory syndrome coronavirus receptor and activates virus entry. J. Virol. 85, 873.

Simmons, G., Gosalia, D.N., Rennekamp, A.J., Reeves, J.D., Diamond, S.L., Bates, P., 2005. Inhibitors of cathepsin L prevent severe acute respiratory syndrome coronavirus entry. Proc. Natl. Acad. Sci. U. S. A. 102, 11876.

Song, W., Gui, M., Wang, X., Xiang, Y., 2018. Cryo-EM structure of the SARS coronavirus spike glycoprotein in complex with its host cell receptor ACE2. PLoS Pathog. 14, e1007236.

Stertz, S., Reichelt, M., Spiegel, M., Kuri, T., Martínez-Sobrido, L., García-Sastre, A., Weber, F., Kochs, G., 2007. The intracellular sites of early replication and budding of SARS-coronavirus. Virology 361, 304.

Su, S., Wong, G., Shi, W., Liu, J., Lai, A.C.K., Zhou, J., Liu, W., Bi, Y., Gao, G.F., 2016. Epidemiology, genetic recombination, and pathogenesis of coronaviruses. Trends Microbiol. 24, 490.

Supekar, V.M., Bruckmann, C., Ingallinella, P., Bianchi, E., Pessi, A., Carfí, A., 2004. Structure of a proteolytically resistant core from the severe acute respiratory syndrome coronavirus S2 fusion protein. Proc. Natl. Acad. Sci. U. S. A. 101, 17958.

Swanson, K., Wen, X., Leser, G.P., Paterson, R.G., Lamb, R.A., Jardetzky, T.S., 2010. Structure of the Newcastle disease virus F protein in the post-fusion conformation. Virology 402, 372.

Swanson, K.A., Settembre, E.C., Shaw, C.A., Dey, A.K., Rappuoli, R., Mandl, C.W., Dormitzer, P.R., Carfi, A., 2011. Structural basis for immunization with postfusion respiratory syncytial virus fusion F glycoprotein (RSV F) to elicit high neutralizing antibody titers. Proc. Natl. Acad. Sci. U. S. A. 108, 9619.

Thorp, E.B., Boscarino, J.A., Logan, H.L., Goletz, J.T., Gallagher, T.M., 2006. Palmitoylations on murine coronavirus spike proteins are essential for virion assembly and infectivity. J. Virol. 80, 1280.

Tortorici, M.A., Walls, A.C., Lang, Y., Wang, C., Li, Z., Koerhuis, D., Boons, G.J., Bosch, B.J., Rey, F.A., de Groot, R.J., Veesler, D., 2019. Structural basis for human coronavirus attachment to sialic acid receptors. Nat. Struct. Mol. Biol. 26, 481.

Traggiai, E., Becker, S., Subbarao, K., Kolesnikova, L., Uematsu, Y., Gismondo, M.R., Murphy, B.R., Rappuoli, R., Lanzavecchia, A., 2004. An efficient method to make human monoclonal antibodies from memory B cells: potent neutralization of SARS coronavirus. Nat. Med. 10, 871.

Van Hamme, E., Dewerchin, H.L., Cornelissen, E., Verhasselt, B., Nauwynck, H.J., 2008. Clathrin- and caveolae-independent entry of feline infectious peritonitis virus in monocytes depends on dynamin. J. Gen. Virol. 89, 2147.

Vlasak, R., Luytjes, W., Spaan, W., Palese, P., 1988. Human and bovine coronaviruses recognize sialic acid-containing receptors similar to those of influenza C viruses. Proc. Natl. Acad. Sci. U. S. A. 85, 4526.

Walls, A., Tortorici, M.A., Bosch, B.J., Frenz, B., Rottier, P.J., DiMaio, F., Rey, F.A., Veesler, D., 2017a. Crucial steps in the structure determination of a coronavirus spike glycoprotein using cryo-electron microscopy. Protein Sci. 26, 113.

Walls, A.C., Tortorici, M.A., Bosch, B.J., Frenz, B., Rottier, P.J.M., DiMaio, F., Rey, F.A., Veesler, D., 2016a. Cryo-electron microscopy structure of a coronavirus spike glycoprotein trimer. Nature 531, 114.

Walls, A.C., Tortorici, M.A., Frenz, B., Snijder, J., Li, W., Rey, F.A., DiMaio, F., Bosch, B.J., Veesler, D., 2016b. Glycan shield and epitope masking of a coronavirus spike protein observed by cryo-electron microscopy. Nat. Struct. Mol. Biol. 23, 899.

Walls, A.C., Tortorici, M.A., Snijder, J., Xiong, X., Bosch, B.J., Rey, F.A., Veesler, D., 2017b. Tectonic conformational changes of a coronavirus spike glycoprotein promote membrane fusion. Proc. Natl. Acad. Sci. U. S. A. 114, 11157.

Walls, A.C., Xiong, X., Park, Y.J., Tortorici, M.A., Snijder, J., Quispe, J., Cameroni, E., Gopal, R., Dai, M., Lanzavecchia, A., Zambon, M., Rey, F.A., Corti, D., Veesler, D., 2019. Unexpected receptor functional mimicry elucidates activation of coronavirus fusion. Cell 176, 1026.

Wang, N., Shi, X., Jiang, L., Zhang, S., Wang, D., Tong, P., Guo, D., Fu, L., Cui, Y., Liu, X., Arledge, K.C., Chen, Y.H., Zhang, L., Wang, X., 2013. Structure of MERS-CoV spike receptor-binding domain complexed with human receptor DPP4. Cell Res. 23, 986.

Wang, Q., Qi, J., Yuan, Y., Xuan, Y., Han, P., Wan, Y., Ji, W., Li, Y., Wu, Y., Wang, J., Iwamoto, A., Woo, P.C., Yuen, K.Y., Yan, J., Lu, G., Gao, G.F., 2014. Bat origins of MERS-CoV supported by bat coronavirus HKU4 usage of human receptor CD26. Cell Host Microbe 16, 328.

Wang, S., Guo, F., Liu, K., Wang, H., Rao, S., Yang, P., Jiang, C., 2008. Endocytosis of the receptor-binding domain of SARS-CoV spike protein together with virus receptor ACE2. Virus Res. 136, 8.

Wicht, O., Li, W., Willems, L., Meuleman, T.J., Wubbolts, R.W., van Kuppeveld, F.J., Rottier, P.J., Bosch, B.J., 2014. Proteolytic activation of the porcine epidemic diarrhea coronavirus spike fusion protein by trypsin in cell culture. J. Virol. 88, 7952.

Wickramasinghe, I.N., de Vries, R.P., Grone, A., de Haan, C.A., Verheije, M.H., 2011. Binding of avian coronavirus spike proteins to host factors reflects virus tropism and pathogenicity. J. Virol. 85, 8903.

Williams, R.K., Jiang, G.S., Holmes, K.V., 1991. Receptor for mouse hepatitis virus is a member of the carcinoembryonic antigen family of glycoproteins. Proc. Natl. Acad. Sci. U. S. A. 88, 5533.

Wong, A.H.M., Tomlinson, A.C.A., Zhou, D., Satkunarajah, M., Chen, K., Sharon, C., Desforges, M., Talbot, P.J., Rini, J.M., 2017. Receptor-binding loops in alphacoronavirus adaptation and evolution. Nat. Commun. 8, 1735.

Wong, J.J., Paterson, R.G., Lamb, R.A., Jardetzky, T.S., 2016. Structure and stabilization of the Hendra virus F glycoprotein in its prefusion form. Proc. Natl. Acad. Sci. U. S. A. 113, 1056.

Wu, K., Li, W., Peng, G., Li, F., 2009. Crystal structure of NL63 respiratory coronavirus receptor-binding domain complexed with its human receptor. Proc. Natl. Acad. Sci. U. S. A. 106, 19970.

Xiong, X., Coombs, P.J., Martin, S.R., Liu, J., Xiao, H., McCauley, J.W., Locher, K., Walker, P.A., Collins, P.J., Kawaoka, Y., Skehel, J.J., Gamblin, S.J., 2013. Receptor binding by a ferret-transmissible H5 avian influenza virus. Nature 497, 392.

Xiong, X., Tortorici, M.A., Snijder, J., Yoshioka, C., Walls, A.C., Li, W., McGuire, A.T., Rey, F.A., Bosch, B.J., Veesler, D., 2018. Glycan shield and fusion activation of a deltacoronavirus spike glycoprotein fine-tuned for enteric infections. J. Virol. 92 (4), pii: e01628-17. https://doi.org/10.1128/JVI.01628-17.
Xu, K., Chan, Y.P., Bradel-Tretheway, B., Akyol-Ataman, Z., Zhu, Y., Dutta, S., Yan, L., Feng, Y., Wang, L.F., Skiniotis, G., Lee, B., Zhou, Z.H., Broder, C.C., Aguilar, H.C., Nikolov, D.B., 2015. Crystal structure of the pre-fusion nipah virus fusion glycoprotein reveals a novel hexamer-of-trimers assembly. PLoS Pathog. 11, e1005322.
Xu, Y., Liu, Y., Lou, Z., Qin, L., Li, X., Bai, Z., Pang, H., Tien, P., Gao, G.F., Rao, Z., 2004a. Structural basis for coronavirus-mediated membrane fusion. Crystal structure of mouse hepatitis virus spike protein fusion core. J. Biol. Chem. 279, 30514.
Xu, Y., Su, N., Qin, L., Bai, Z., Gao, G.F., Rao, Z., 2004b. Crystallization and preliminary crystallographic analysis of the heptad-repeat complex of SARS coronavirus spike protein. Acta Crystallogr. D Biol. Crystallogr. 60, 2377.
Yamada, Y., Liu, D.X., 2009. Proteolytic activation of the spike protein at a novel RRRR/S motif is implicated in furin-dependent entry, syncytium formation, and infectivity of coronavirus infectious bronchitis virus in cultured cells. J. Virol. 83, 8744.
Yang, Y., Liu, C., Du, L., Jiang, S., Shi, Z., Baric, R.S., Li, F., 2015. Two mutations were critical for bat-to-human transmission of Middle East respiratory syndrome coronavirus. J. Virol. 89, 9119.
Ye, R., Montalto-Morrison, C., Masters, P.S., 2004. Genetic analysis of determinants for spike glycoprotein assembly into murine coronavirus virions: distinct roles for charge-rich and cysteine-rich regions of the endodomain. J. Virol. 78, 9904.
Yeager, C.L., Ashmun, R.A., Williams, R.K., Cardellichio, C.B., Shapiro, L.H., Look, A.T., Holmes, K.V., 1992. Human aminopeptidase N is a receptor for human coronavirus 229E. Nature 357, 420.
Yin, H.S., Paterson, R.G., Wen, X., Lamb, R.A., Jardetzky, T.S., 2005. Structure of the uncleaved ectodomain of the paramyxovirus (hPIV3) fusion protein. Proc. Natl. Acad. Sci. U. S. A. 102, 9288.
Yin, H.S., Wen, X., Paterson, R.G., Lamb, R.A., Jardetzky, T.S., 2006. Structure of the parainfluenza virus 5 F protein in its metastable, prefusion conformation. Nature 439, 38.
Youn, S., Collisson, E.W., Machamer, C.E., 2005. Contribution of trafficking signals in the cytoplasmic tail of the infectious bronchitis virus spike protein to virus infection. J. Virol. 79, 13209.
Yu, X., Zhang, S., Jiang, L., Cui, Y., Li, D., Wang, D., Wang, N., Fu, L., Shi, X., Li, Z., Zhang, L., Wang, X., 2015. Structural basis for the neutralization of MERS-CoV by a human monoclonal antibody MERS-27. Sci. Rep. 5, 13133.
Yuan, Y., Cao, D., Zhang, Y., Ma, J., Qi, J., Wang, Q., Lu, G., Wu, Y., Yan, J., Shi, Y., Zhang, X., Gao, G.F., 2017. Cryo-EM structures of MERS-CoV and SARS-CoV spike glycoproteins reveal the dynamic receptor binding domains. Nat. Commun. 8, 15092.
Zhang, H., Wang, G., Li, J., Nie, Y., Shi, X., Lian, G., Wang, W., Yin, X., Zhao, Y., Qu, X., Ding, M., Deng, H., 2004. Identification of an antigenic determinant on the S2 domain of the severe acute respiratory syndrome coronavirus spike glycoprotein capable of inducing neutralizing antibodies. J. Virol. 78, 6938.
Zheng, Q., Deng, Y., Liu, J., van der Hoek, L., Berkhout, B., Lu, M., 2006. Core structure of S2 from the human coronavirus NL63 spike glycoprotein. Biochemistry 45, 15205.

CHAPTER FIVE

Structural analysis of pleomorphic and asymmetric viruses using cryo-electron tomography and subtomogram averaging

Martin Obr, Florian K.M. Schur*
Institute of Science and Technology Austria (IST Austria), Klosterneuburg, Austria
*Corresponding author: e-mail address: florian.schur@ist.ac.at

Contents

Advances in Virus Research, Volume 105
ISSN 0065-3527
https://doi.org/10.1016/bs.aivir.2019.07.008

Abstract

Describing the protein interactions that form pleomorphic and asymmetric viruses represents a considerable challenge to most structural biology techniques, including X-ray crystallography and single particle cryo-electron microscopy. Obtaining a detailed understanding of these interactions is nevertheless important, considering the number of relevant human pathogens that do not follow strict icosahedral or helical symmetry. Cryo-electron tomography and subtomogram averaging methods provide structural insights into complex biological environments and are well suited to go beyond structures of perfectly symmetric viruses. This chapter discusses recent developments showing that cryo-ET and subtomogram averaging can provide high-resolution insights into hitherto unknown structural features of pleomorphic and asymmetric virus particles. It also describes how these methods have significantly added to our understanding of retrovirus capsid assemblies in immature and mature viruses. Additional examples of irregular viruses and their associated proteins, whose structures have been studied via cryo-ET and subtomogram averaging, further support the versatility of these methods.

Abbreviations

6HB	6-helix bundle
AIDS	acquired immunodeficiency syndrome
ATLL	adult T-cell leukemia/lymphoma
BSL	biosafety level
BVM	bevirimat
CA	capsid protein
CCD	charge-coupled device
cryo-EM	cryo-electron microscopy
cryo-ET	cryo-electron tomography
cryo-FIBSEM	cryo-focused ion beam milling scanning electron microscopy
CTD	C-terminal domain
CTF	contrast transfer function
DED	direct electron detector
DNA	deoxyribonucleic acid
ds	double stranded
EFTEM	energy-filtered transmission electron microscopy
EIAV	equine infectious anemia virus
FHV	Flock house nodavirus
HIV	human immunodeficiency virus
HTLV	human T-cell lymphotropic virus
MA	matrix protein
MI	maturation inhibitor
MLV	murine leukemia virus
M-PMV	Mason-Pfizer monkey virus
NC	nucleocapsid protein
NMR	nuclear magnetic resonance
NTD	N-terminal domain
PVAP	protein forming virus-associated pyramids
RNA	ribonucleic acid

RSV	Rous sarcoma virus
RVFV	Rift valley fever virus
SAXS	small angle X-ray scattering
SIRT	simultaneous iterative reconstruction technique
SIRV	*Sulfolobus islandicus* rod-shaped virus
SNR	signal to noise ratio
SPA	single particle analysis
ss	single-stranded
TMV	tobacco mosaic virus
VAP	virus-associated pyramid
VLP	virus-like particle
WBP	weighted back-projection
XMRV	xenotropic murine leukemia virus-related virus

1. Introduction: Structural biology of pleomorphic and asymmetric viruses

Structural biology of viruses covers a wide range of complexity, from monomeric viral proteins to complete virus particles that can be highly regular (e.g., icosahedral) or pleomorphic and asymmetric. In order to visualize virus structures at these different levels of complexity, numerous structural biology techniques are available, including X-ray crystallography, nuclear magnetic resonance (NMR) spectroscopy and cryo-electron microscopy (cryo-EM).

Many non-structural proteins of viruses can be successfully studied on a monomeric or oligomeric level (e.g., after their purification) using X-ray crystallography or NMR. Structural virus proteins, however, need to establish protein-protein, protein-membrane and protein-nucleic acid interactions to assemble a virus particle. Observing them in isolation is, therefore, often insufficient to describe the intricate interactions that provide viruses with the meta-stability that is required for efficient assembly, as well as disassembly taking place during the infection of new cells. Placing the information of the individual structural units into the context of the entire virus particle often requires the virus to be studied *in toto*.

X-ray crystallography and single-particle analysis (SPA) cryo-EM are well established methods to determine structures of complete regular icosahedral or helical virus particles at near-atomic resolution (for examples, see Fromm et al., 2015; Grigorieff and Harrison, 2011; Harrison et al., 1978; Liu et al., 2010). A new frontier in structural virology is to study complex and pleomorphic viruses with less symmetry at high resolution. The

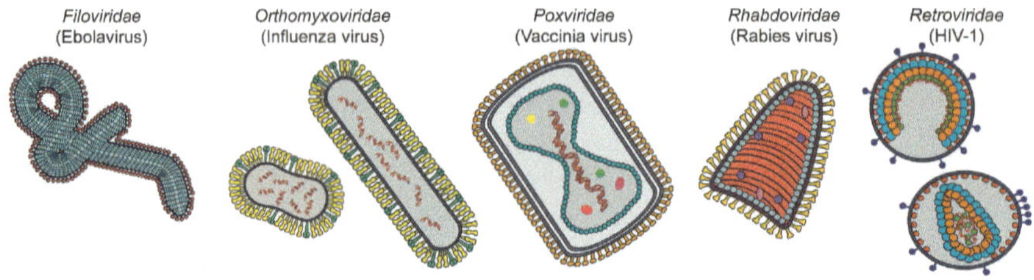

Fig. 1 Examples of pleomorphic and asymmetric viruses. A number of relevant human pathogens or extensively used viral model systems form pleomorphic and asymmetric virus assemblies. Virus morphologies can range from filamentous or bulletshaped viruses with changing helical symmetry (*Filoviridae* and *Rhabdoviridae*, respectively), to block-shaped viruses with different layers adopting different structural conformations (Poxviridae). Influenza virus provides a prominent example of particles that can adopt both roughly isometric and filamentous anisometric morphologies, depending on, among other factors, whether the virus is obtained from laboratory strains or clinical isolates. Retroviruses (see below for more detail) undergo significant structural rearrangements in the course of their life cycle, resulting in immature (top) and mature (bottom) morphologies.

importance of obtaining a detailed structural description of pleomorphic viruses becomes apparent when considering the large number of human pathogens or prominently used viral model systems that display variable assemblies, such as *Retroviridae, Filoviridae, Orthomyxoviridae, Rhabdoviridae* or *Poxviridae* (Fig. 1). Members of the *Retroviridae* family are a prime example of the large degree of morphological variance that can occur within viruses of a single species. Every virus particle is structurally unique, with most of them displaying a diameter ranging from approximately 100–200 nm and a differing number of proteins forming the virus shell. In addition, every retrovirus undergoes a dramatic structural rearrangement in the course of maturation, a process required for the virus particle to become infectious (Freed, 2015). This lack of global regularity poses a considerable challenge for both X-ray crystallography and SPA cryo-EM.

A specialized application of cryo-EM, termed cryo-electron tomography (cryo-ET) combined with an image processing method called subtomogram averaging allows for the determination of structures from non-symmetric viral capsids (Stass et al., 2018) or viral envelope proteins on the virus surface (Subramaniam et al., 2007).

Over the years, significant advancements in cryo-ET and subtomogram averaging have been made that allow for more and more detailed insights at higher resolutions of complex macromolecular assemblies. While our understanding of retroviruses has significantly benefited from these advances, which include more sensitive instrumentation, data acquisition automation and sophisticated image processing algorithms, retroviruses as a model were at the same time also instrumental in improving existing methodological workflows in cryo-ET and subtomogram averaging. Retrovirus capsid (CA) assemblies were the first to be solved to subnanometer and later sub-4Å resolutions using cryo-ET and subtomogram averaging (Schur et al., 2015b, 2016). These structures revealed important features in the conservation and diversity of immature and mature assemblies and also key regulatory aspects of virus maturation.

These results prove that cryo-ET and subtomogram averaging are valid high-resolution structure determination techniques that will potentially allow for the determination of high-resolution structures from additional pleomorphic virus assemblies. This is supported by recent work on other pleomorphic specimens, including filoviruses (Wan et al., 2017).

This chapter summarizes how cryo-ET, subtomogram averaging and associated hybrid image processing techniques are ideally suited to provide new insights into hitherto unknown structural features of pleomorphic and asymmetric virus particles. To this end, first a short definition of pleomorphism and asymmetricity in the context of viruses will be provided in Section 2 of this chapter. Section 3 summarizes the challenges in determining structures in pleomorphic and asymmetric viruses and also introduces the principles and the potential of cryo-ET and subtomogram averaging. Section 4 discusses how cryo-ET and subtomogram averaging have been vital in extending our knowledge on the structural interactions of retroviral CA in immature and mature viruses.

In the fifth part, further examples of cryo-ET and subtomogram averaging studies of pleomorphic and asymmetric viruses will be summarized. The chapter will conclude with future perspectives on how further methodological developments can continue to revolutionize structural characterizations of pleomorphic viruses.

2. Viral asymmetry and pleomorphism

The physical principles governing the construction of regular viruses have been laid out in the seminal work by Caspar and Klug, which describes the dominant assembly morphologies in helical and icosahedral viruses (Caspar and Klug, 1962). The principles that govern the construction of pleomorphic and asymmetric viruses are more complicated due to their morphological variability (Fig. 1). Virus pleomorphism can be defined by the irregularity of particles from the same virus species, i.e., the variation of size and shape as well as the protein copy number that constitute the virus particle. Asymmetry in the context of virus structures does not automatically imply a complete absence of symmetry. Pleomorphic viruses often contain lattice forming proteins, such as capsid or matrix proteins, which display local symmetry. This is, for example, evident in CA assemblies of retroviruses having local six- or fivefold symmetry (Mattei et al., 2016b). Helical symmetry is very common among viruses of different families, some of which have been extensively studied structurally. Tobacco mosaic virus (TMV), for example, displays uniform helical symmetry that does not deviate within individual virus particles (Fromm et al., 2015). Other filamentous helical viruses instead can display changing helical symmetries within the same particle. This includes filoviruses, such as Ebola and Marburg virus, but also members of the *Paramyxoviridae*, and *Rhabdoviridae* family (Ge et al., 2010; Ke et al., 2018; Wan et al., 2017) (Fig. 1).

Such a lack of global symmetry can often be observed in enveloped viruses. Since the membrane of enveloped viruses separates viral proteins from the environment, a closed capsid shell of precisely defined shape is often not absolutely required to protect the genome and viral cofactors, as it is in the case of naked viruses. The definition of viral symmetry is further complicated by the fact that even viruses with a high degree of global symmetry (such as icosahedral viruses) might display local disorder. For example, viral proteins can be unevenly distributed on the surface of enveloped icosahedral particles, e.g., in Hepatitis B virus or Herpes simplex virus (Grünewald et al., 2003; Seitz et al., 2007). Viral genomes can also be asymmetrically located within an otherwise icosahedral particle (Stass et al., 2018).

The abovementioned factors demonstrate that asymmetricity and pleomorphism refer to a wide spectrum of virus particle shape and size. Therefore, in pleomorphic and asymmetric virus species every virus particle needs to be considered as a unique structural entity. This increases the experimental challenges in determining the structure of such viruses.

3. Methodological challenges for determining the structure of pleomorphic and asymmetric viruses

3.1 Virus preparation

Naturally, the most accurate analysis would be to study viruses in the context of their host or in a cellular system that does not result in perturbations of viral structure, irrespective of whether the viruses under study are regular or pleomorphic. Such experimental conditions, however, are often not reconcilable with the available structural biology techniques which typically require purification of specimens. Nevertheless, purified viruses do still allow for accurate descriptions of selected viral proteins in their native environment, i.e., structural proteins protected by the viral membrane. Viral glycoproteins, on the other hand, might be affected by the purification procedure, and a controlled experimental setup is therefore necessary to allow precise interpretation of structural and functional states (Stass et al., 2018). Viruses within cells or upon budding can be studied by cryo-ET (for examples, see Carlson et al., 2008; Cyrklaff et al., 2005; Hampton et al., 2017), but this suffers from other experimental limitations as outlined in Section 3.3.

Worldwide only few laboratories or facilities are equipped with cryo-electron microscopes or beam lines operated under biosafety level (BSL) 2 or 3. Therefore, many pathogenic viruses need to be chemically inactivated prior to further structural analysis, in order to comply with safety regulations. The use of virus-like particles (VLPs) is one approach to overcome the increased structural complexity of viruses and potentially related biosafety issues for human pathogens (Zeltins, 2013). VLPs can closely reflect the organization of authentic virus particles, but lack one or more viral components. Depending on the used VLP system, such particles can be assembled *in vitro* using purified components. Alternatively, VLPs can be produced via cell culture, where only a subset of viral proteins, sufficient to produce a non-infectious particle, are expressed.

In vitro assembled VLPs are particularly beneficial for analysis of structural aspects of viruses, as such systems generally offer more homogeneous particles resulting in more efficient data acquisition and image processing. VLP systems can also be used to elucidate molecular mechanisms that are critical to assembly and morphological changes in viruses that are otherwise difficult to study *in situ*. Careful verification of structures from *in vitro* assembled VLPs is necessary to show that they are representative of the structures found in authentic virus particles *in situ*. The term *in situ* is used variably across and even within scientific disciplines. In the context of this chapter, it is used for proteins or macromolecular complexes that are not extracted from their native environment. This therefore also includes proteins within purified virus particles.

3.2 Cryo-electron microscopy of regular and pleomorphic viruses

X-ray crystallography requires the specimen to form a regular three-dimensional lattice. This approach is therefore not compatible with all the conformational states that a viral lattice in pleomorphic viruses may adopt. Cryo-EM instead allows for studying specimens in a native hydrated state via a process called vitrification, i.e., the rapid cooling of a specimen without formation of ice crystals. Samples can then be studied "in solution" in a layer of vitrified aqueous buffer without the prerequisite to form a crystalline lattice.

In SPA cryo-EM (for a detailed review see Cheng et al., 2015), 2D electron micrographs are acquired containing individual particles of the specimen in random orientations. In order to obtain a 3D structure of these particles, the orientational relation between them needs to be determined. Commonly, this is achieved by aligning the individual particles against a reference. In order to compensate for specimen heterogeneity (i.e., in shape and molecular composition), particles can be computationally sorted and classified. This is ultimately limited by the ability to distinguish individual particles and different particle conformations, which in turn is also related to the signal to noise ratio (SNR) of the acquired data.

Therefore, the ultimate resolution-limiting factor in imaging vitrified specimens is the cumulative electron dose applied to the sample and also the radiation damage caused by it. Hence, the dose that is used to image a specimen needs to be kept as low as possible, which naturally negatively impacts the SNR. One other limitation in SPA, as the name implies, is that it requires single particles that are not overlapping within the vitrified layer. The orientation of superimposing particles cannot be unambiguously distinguished in 2D micrographs, which prevents successful alignment. Virus particles following a strict icosahedral or helical symmetry, if considered *in toto*, represent an exception. Although the individual proteins within the virus particle shell superimpose, successful alignment is still possible due to the highly regular geometry of the virus particle. The power of regular global symmetry lies in the fact that a single view of a virus can be extrapolated onto the entire virus particle, since the relationship of different views of a particle is known. The final average of the virus structure at higher resolution can then be generated by taking this symmetry into account. However, this also implies that the successful application of SPA cryo-EM to viruses requires them to be structurally indistinguishable from one another. This is why SPA cryo-EM cannot be applied to pleomorphic and asymmetric viruses.

As already explained above, viral proteins forming an asymmetric particle are not related by a defined geometry, and each virus particle can be considered to have a unique structure due to the variation in virus size and shape (Fig. 1). Hence, the views of such particles, obtained from different orientations, are not identical and averaging of entire virus particles will not yield a high-resolution structure. Aligning the individual proteins within an asymmetric virus (and not the entire virus) is also aggravated in SPA cryo-EM due to the above explained problem of superimposing densities in 2D images. Exceptions for specialized applications exist, where SPA cryo-EM has been applied successfully for asymmetric reconstructions of viral components, such as of the asymmetrically located genome within an icosahedral particle (Dai et al., 2017; Stass et al., 2018).

For most pleomorphic viruses, however, it is necessary to obtain 3D information to separate superimposing densities. Such information can be obtained by cryo-ET and subtomogram averaging.

3.3 Principles of cryo-electron tomography and subtomogram averaging

In the following paragraphs we will give a comprehensive summary of cryo-ET and subtomogram averaging, including the potential and limitations of these methods. For the purpose of this chapter, virological applications are used to explain these methods. However, the potential of cryo-ET and subtomogram averaging extends far beyond viral specimens (Hutchings and Zanetti, 2018; Lucic et al., 2005). For additional details the reader is also referred to recent reviews discussing these methods (Schur, 2019; Wan and Briggs, 2016).

A complete cryo-ET and subtomogram averaging workflow for studying retroviral VLPs is depicted in Fig. 2. In cryo-ET, first a so-called tilt series is acquired, in which projection images containing always the same objects (i.e., several VLPs) are acquired from different angles by rotating the specimen stage within the electron microscope. This tilt series can then be computationally reconstructed into a 3D volume (a tomogram) (Fig. 2A).

Cryo-ET does not rely on the repetitive occurrence of structures for 3D reconstruction and detailed information on the structural arrangement of proteins within a virus can be already derived from single tomograms. Hence, cryo-ET allows describing individual macromolecular complexes or viruses to a resolution of several nanometers without the need for further computational averaging. Higher resolution structures, optimally beyond the subnanometer barrier, can then be obtained by subtomogram averaging (as detailed below) when the feature of interest is present multiple times within a tomogram, i.e., many copies of a virus structural protein forming a spherical or tubular lattice (Fig. 2B). Here, copies of the individual proteins

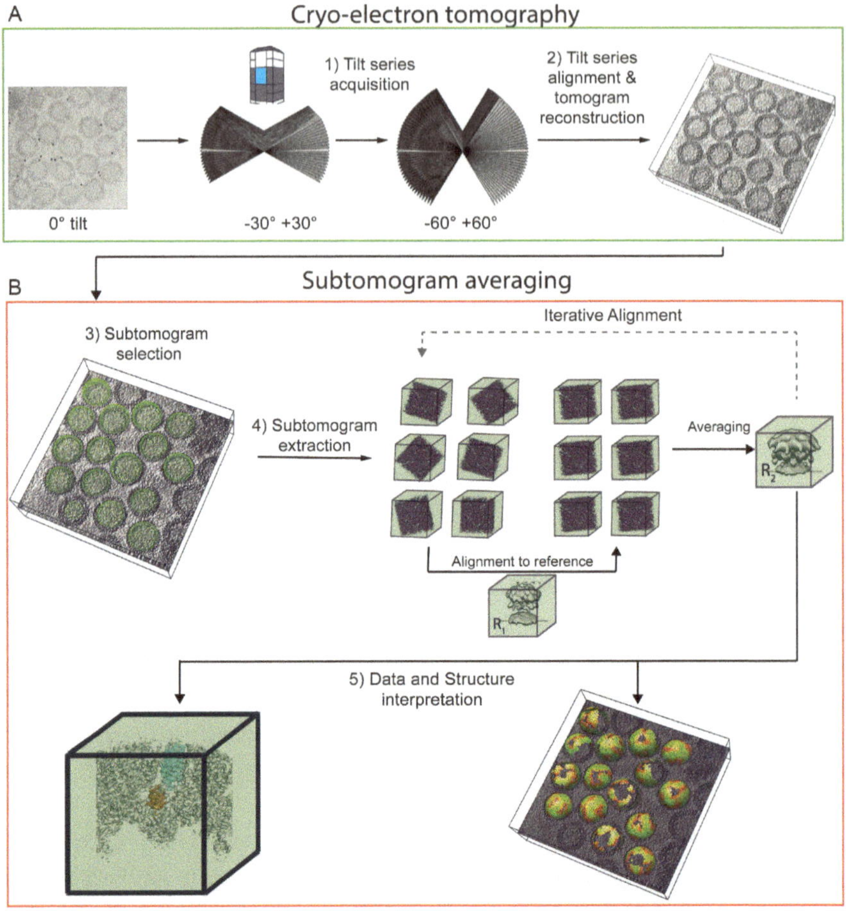

Fig. 2 Cryo-ET and subtomogram averaging workflow applied to retroviral VLPs. (A) Cryo-ET data acquisition and tomogram reconstruction: Retrovirus VLPs (in the depicted case derived from a truncated HIV-1 Gag variant, see Section 4.2 for more information) are vitrified in a thin layer of buffer. A tilt series is then acquired (1), wherein the tilt scheme depends on the desired outcome of data acquisition and image processing (i.e., dataset size vs. achievable resolution). In order to generate a 3D tomogram (2) of the imaged specimen, the individual micrographs within a tilt series need to be precisely aligned. (B) Subtomogram averaging: Initially, the extraction positions of the subtomograms containing the protein of interest need to be determined (3). The extracted subtomograms are then subjected to an iterative alignment against a reference, until the structure converges (4). By putting the aligned subtomogram translations and rotations back into the tomogram, contextual meta-information of the protein of interest can be obtained (5). In the depicted case, the HIV-1 CA proteins form an incomplete hexameric lattice with gaps and cracks. *Data shown in this figure are derived from Schur, F.K.M., Obr, M., Hagen, W.J.H., Wan, W., Jakobi, A.J., Kirkpatrick, J.M., Sachse, C., Kräusslich, H.G., Briggs, J.A.G., 2016. An atomic model of HIV-1 capsid-SP1 reveals structures regulating assembly and maturation. Science (80-.) 353, 506–508, available in the Electron Microscopy databank under accession codes EMD-4015 and EMPIAR-10164.*

are extracted into smaller subvolumes (i.e., subtomograms), which are then subjected to iterative alignment and averaging.

Subtomogram averaging can also provide important contextual information on the orientation and location of the aligned proteins in their native environment. In the case of retroviruses, for example, subtomogram averaging has been central in confirming the global arrangement of the retroviral Gag protein lattice forming an incomplete hexameric layer underneath the membrane of virus particles (Briggs et al., 2009; Carlson et al., 2008; Wright et al., 2007) (such an example for an incomplete lattice in retroviral VLPs is given in Fig. 2B). Subtomogram averaging can also be used to render the structure of purified virus particles, by placing the determined structures of the individual viral proteins back into a 3D volume at their respective aligned positions (Halldorsson et al., 2018; Mattei et al., 2016a).

Despite their potential in visualizing the constituents of viral assemblies at ever increasing resolutions, cryo-ET and subtomogram averaging also suffer from various limitations that ultimately limit data quality and the resulting resolution (see also review Schur, 2019). In order to put the recent advances in describing viral assemblies at high resolutions using cryo-ET and subtomogram averaging into the context of present methodological limitations, we will provide a short summary of the current frontiers in cryo-ET and subtomogram averaging. To acknowledge the limited resolution that commonly can be obtained with cryo-ET and subtomogram averaging, the term high-resolution is used for subnanometer resolutions throughout this text and therefore might deviate from its use in X-ray crystallography or SPA cryo-EM.

As outlined in Section 3.2, the reduced SNR in cryo-EM micrographs impacts all further image processing steps. In cryo-ET, the cumulative dose needs to be distributed over an entire tilt series resulting in individual projections with a low SNR. The amount of electron dose that can be used strongly depends on the specimen itself. The different susceptibility of biological molecules to radiation can be exploited experimentally, as demonstrated by the development of so-called "bubblegrams," which for example allows visualization of the location of the viral genome within virus particles (Fontana et al., 2015).

The rise of cryo-EM in recent years was predominantly caused by the introduction of direct-electron detectors (DEDs), which provide superior SNR compared to the previously used CCD cameras or photographic film (Kühlbrandt, 2014). One further advantage of DEDs is their ability to acquire movies. Individual movie frames can be aligned to compensate for beam induced motion and specimen drift (Brilot et al., 2012; Li et al., 2013).

Depending on the desired resolution, the selection of the tilt scheme (i.e., the number and sequence in which the images are acquired) is of importance. The type of the used tilt scheme results in a trade-off between dataset size and high-resolution content in tomograms. In the optimal case, views of a specimen should be acquired over the entire range of 180 degrees, which would provide a complete description of its three-dimensional architecture. Due to mechanical limitations of microscope stages and specimen holders, and due to tilting of the specimen in discrete angular steps (i.e., resulting in tilt series from −60 to +60 degrees with 3 degrees increment between individual acquisitions), not all views of the sample can be obtained. This results in missing information in real and Fourier space and causes anisotropic information in reconstructed tomograms (commonly referred to in tomography as 'missing wedge'). While the resolution in the x and y dimensions can be more evenly sampled, the resolution along the optical axis (Z-axis) cannot. In cases where the proteins of interest are evenly oriented within a tomogram (such as viral proteins that make up the viral protein lattice), subtomogram averaging can partially compensate for the incomplete sampling of data in tomograms by effectively filling in the missing wedge. Local symmetry in the subvolumes, for example, as seen in retroviral CA assemblies that form hexameric or pentameric structures, significantly facilitate subtomogram averaging and structure determination.

Cryo-EM micrographs are usually acquired out-of-focus to obtain phase-contrast. While this increases the visibility of the imaged specimen, it also results in the loss of high-resolution information in micrographs due to the defocus-dependent contrast transfer function (CTF). Phase plates can be used to avoid the requirement of defocusing to gain phase contrast (Danev and Nagayama, 2010). They modulate the electron wave by introducing a phase shift, resulting in improved contrast at low and high spatial frequencies. Phase plates can also be beneficial to compensate for the reduced contrast in cryo-ET, which is due to the stronger scattering of electrons, caused by the increased sample thickness of specimens (as larger objects are imaged). In particular the elevated apparent sample thickness in cryo-ET upon tilting of specimens considerably reduces the interpretable image content. Energy-filtered transmission electron microscopy (EFTEM) is commonly employed, as it removes scattered electrons that otherwise would not contribute constructively to the acquired image (Lucic et al., 2005).

Defocusing to increase phase contrast leads to contrast inversion at certain spatial frequencies and loss of high-resolution information. This information therefore needs to be restored via correction for the CTF.

This correction depends on an accurate estimation of the defocus value of the acquired micrograph. In cryo-ET, CTF correction does not only suffer from the reduced SNR in projections, but also from the defocus change across the tilted sample. Stable acquisitions, where the specimen remains at a defined eucentric height throughout the entire tilt series allows averaging of the signal of all tilts in a series and hence boosting of the SNR for estimating the defocus (Schur et al., 2013; Zanetti et al., 2009). If the SNR of the data is sufficiently high, defocus estimation can also be attempted directly on individual tilts. This approach is becoming more and more feasible with the advent of new, more sensitive direct electron detectors (Hutchings et al., 2018; Schur et al., 2016). Approaches to correct for the defocus gradient within thick samples and therefore allowing even more precise CTF correction are discussed in Section 4.5.

When aiming for high-resolution structure determination in cryo-ET, high magnifications need to be used for data acquisition, resulting in a small pixel size, and hence, a restricted field of view. This negatively affects the dataset size as fewer virus particles will be present in the individual micrographs. Moreover, the limited stage accuracy requires iterative tracking and focusing during acquisition to keep the field of view precisely centered on the area of interest. Fully automated data acquisition is available in several commercial (Thermo Fisher Tomography, Gatan Latitude S, TVIPS EM-Tomo) and free (Mastronarde, 2005; Suloway et al., 2009) software packages. However, few software packages contain advanced scripting functionalities that allow adapting the data acquisition protocol to specific needs, i.e., to develop new tilt schemes tailored for high-resolution (see Hagen et al. (2017) for an example of such a tilt scheme). More eucentric specimen holders have led to the development of faster acquisition schemes without the need of iterative tracking (Chreifi et al., 2019). These new holders may result in dataset sizes in cryo-ET that are comparable to what is currently possible in SPA cryo-EM.

Precise tilt series alignment (which ultimately influences the final resolution) relies on the ability to correlate individual tilts. For viral samples, as they are often obtained after purification, electron dense fiducial markers (e.g., colloidal gold) are commonly added to the viral solution prior to vitrification and can serve as registration points present in all projections.

Different reconstruction methods are used together with deconvolution and denoising approaches in order to improve contrast of reconstructed tomograms, which in turn facilitates data interpretation and also subtomogram averaging (Lucic et al., 2005). The most widely used tomogram reconstruction algorithm in subtomogram averaging studies is weighted backprojection (WBP), which sufficiently retains spatial information during the reconstruction process, as evidenced by high-resolution cryo-ET and subtomogram averaging experiments (Bharat et al., 2015; Hutchings et al., 2018; Schur et al., 2016). Structure determination in cryo-ET requires reliable identification of the proteins of interest. Here, alternative tomographic reconstruction methods (such as algebraic reconstruction techniques, e.g., the simultaneous iterative reconstruction technique (SIRT)) can be advantageous as they can provide superior contrast of tomograms compared to WBP, potentially at the cost of losing high-resolution information.

Virus particles often allow for a straightforward assignment of initial subtomogram positions of proteins of interest, since they either form large assemblies that are easily discernible or since their lattices might be arranged in a geometrically defined way (e.g., radially underneath the viral membrane). This *a priori* information can be used to initiate subtomogram averaging experiments. In the case of spherical pleomorphic viruses, such as retroviruses, initial subtomogram positions can be assigned by defining the radial distance from the particle center, assuming that viruses do not excessively deviate from a perfect sphere (Briggs et al., 2009; de Marco et al., 2010a) (Fig. 2B). An initial angular assignment is then possible by taking the normal orientation of the viral capsid proteins with respect to the membrane into account. Deviations of virus particles from a perfect sphere can be compensated for by allowing larger translational and rotational freedom in the initial iterations of subtomogram alignment.

In tubular virus assemblies, such as filoviruses, initial positions can be assigned by fitting a spline curve along the central axis of the virus, and again using the radial distance from the center of the virus to define positions of proteins close to the membrane (Bharat et al., 2011, 2012b). Given the assumption that viral proteins are again radially underlying the membrane, their orientation can be then assumed to be normal in relation to the virus surface. Alternatively, manual assignment of extraction positions is possible, though very laborious. Template matching can be used, if a template (such as a previously obtained structure) resembling the protein of interest is available (Beck and Baumeister, 2016; Nickell et al., 2006).

Subtomogram averaging can be performed in several software packages, such as Dynamo (Castano-Diez et al., 2012), TOM (Nickell et al., 2005), PyTOM (Hrabe et al., 2012), AV3 (Förster et al., 2005), PEET (Nicastro et al., 2006), Relion (Bharat et al., 2015) or EMAN (Galaz-Montoya et al., 2015), which provide different sets of functionalities. These can include in addition to alignment and averaging, data management and particle picking functionalities. Currently, high-resolution structures are obtained most often via a combination of different software packages, where their specific functionalities are chosen for defined tasks in the processing workflow (Himes and Zhang, 2018; Schur et al., 2016).

The increased number of interpolations in image processing in cryo-ET and subtomogram averaging can further attenuate the obtainable resolution (Schur, 2019). Subtomogram averaging can therefore also be combined with single-particle reconstruction approaches to form hybrid methods partially compensating for this increased number of interpolations. In such hybrid approaches subtomogram averaging is used to provide starting models or angular restraints for subsequent single-particle refinement and reconstruction from the same dataset (Bartesaghi et al., 2012; Bharat et al., 2012a; Himes and Zhang, 2018) (see Section 4.3 for detailed examples applied to retroviruses). For many biological molecules that display different conformational states, computational sorting (or classification) is necessary to obtain high resolution structures. However, as previously discussed, this is aggravated by the low SNR and the missing wedge in cryo-ET.

This section details that several factors must be met to ensure that the highest-possible resolution can be obtained. Choosing the correct experimental parameters for data acquisition and image processing is critical, but also the specimen will ultimately determine the result that can be achieved.

4. Elucidating retrovirus assemblies with cryo-ET and subtomogram averaging

Virus capsids are in many cases ideal samples for cryo-ET and subtomogram averaging as they might display low local conformational variability. They often form paracrystalline lattices *in situ* that vary only in size and curvature, but not in the conformation of the individual proteins.

This is also one of the several reasons why cryo-ET and subtomogram averaging have been so successful in providing unprecedented insights into the assembly of retroviral CA in both immature and mature viruses (Mak and de Marco, 2018). In the next section we will give a brief overview of retroviruses, followed by a short historical perspective of how cryo-EM and later high-resolution cryo-ET and subtomogram averaging helped to elucidate important structural features of these viruses. Providing an exhaustive description of the retroviral life cycle would go beyond the scope of this chapter. The reader is therefore also referred to existing literature (Craigie and Bushman, 2012; Sundquist and Krausslich, 2012; Wilen et al., 2012).

4.1 Retroviruses: HIV-1 and beyond

Retroviruses are positive-sense single-stranded (ss)RNA viruses that replicate through a DNA intermediate via reverse transcription, and integrate into the genome of the host. This gives rise to a lifelong infection, for which no cure exists.

Retrovirus genera include α-, β-, γ-, δ-, ε- retroviruses and lentiviruses (*Orthoretrovirinae*), and spumaviruses (*Spumavirinae*). Among them, two retroviruses are important human pathogens. The lentivirus human immunodeficiency virus 1 (HIV-1) is the causative agent of the acquired immunodeficiency syndrome (AIDS), responsible for more than 35 million infected individuals worldwide and approximately one million deaths in 2017. Infection with the δ-retrovirus human T-cell lymphotropic virus (HTLV) mostly remains asymptomatic, but in a few percent of cases it can lead to severe diseases, such as tropical spastic paraparesis (also known as HTLV-1 associated myelopathy) or Adult T-cell leukemia/lymphoma (ATLL), an aggressive type of cancer with a median survival of less than one year. For several decades, retroviruses have been the focus of extensive research into understanding their structural hallmarks to identify targets for therapeutic intervention. To this end, not only medically relevant viruses, but also animal retroviruses are studied, where the latter serve as model systems that offer insights into conserved and divergent features of retroviral assembly, maturation and infection.

Retroviral assembly is driven by the major structural polyprotein Gag, which is found in all retroviruses (Briggs and Krausslich, 2011; Sundquist and Krausslich, 2012) (Fig. 3A). Gag forms the protein shell of the nascent immature virus particle and is composed of a varying number of domains, which are connected by flexible linkers. Three canonical Gag domains are present in all retroviruses: the membrane-binding matrix (MA) domain; the main protein-protein interaction capsid (CA) domain, consisting of the independently folded N-terminal CA (CA-NTD) and C-terminal CA (CA-CTD); and the RNA binding nucleocapsid (NC) domain. Outside of the conserved Gag regions, different retroviruses can have various other domains or short peptides (Fig. 3A). These non-canonical domains serve diverse functions. HIV-1 spacer peptide 1 (SP1) plays a role in the assembly and stabilization of the immature lattice and the regulation of maturation (Datta et al., 2011a; Gross et al., 2000; Kräusslich et al., 1995). Peptide stretches with similar function (either as individual domain or part of CA) exist also in other retroviruses, such as Rous Sarcoma virus (RSV, *α-retrovirus*) (Bush et al., 2014) or Murine Leukemia virus (MLV, *γ-retrovirus*) (Cheslock et al., 2003).

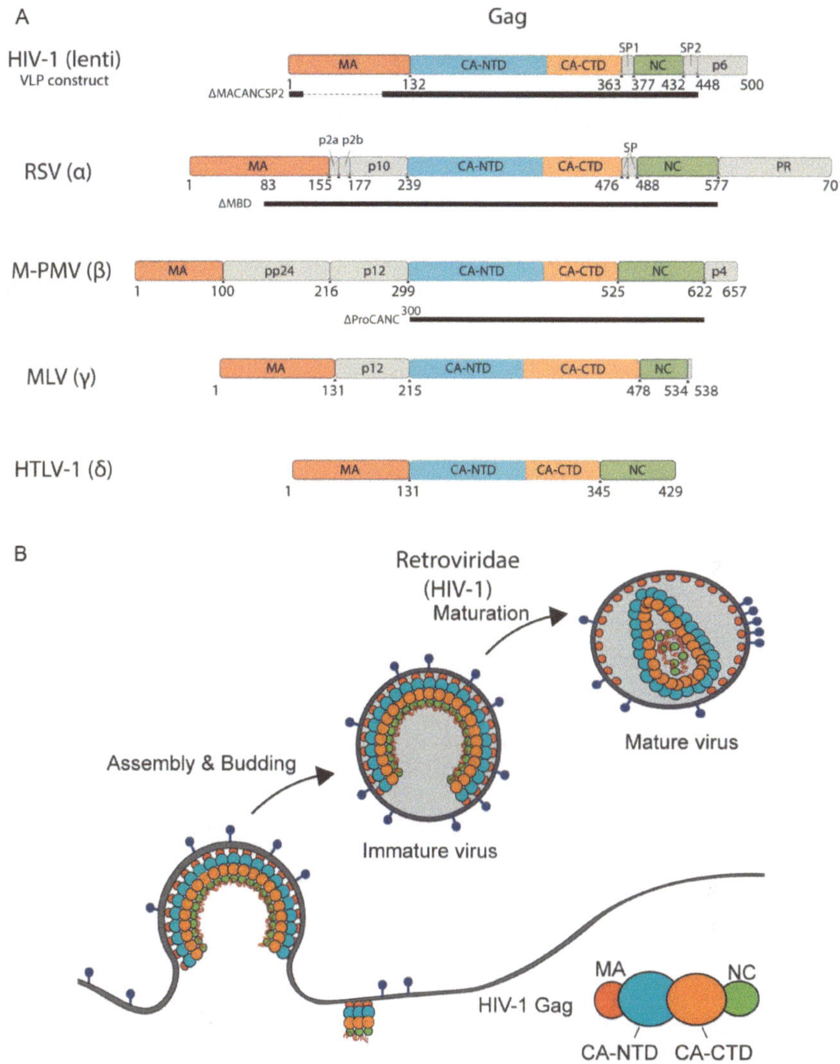

Fig. 3 Retrovirus assembly and maturation. (A) Schematic depictions of the Gag domain architecture in retroviruses from different retroviral genera. All viruses share the conserved Gag domains MA, CA and NC, but can harbor additional non-canonical domains or short peptide stretches. For HIV-1, M-PMV and RSV, truncated Gag variants are depicted, used for *in vitro* assembly and which are also discussed in this chapter. The cleavage sites (and the corresponding residues within Gag) are indicated with black arrowheads. The color coding for the conserved domains is as in (B). (B) Retroviruses undergo significant conformational changes during their life cycle. In the case of HIV-1, for example, Gag proteins assemble at the plasma membrane into an incomplete hexameric lattice. The main lateral structural interactions are established by the CA domain within Gag. During maturation, the viral protease cleaves the polyprotein Gag at selected positions to free the individual domains. These domains then rearrange and multimerize into the mature lattice, resulting in an infectious virus. Released CA forms fullerene shaped structures that encapsulate the viral RNA genome. During this structural transition CA forms both hexameric and pentameric assemblies that have unique sets of interactions that are different from those in the immature Gag lattice.

Depending on the retrovirus species, assembly of retrovirus particles has been reported to occur either in the cytoplasm (e.g., Mason-Pfizer Monkey virus, M-PMV, *β-retrovirus* (Rhee and Hunter, 1990)) or at internal membranes (e.g., Equine infectious anemia virus, EIAV, *lentivirus* (Zhang et al., 2016)) or at the plasma membrane of the infected cell (such as HIV-1, Fig. 3B) (Sundquist and Krausslich, 2012). In newly released immature retroviruses Gag radially underlines the viral membrane. Concomitantly with or shortly after budding the viral protease becomes active and cleaves Gag at defined positions (Fig. 3A), releasing the individual domains. This leads to a dramatic structural rearrangement of the virion interior and renders the virus infectious (Freed, 2015).

While MA stays attached to the membrane, a subset of CA reassembles into a fullerene-like structure consisting of a hexameric lattice that is closed by the incorporation of 12 CA pentamers (Briggs and Krausslich, 2011; Mattei et al., 2016b). This core encapsulates the condensed NC-RNA complex. HIV-1 predominantly forms cone-shaped cores, where seven CA pentamers are located on the broad end and five pentamers on the narrow end. Other retroviruses form different fullerene-like morphologies, resembling cylinders (e.g., M-PMV) or spheres (e.g., MLV) (Mattei et al., 2016b).

Retroviral glycoproteins, due to their exposure to the host immune system are also of high structural relevance. SPA cryo-EM, as well as cryo-ET and subtomogram averaging have provided insights into the trimeric structure of retroviral glycoproteins *in vitro* and on viral membranes (Lee et al., 2016; Liu et al., 2008; White et al., 2010; Zanetti et al., 2006; Zhu et al., 2006). The reader is referred to existing literature (Mak and de Marco, 2018) and Chapter "Structures of enveloped virions determined by cryogenic electron microscopy and tomography" by Stass et al. of this book for a more detailed discussion of cryo-EM and cryo-ET of enveloped virions.

For many retroviruses it is possible to spontaneously assemble purified full-length or truncated Gag protein, in the absence of membrane, into VLPs *in vitro* (Bush and Vogt, 2014) (see Fig. 4A for a comparison of authentic immature HIV-1 virus particles and *in vitro* assembled HIV-1 VLPs). *In vitro* assemblies that predominantly consist of the CA domain with surrounding Gag stretches are also often referred to as capsid-like particles. For simplicity, we will refer here to all particles as VLPs. *In vitro* conditions allow a much more regulated experimental setup in comparison to cell culture derived virus particles. Parameters such as the chosen Gag/CA construct and buffer conditions (pH and ionic strength) can be tuned to produce particles representing either immature or mature assemblies. VLP systems

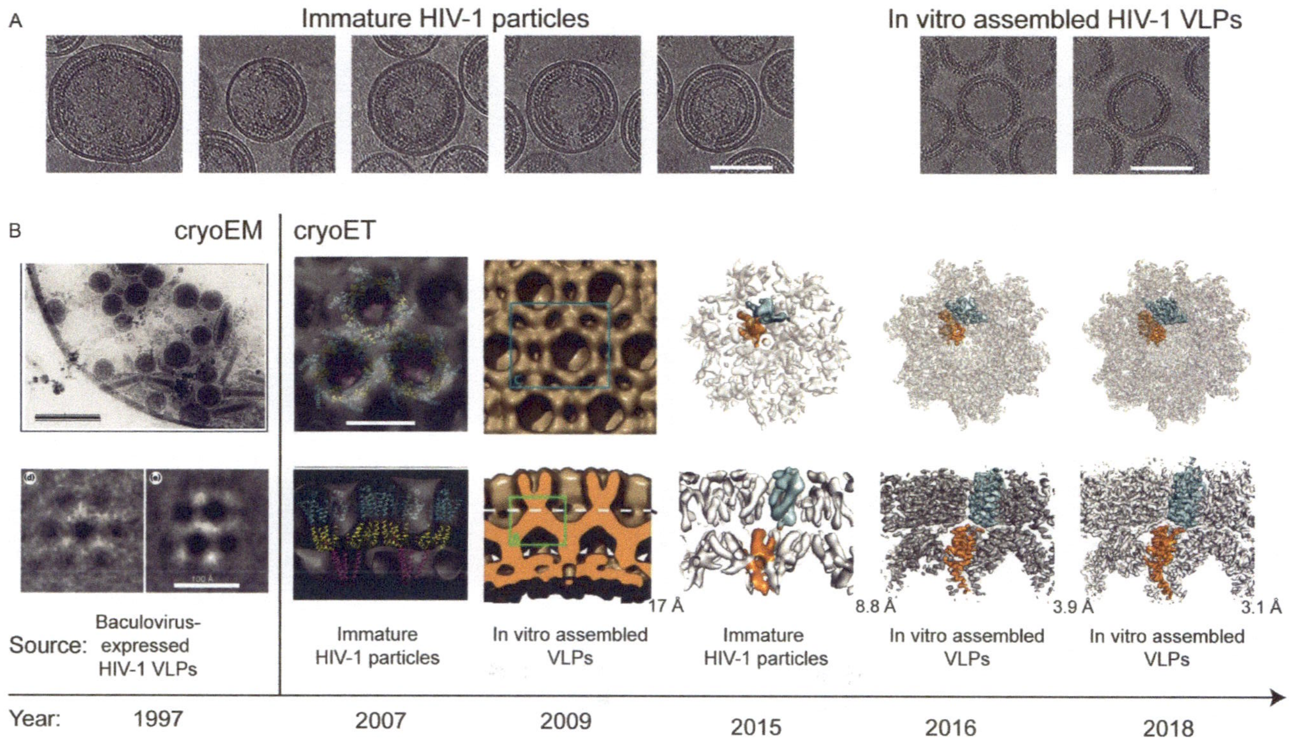

Fig. 4 See legend on next page.

also help in elucidating molecular mechanisms underlying assembly and morphology in retroviruses that are difficult to propagate in cell cultures (such as RSV, EIAV or HTLV-1).

4.2 2D Cryo-EM of retroviruses

For a long time, available structures of individual Gag domains were mostly relevant for the mature form of retroviruses, as X-ray crystallography and NMR provided only insights into the structure of the isolated MA, CA and NC domains of several different retroviruses (for examples, see Cornilescu et al., 2001; Gamble et al., 1997; Hill et al., 1996; Kingston et al., 2000; Macek et al., 2009; Mortuza et al., 2004; Tang et al., 2002; Worthylake et al., 1999). X-ray crystallography studies of CA in its mature pentameric and hexameric oligomeric state, as well as 2D cryo-EM experiments of *in vitro* assembled CA tubes helped to resolve how retroviral CA forms the building blocks of the mature infectious virus particle (Kingston et al., 2000; Li et al., 2000; Pornillos et al., 2009, 2011). One

Fig. 4 Cryo-EM and cryo-ET of immature HIV-1 viruses and VLPs. (A) A gallery of authentic immature HIV-1 virus particles isolated from cells and immature-like HIV-1 VLPs assembled *in vitro*, imaged by cryo-ET. See Fig. 3A for a schematic depiction of the truncated Gag construct used to assemble VLPs. The variability in shape and size of the native viruses is clearly visible. In contrast, VLPs display a more regular size distribution, but nevertheless still vary in shape, deviating from a perfect sphere. The strong regular protein density visible in both viruses (here underlying the lipid bilayer) and VLPs is the CA-SP1 layer. The NC appears disordered. Protein density is black. Scale bar is 100 nm. Viruses and VLPs are derived from a dataset published in Schur et al. (2016). (B) Timeline of the obtained structures describing the immature Gag assembly in HIV-1. Initial results were obtained by 2D cryo-EM. With the advent of cryo-ET and subtomogram averaging, it became then possible to describe the Gag assembly in viruses or *in vitro* assembled VLPs (as shown in A) and to fit previously solved molecular models for the individual domains. Over the years, with improving methodology, refined models of the HIV-1 CA-SP1 assembly became available. Coloring of the three structures to the right corresponds to CA-NTD (cyan), CA-CTD (orange) and SP1 (orange). The subnanometer resolution structures on the right were generated from their respective EMDB-files (EMD-2706, EMD-4015, EMD-8986). *The figures in (B), originally published in Fuller, S.D., Wilk, T., Gowen, B.E., Kräusslich, H.-G., Vogt, V.M., 1997. Cryo-electron microscopy reveals ordered domains in the immature HIV-1 particle. Curr. Biol. 7, 729–738, reproduced with permission from Elsevier; Wright, E.R., Schooler, J.B., Ding, H.J., Kieffer, C., Fillmore, C., Sundquist, W.I., Jensen, G.J., 2007. Electron cryotomography of immature HIV-1 virions reveals the structure of the CA and SP1 Gag shells. EMBO J. 26, 2218–2226 reproduced with permission from Wiley and Sons; Briggs, J.A., Riches, J.D., Glass, B., Bartonova, V., Zanetti, G., Krausslich, H.G., 2009. Structure and assembly of immature HIV. Proc. Natl. Acad. Sci. U.S.A. 106, 11090–11095, reproduced with permission of Dr. John Briggs.*

difficulty was how to extrapolate the information of the structure of the individual domains (or oligomeric building blocks) to describe the full CA lattice of the mature viral particle. Subnanometer resolution 2D cryo-electron crystallography experiments of HIV-1 CA mutants and helical reconstruction cryo-EM of *in vitro* assembled HIV-1 CA tubes, combined with all-atom molecular dynamics simulation improved our understanding of the complete HIV-1 core (Ganser-Pornillos et al., 2007; Zhao et al., 2013).

Low-resolution structures of soluble full-length Gag were determined for HIV-1, MLV and BLV using small angle X-ray scattering (SAXS) (Datta et al., 2007, 2011b; Qualley et al., 2019). Information on the Gag-Gag interactions in the assembled state in the immature viruses, however, remained more elusive as the flexibility of full-length Gag interfered with high-resolution structure determination by X-ray crystallography.

Early attempts to describe the arrangement of retroviral Gag in immature viruses were done via negative stain and thin-section electron microscopy experiments. This early work was based on the commonly accepted dogma that many retroviruses, including HIV-1, adopt a regular icosahedral symmetry (reviewed in Wilk and Fuller, 1999). Variations in particle size were assumed to be caused by the ability of retroviruses to adopt different triangulation numbers to form icosahedral symmetry (Nermut et al., 1994). Cryo-EM later revealed that retroviruses do not have icosahedral symmetry, as a meaningful structure from averaged views of individual particles could not be obtained. It was also shown that Gag aligns radially underneath the viral membrane in immature viruses with separately folded domains and that upon maturation a conformational change occurs that is dramatic enough to be visualized within standard 2D cryo-EM micrographs (Briggs et al., 2003; Fuller et al., 1997; Yeager et al., 1998). The understanding of retrovirus assembly and maturation was enhanced by using image analysis tools on 2D cryo-EM images, revealing the local hexameric symmetry in native immature and mature virus particles and the spacing of immature and mature hexameric CA lattices of approximately 8 and 10 nm, respectively (Briggs et al., 2003; Fuller et al., 1997). Although almost all CA cores in mature HIV-1 appeared conical, a minor fraction appeared tubular, which provided evidence of the versatility of CA proteins to form assemblies of varying shape and size.

While these 2D cryo-EM studies provided important insights into structural assembly features of immature and mature viruses, it was not possible to describe the protein-protein interfaces that Gag forms during assembly of the immature virus. It was also not possible to unambiguously assign the observed densities to the individual Gag domains.

4.3 Cryo-ET and subtomogram averaging of immature and mature CA assemblies

3D visualizations of immature HIV-1 virus particles were provided by the first cryo-ET and subtomogram averaging experiments, which showed that the Gag lattice is a hexameric array containing holes and defects (referred to as an incomplete lattice) (Briggs et al., 2009; Wright et al., 2007) (Fig. 4B). These experiments described the individual hexameric building blocks as resembling a goblet with the CA-NTD forming the walls and the CA-CTD region forming the base of the cup and SP1 forming the stem. These studies also performed fits of the available molecular models into the cryo-EM densities, thus describing how the immature Gag lattice within viruses might be stabilized. Cryo-ET of native HIV-1 budding sites and subsequent subtomogram averaging of the immature Gag lattice in budding viruses revealed that their lattices are indistinguishable from those of released immature virions (Carlson et al., 2008). This suggested that the organization of Gag in immature HIV-1 virions is determined during assembly at the plasma membrane in the cell and that no major rearrangements occur until the proteolytic processing of Gag leads to maturation.

Immature virions from different retroviral genera were predicted to form similar Gag lattice arrangements because of the strong tertiary structure conservation of their CA domains (Campos-Olivas et al., 2000; Cornilescu et al., 2001; Khorasanizadeh et al., 1999; Kingston et al., 2000; Macek et al., 2009; Mortuza et al., 2004; Tang et al., 2002). This idea was also supported by comparative low-resolution cryo-ET and subtomogram averaging experiments on *in vitro* assembled Gag particles of HIV-1, M-PMV, RSV and xenotropic murine leukemia virus-related virus (XMRV) (de Marco et al., 2010a; Hadravová et al., 2012). Gag-derived constructs from these retroviruses formed immature hexameric lattices with similar inter-hexamer spacing of ~80 Å. The CA-NTD in HIV-1, M-PMV and RSV formed a ring around the hexameric center, with what appeared to be extensive inter-hexamer interactions. The CA-NTD of the *in vitro* assembled XMRV particles appeared disordered, presumably caused by the *in vitro* assembly system of XMRV (Hadravová et al., 2012).

At the CA-CTD level of the immature lattice of these retroviruses, a dimeric interface linking adjacent hexamers was identified, in agreement with the previously published cryo-ET data of HIV-1 and available crystal structures of the dimeric CA-CTD (Bartonova et al., 2008; Briggs et al., 2009). More differences in the immature assemblies of the retroviruses were observed in the region immediately downstream of CA, where the spacer peptides in HIV-1 and RSV formed pillar-like structures resembling a 6-helix bundle (6HB) (Accola et al., 1998; Bush et al., 2014). In XMRV, despite the lack of a spacer peptide stretch, the region between CA and NC also formed a rod-like structure similar to what was observed in HIV-1 and RSV.

Cryo-ET and subtomogram averaging have also been used to better understand the mode of action of the prototypic maturation inhibitor (MI) Bevirimat (BVM) (Keller et al., 2011). Small remnants of uncleaved CA-SP1 have a dominant-negative effect on infectivity and MI's interfere with the cleavage at the CA-SP1 site in HIV-1, rendering inhibited viruses non-infectious. Tomograms and structures of this first structural study showed that upon BVM inhibition HIV-1 particles were not able to complete maturation as the inhibitor stabilized the immature lattice and prevented mature core formation. The resolution of this study, however, was not yet sufficient to describe the binding site of BVM.

In order to obtain a subnanometer structure of immature Gag interactions, a hybrid approach was developed that combined the advantages of cryo-ET, subtomogram averaging, and single-particle cryo-EM. Bharat et al. (2012a) used a truncated M-PMV Gag construct (Fig. 3A) that assembled tubes with helical symmetry and immature lattice spacing. The tubular VLP assemblies did not show homogenous helical parameters, interfering with conventional 2D helical analysis. Subtomogram averaging was therefore employed to derive the exact helical parameters for the individual tubes, which were then used to calculate the 3D structure of the CA assembly using iterative real-space helical reconstruction (Egelman, 2007). The structures of the immature CA hexamer obtained from individual tubes were then averaged together in order to derive an 8 Å structure. At this resolution alpha-helices could be clearly visualized and hence allowed for unambiguous fitting of available HIV-1 CA crystal structures into the electron microscopy densities, providing the first description of immature retroviral CA interactions. Whether this structural assembly represented a general immature retroviral assembly conserved for all retrovirus genera remained unclear (Bush and Vogt, 2014).

In order to further validate the structure derived from the tubular *in vitro* M-PMV Gag assembly, HIV-1 CANC tubes bearing a mutation in the CA-CTD, which induces an immature-like assembly, were subjected to the same 3D-2D approach (Bharat et al., 2014). The obtained subnanometer reconstruction revealed that the CA-CTD formed a similar immature arrangement as in M-PMV, while the CA-NTD layer arranged in a way that was not reconcilable with the assembly within real viruses.

Another example of a hybrid structure determination was described in Effantin et al. (2016). The authors used a combination of subtomogram averaging and SPA to determine the structure of the foamy virus (a spumavirus) viral glycoprotein. At first subtomogram averaging was used to determine the shape of the glycoprotein trimer. This low-resolution structure was then used to pick particles from 2D micrographs, which were then refined using SPA to 9 Å resolution.

4.4 Toward high-resolution cryo-electron tomography and subtomogram averaging of retroviral CA assemblies

While these early insights by cryo-ET and subtomogram averaging substantially expanded the understanding of retrovirus Gag and CA assemblies, the best resolutions that were obtained by conventional subtomogram averaging were limited to around 17 Å (Briggs et al., 2009). Although there is no inherent limitation restricting cryo-ET and subtomogram averaging methods from obtaining higher resolutions, several resolution-limiting factors, as explained above, need to be mitigated to obtain results comparable to SPA cryo-EM.

The first cryo-ET and subtomogram averaging subnanometer reconstruction was achieved by systematically addressing some of these factors, e.g., by including an optimized defocus determination in order to correct for the CTF, and by increasing the dataset size via automated data acquisition for high-resolution cryo-ET (Schur et al., 2013). Due to the low SNR in micrographs acquired with an energy-filtered CCD camera, optimization of data acquisition for increased stability was necessary to allow for averaging of the signal from several tilts within the same tilt series to increase the SNR for a more precise defocus determination and hence CTF correction. The study revisited the *in vitro* assembled M-PMV CANC tubes described earlier (Bharat et al., 2012a) and subjected them to the optimized cryo-ET data collection and subtomogram averaging image processing workflow. This resulted in a reconstruction of the M-PMV CANC assembly at 8.5 Å resolution and represented a substantial improvement in the resolution obtained using cryo-ET and subtomogram averaging. Moreover, this result also served as an important proof-of-principle that these methods can be used to visualize protein secondary structure. Achieving high resolution with subtomogram averaging alone, and not using a hybrid 3D/2D helical reconstruction approach had advantages. Foremost it did not require precise determination of the helical symmetry parameters in advance, since each unit cell in the specimen could be aligned independently.

The same optimized cryo-ET and subtomogram averaging workflow was then applied to derive subnanometer resolution structures of the immature CA arrangement in purified authentic immature HIV-1 and M-PMV virus particles (Schur et al., 2015b). This allowed for an unambiguous description of the tertiary and quaternary CA interactions that mediate assembly in these retroviruses. Strikingly, the immature HIV-1 CA arrangement differed from the immature CA arrangement in truncated *in vitro* M-PMV Gag assemblies and authentic immature M-PMV viruses. This finding showed that retroviral CA proteins, while having conserved tertiary structures, adopt different quaternary arrangements on the CA-NTD level during virus assembly. This study also confirmed previous observations that the regions downstream of HIV-1 CA form a 6HB (Bharat et al., 2014).

In order to determine whether different immature CA arrangements do also occur in retroviruses of other genera, cryo-ET and subtomogram averaging was then performed on *in vitro* assembled RSV Gag assemblies lacking the Gag membrane binding domain (Schur et al., 2015a). The obtained 7.7 Å structure revealed that RSV again adopts a different quaternary CA arrangement at the CA-NTD, and also that RSV SP likely adopts a similar 6HB as compared to HIV-1. Although the resolution of the density in this region was too low to unambiguously visualize the individual helices, a simultaneous publication performing all-atom simulation and mutagenesis experiments on an immature RSV CA-SP assembly further corroborated this observation (Goh et al., 2015). The structure of immature RSV obtained by subtomogram averaging also confirmed a previous study that described that the non-canonical RSV Gag domain p10 (see Gag schematics in Fig. 3A) folds back into the CA-NTD hexamer to stabilize the immature assembly, and that this potentially regulates proteolytic maturation (Phillips et al., 2008).

Similarly, a new immature CA arrangement was observed by studying immature and mature MLV virus particles by cryo-ET and subtomogram averaging (Qu et al., 2018). The differences between immature and mature MLV CA arrangements were significantly smaller than what was observed for other viruses, allowing for potential alternative mechanisms for CA disassembly and reassembly during maturation.

These subnanometer resolution structures from native viruses already provided a new level of understanding of conserved and divergent features in retrovirus assembly and maturation. At this resolution, however, only secondary structure features like alpha-helices could be visualized and a description of the exact interaction interfaces at the level of individual amino acids was not possible. This requires resolutions at approximately 4 Å to allow either *de novo* model building or refinement of available high-resolution structures obtained from X-ray crystallography or NMR experiments in cryo-electron microscopy densities.

Higher resolutions were first obtained for *in vitro* HIV-1 Gag assemblies at 3.9 Å, and authentic immature HIV-1 viruses at 4.2 Å, by further optimization of cryo-ET data acquisition and subtomogram averaging protocols (Schur et al., 2016). The workflow made use of energy-filtered DEDs, movie-mode processing and weighting of tilts for cumulative dose to computationally increase the SNR in the reconstructed tilt series and to account for the dose-dependent loss of high spatial frequencies (Grant and Grigorieff, 2015). It also implemented a novel dose-symmetric tilt scheme that ensured optimal dose-distribution over the entire tilt series (Hagen et al., 2017). The higher SNR achieved by using DEDs improved alignment during subtomogram averaging and also allowed reliable defocus estimation from individual projections, resulting in a more precise CTF-correction. Both structures from *in vitro* assemblies and native viruses appeared identical, confirming that VLPs can be used to generate structural models of retroviral assemblies. The 3.9 Å structure of HIV-1 CA-SP1 revealed the network of intra- and interhexameric interactions mediating immature HIV-1 assembly. It also allowed *de novo* modeling of the helix formed by the last and first residues of CA and SP1, respectively, that form the 6HB. This showed that cleavage between CA and SP1 is not possible in the assembled immature lattice without destabilization of the 6HB, which is required for the viral protease to gain access to the cleavage site to complete maturation. In this study the truncated Gag construct was also assembled in the presence of the maturation inhibitor BVM. A clear density corresponding to BVM in the center of the 6HB allowed for the first description of the compound's mode of action. It confirmed previous biochemical results that indicated that BVM prevents cleavage by stabilizing the immature lattice (Keller et al., 2011) and further showed that resistance mutations and naturally occurring polymorphisms in Gag, conferring resistance to BVM, counteract the MI by weakening the immature CA-SP1 interactions, thus destabilizing the lattice.

A similar approach exploiting developments of DEDs and image processing was then performed to describe the native CA arrangement within purified mature HIV-1 particles (Mattei et al., 2016a). Despite the higher conformational variability of mature CA cores compared to the CA lattice formed in immature viruses, it was possible to visualize mature

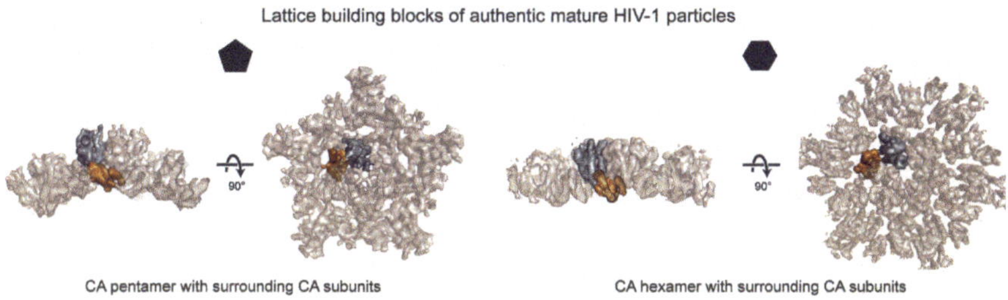

Fig. 5 Structures of the lattice building blocks in the authentic mature HIV-1 particles. The structures of the pentameric (left) and hexameric (right) CA building blocks of mature cores within authentic HIV-1 particles (Mattei et al., 2016a). The CA pentamer is shown at 8.8Å resolution (image generated using the structure deposited in the EMDB under accession code EMD-3466), and the CA hexamer at 6.8Å resolution (EMD-3465). A single CA monomer is colored as in Fig. 4B. Side and top views are shown. Note the difference between the curvature of the pentamer and hexamer (apparent from the side view), due to the pentamer location at the wide and broad ends of the fullerene cores.

CA hexamers and pentamers within HIV-1 viruses at resolutions of approximately 7 and 9Å, respectively (Fig. 5). By classifying CA hexamers according to their varying curvature in the mature core, this approach also allowed determining how CA flexes to form the mature fullerene assembly. It also revealed that the mature CA pentamer forms different interfaces than in previously reported crystal structures of the cross-linked CA pentamer (Pornillos et al., 2011).

HIV-1 Gag maturation intermediates were also studied using cryo-ET and subtomogram averaging by imaging virus particles with point mutations abolishing the individual cleavage sites between the Gag domains. In an early study the obtained results suggested that disassembly of the immature Gag lattice requires cleavage on both sides of CA-SP1 (de Marco et al., 2010b). Exploiting recent advances in high-resolution cryo-ET and subtomogram averaging, Mattei et al. (2018) determined structures of the same cleavage mutant viruses studied earlier, but this time at subnanometer resolutions and beyond. These structures showed that cleavage on only one side of CA-SP1 was in principle allowing for the formation of a mature lattice. In order to form a complete conical mature core, complete proteolytic processing was required.

The structural understanding of retroviral assembly and maturation we have today would not have been possible without continuous developments in cryo-ET and subtomogram averaging methods. Several fundamental questions, however, are still open (see Mattei et al. (2016b) for more information). These questions include, for example: How (or if) the CA lattice disassembles into small oligomers to form the mature core? How does mature core growth precisely initiate and how does it proceed? Further optimized cryo-ET and subtomogram averaging can help addressing such (and even more) complex questions.

In the next section, examples will be discussed where retroviruses have been valuable tools to improve available methods in image processing for cryo-ET and subtomogram averaging.

4.5 Retrovirus particles as tools for cryo-ET and subtomogram averaging method development

Several key factors have allowed retrovirus assemblies to be the first pleomorphic structures to be solved to subnanometer (Schur et al., 2013) and later to sub-4 Å resolution (Schur et al., 2016) using cryo-ET and subtomogram averaging. These factors included the large copy number of assembled retroviral CA in virus particles, their inherent local symmetry, their low conformational flexibility, and the availability of VLP systems. This also renders retroviral CA assembly datasets valuable to be used as tools to benchmark and improve novel approaches. Naturally, it is important to mention that not only retroviruses, but also regular viruses, such as TMV or Hepatitis B virus capsids, were used as tools to further increase the methodological portfolio of cryo-ET and subtomogram averaging (Bharat et al., 2015; Kunz and Frangakis, 2017).

As explained above, different resolution limiting factors need to be overcome to achieve the highest-possible resolution with cryo-ET and subtomogram averaging. Among them is the requirement for precise defocus estimation and CTF-correction, and the increased number of interpolations in cryo-ET and subtomogram averaging processing.

The defocus change across a tilted sample is taken into account in 2D-CTF correction by either correcting stripes or tiles of a micrograph with a defocus value that is dependent on the distance from the tilt axis (Fernandez et al., 2006; Xiong et al., 2009; Zanetti et al., 2009). This approach was successful in determining first subnanometer reconstructions and later the 3.9 Å resolution structure of retroviral CA assemblies (Schur et al., 2013, 2015a, 2015b, 2016). Another level of complexity is added by the defocus gradient along the Z-axis of the sample, which is particularly apparent in cryo-ET due to the increased sample thickness. Several 3D CTF-correction methods have been developed to compensate for the defocus variance of objects with a different height within the sample (Kunz and Frangakis, 2017; Voortman et al., 2011). In this regard, it is important to consider that the computational costs in cryo-ET can be significant because of the 3D nature of the volumes that need to be processed. A computationally efficient 3D-CTF correction algorithm was developed using the immature HIV-1 dataset that reached 3.9 Å as benchmark (EMPIAR-10164), and allowed to improve the obtainable resolution to 3.4 Å (Turoňová et al., 2017). Use of this improved CTF-correction also allowed obtaining sub-4 Å reconstructions from considerably smaller datasets than obtained via the above described tile or line based conventional 2D-CTF correction.

The increased number of interpolations can be addressed by again using a hybrid approach combining cryo-ET and subtomogram averaging with SPA cryo-EM.

The authors of the software emClarity (Himes and Zhang, 2018) also benchmarked their software using the publicly available high-resolution cryo-ET dataset containing immature HIV-1 assemblies (EMPIAR-10164). In emClarity, subtomogram averaging is initially performed to obtain refined positions of the proteins of interest. These subtomograms are then used as fiducial markers to iteratively refine tilt series alignment, reducing the negative effects of imprecise tilt series alignment and interpolations in tomogram reconstruction. This allowed to increase the resolution of HIV-1 CA-SP1 even further to 3.1 Å, currently the highest resolution achieved by cryo-ET and subtomogram averaging.

5. Applications beyond retroviruses

Cryo-ET and subtomogram averaging have been instrumental in describing structures in pleomorphic and asymmetric viruses beyond retroviruses. While this chapter emphasizes the high-resolution potential of cryo-ET and subtomogram averaging, some novel mechanisms can be readily unraveled at lower resolution. The biological insights that can be obtained are also highly valuable due to the possibility to combine cryo-ET and subtomogram averaging with studies of transient states (such as described for observing maturation intermediates in HIV-1), or the possibility to visualize viruses and their associated proteins directly within cells.

Integrative structural approaches are also vital, when low-resolution EM structures can be supplemented with high-resolution information obtained via X-ray crystallography or NMR methods. Examples for this include the work of Halldorsson et al. (2018), where X-ray crystallography, SPA cryo-EM,

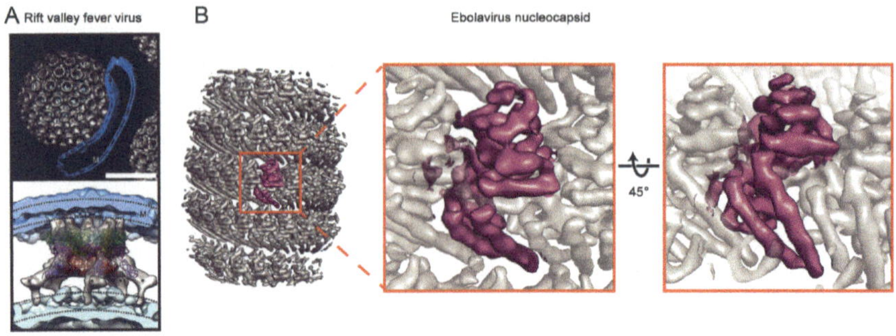

Fig. 6 Structure of the RVFV glycoprotein and Ebola virus nucleocapsid-like assembly obtained by cryo-ET and subtomogram averaging. (A) Fusion of the RVFV glycoproteins at liposomes. Top: segmentation of a tomogram. Scale bar: 100 nm. Bottom: crystal structures of Gn and Gc glycoproteins were fitted into the electron density at the liposome surface, showing the fusion protein inserted into the target membrane, in the first step of the membrane fusion reaction required for cell entry. (B) The electron density of a section from the surface of a helical Ebola NP (containing residues 1–450) *in vitro* assembly is shown. One NP protomer is highlighted in color. The resolution of the structure is 6.6 Å, allowing to clearly identify secondary structure elements such as alpha helices. The individual panels from left to right are the structure of the NP helix; zoom-in showing a single NP; rotated view of NP. The images were generated using the EMDB file: EMD-3869. *Panel (A) was reproduced and modified from Halldorsson, S., Li, S., Li, M., Harlos, K., Bowden, T.A., Huiskonen, J.T., 2018. Shielding and activation of a viral membrane fusion protein. Nat. Commun. 9, 349 and is published under a creative Commons Attribution 4.0 International License (http://creativecommons.org/licenses/by/4.0/).*

cryo-ET and subtomogram averaging were combined to visualize the fusion mechanism of the Rift valley fever virus (RVFV) surface glycoprotein (Fig. 6A). Subtomogram averaging was used to visualize the conformational change of the glycoprotein spike upon fusion between RVFV particles and liposomes. To ensure fusion competency of the surface glycoproteins, cryo-ET experiments had to be performed under BSL-3 conditions using unfixed infectious virus particles.

Other viruses that were studied via cryo-ET and subtomogram averaging included members of the *Paramyxoviridae* family, such as Measles virus or Newcastle Disease virus (Battisti et al., 2012; Liljeroos et al., 2011). The presented structures did not reach high resolutions, but provided valuable information about morphological differences in their respective NC and matrix coating. These examples highlight that with improved cryo-ET and subtomogram averaging workflows exciting structural virology questions should be revisited to obtain higher-resolution insights.

Summarizing all relevant cryo-ET and subtomogram averaging studies on pleomorphic and also regular viruses, or their associated viral proteins, would go beyond the scope of this chapter. To this end, selected examples displaying the potential and versatility of the methods are highlighted below. This also includes structures of vesicle coat proteins, which share similar mechanisms and structural features with pleomorphic viruses.

5.1 Filoviruses: Visualization of asymmetric filamentous viruses at high resolution

The resolution that is regularly achieved with cryo-ET and subtomogram averaging is usually lower than what has been described for retroviruses, mostly due to the increased challenges imposed by most other viral samples. Subnanometer resolution structures of filovirus nucleocapsid assemblies, however, prove that cryo-ET and subtomogram averaging can successfully deliver high-resolutions from even more challenging samples than retroviruses.

The *Filoviridae* family comprises some of the most lethal human pathogens, including Ebola and Marburg virus, which cause severe hemorrhagic fevers. Recent outbreaks confirmed the importance of elucidating the structures of filoviruses as potential targets for medical intervention (Messaoudi et al., 2015). All filoviruses are BSL-4 pathogens, which poses a significant challenge for performing virological and structural studies. The name *Filoviridae* has been derived from the characteristic pleomorphic filamentous shape of the virions, which are approximately 80 nm in diameter and 1 μm in length (Fig. 1). The genome of filoviruses is negative sense ssRNA and encodes for seven proteins. The transcription and replication machinery comprises the polymerase (L), its cofactor VP35, and a transcription factor VP30. The genome is condensed and encapsidated inside a helical shell of nucleoprotein (NP), forming a nucleocapsid (NC). The inner side of the viral membrane is lined with matrix protein (VP40) and VP24, whereas the outer side is decorated with surface glycoproteins (Emanuel et al., 2018).

Due to biosafety concerns when working with native virus particles, simplified VLP systems have been developed in order to address morphological and structural questions. Ebola virus NP, lacking the disordered C-terminus, can assemble with cellular RNA when expressed in *E. coli* (Noda et al., 2010) and also in mammalian cells (Bharat et al., 2011, 2012b), to form loosely coiled nucleocapsid-like structures. This loose coil can be rigidified by binding of VP24 and VP35. When co-expressed in mammalian cells together with VP40, these four proteins form VLPs with a morphology that is indistinguishable from authentic filovirus particles.

Several structural studies have attempted to determine the arrangement of NC using both subtomogram averaging and helical reconstruction, as well as combinations of the two methods.

Native filovirus particles, but also VLPs, have been shown to contain multiple helical symmetries that can switch within individual virus particles, limiting the applicability of helical reconstruction analysis.

The first low-resolution structures of Ebola and Marburg virus assemblies used the identical 3D/2D approach as described above for M-PMV tubes in Section 4.3. Cryo-ET and subtomogram averaging were first used to determine helical symmetry and handedness parameters of sections of the native viruses. This information was then exploited for iterative helical real space reconstruction (Bharat et al., 2011, 2012b). This allowed deriving structures of the Ebola and Marburg virus NC at 3.6 and 3.4 nm resolution, respectively.

High-resolution structures of Ebola and Marburg virus nucleocapsid were then solved via conventional cryo-ET and subtomogram averaging (Fig. 6B) using the methodological improvements described for obtaining the sub-4Å resolution structures of immature HIV-1 CA. Wan et al. (2017) determined the structure of recombinant Ebolavirus nucleocapsid (NC)-like particles at 6.6Å resolution and then validated these *in vitro* structures with subnanometer resolution reconstructions of NC within intact Ebola and Marburg virus particles. This revealed the arrangement of the nucleoprotein components and how subunit polymerization, RNA recruitment and co-factor attachment are coordinated.

Recent SPA cryo-EM work has further improved our understanding of the Ebola virus nucleoprotein-RNA complex by analyzing *in vitro* assembled NC tubes via helical reconstruction, yielding 3.6Å resolution (Sugita et al., 2018). This work shows that certain pleomorphic *in vitro* specimens are permissible also to SPA cryo-EM, if proper classification is possible.

5.2 Cryo-ET and subtomogram averaging of virus associated proteins

Viruses often establish specialized compartments and virus assemblies in infected cells to support their own replication. In case these compartments cannot be reproduced *in vitro*, or if visualization of the viral proteins in a native context is desired, the complexes have to be studied inside the infected cells, or fractions thereof. This increases the experimental complexity and therefore imposes further limitations on the resolution that can be achieved. However, the insights that can be obtained at low to medium resolution are often already sufficient to describe completely novel and important features of the studied viruses and their associated proteins.

Most positive-sense ssRNA and dsRNA viruses adopt an icosahedral shape. This geometry is ideally suited for structural characterization by X-ray crystallography or SPA cryo-EM. Nonetheless, subtomogram averaging also finds its place in structural biology of these viruses, as it is the go-to method for low-copy viral components and events in the life-cycle that need to be observed in complex environments.

Novel insights into the replication of nodaviruses were obtained by the work of Ertel et al. (2017), in which the authors analyzed mitochondria from S2 insect cells infected with Flock house nodavirus (FHV, an insect virus and common model system for the study of non-enveloped viruses). Cryo-ET revealed the structures of mitochondrion associated FHV replication compartments, named spherules. At the neck of each of these spherules a cupped ring structure containing multifunctional FHV RNA replication protein was visualized (Fig. 7A). This so-called crown structure was then subjected to subtomogram averaging. Although at low resolution, the structure revealed novel features, such as a 12-fold symmetry, and the crown association with long cytoplasmic fibrils, presumed to be exported progeny RNA.

Sulfolobus islandicus rod-shaped virus 2 (SIRV2) is a non-enveloped dsDNA phage, which infects thermophilic archaeans. Similar to the majority of archean viruses, SIRV2 does not exit the host cell via lysis. Daum et al. (2014) employed room temperature TEM and also cryo-ET to visualize virus-associated pyramid (VAP) structures formed in *E. coli* expressing the virus-encoded protein PVAP (Fig. 7B), which is required for SIRV2 to exit its host. PVAP forms a sevenfold symmetric pyramid that can form an iris-like opening at the lipid membrane, extending through the S-layer of the host. Only 57 subtomograms were averaged to obtain a final structure with a resolution of approximately 43 Å. Future experiments using novel technological advances in cryo-ET imaging and subtomogram averaging could increase the obtainable resolution from such low-copy events, providing more detailed explanations on the mechanisms of PVAP formation. The authors also showed that virus-associated pyramids are not host-specific and can insert into any kind of biological membrane, potentially allowing the exploitation of this mechanism for specific biotechnological applications.

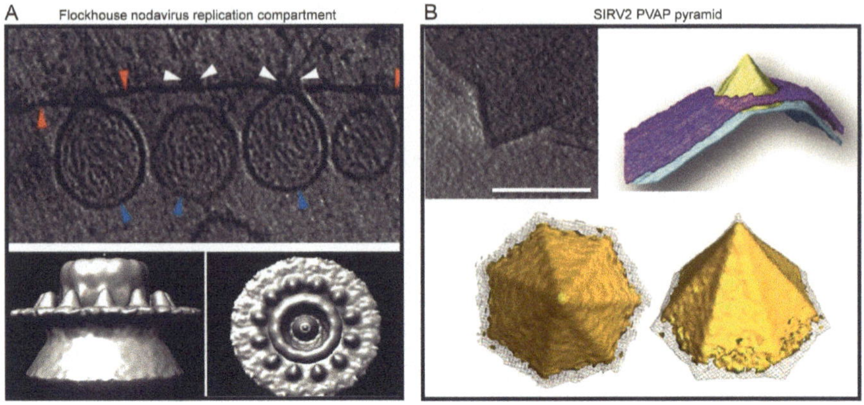

Fig. 7 Virus-related structures in complex environments solved by subtomogram averaging. (A) Structural analysis of the crown structure of the Flockhouse nodavirus replication compartment. Upper panel: a slice through a tomogram of mitochondria from S2 cells, showing the spherules—viral structures forming invaginations in the mitochondria. Scale bar: 100 nm. Bottom panel: side and top view of the electron density of the crown structure. (B) VAP structure at the surface of *E. coli* cells expressing a protein called PVAP from *Sulfolobus islandicus* rod-shaped virus 2. Top left: slice through tomogram showing the cross-section through the VAP. Scale bar: 200 nm. Top right: segmentation of the VAP structure showing the pyramid in yellow, plasma membrane in cyan and cell wall in purple. Bottom: isosurface representation of VAP. *Panel (A) was reproduced and modified from Ertel, K.J., Benefield, D., Castaño-Diez, D., Pennington, J.G., Horswill, M., den Boon, J.A., Otegui, M.S., Ahlquist, P., 2017. Cryo-electron tomography reveals novel features of a viral RNA replication compartment. Elife 6, 1–24, respectively, and is published under a creative Commons Attribution 4.0 International License (http://creativecommons.org/licenses/by/4.0/). Panel (B) was modified with permission from Daum, B., Quax, T.E.F., Sachse, M., Mills, D.J., Reimann, J., Yildiz, Ö., Häder, S., Saveanu, C., Forterre, P., Albers, S.-V., Kühlbrandt, W., Prangishvili, D., 2014. Self-assembly of the general membrane-remodeling protein PVAP into sevenfold virus-associated pyramids. Proc. Natl. Acad. Sci. U.S.A. 111, 3829–34.*

5.3 Structural studies of intracellular transport vesicles by cryo-ET and subtomogram averaging

Coated vesicles and enveloped viruses share certain similar mechanisms of structural assembly and can be considered to perform homologous processes. Both are higher-order pleomorphic structures, assembled to transport their selectively recruited cargo to distinct target locations. Hence, results obtained via cryo-ET and subtomogram averaging for coated vesicles demonstrate the possibilities that these methods also have for pleomorphic viruses.

Cargo-selective trafficking vesicles can derive from the plasma membrane, Golgi, ER or endosomal membranes and their formation is driven by assembly of different classes of coat proteins, i.e., Clathrin, COPI,

COPII, or retromer complex, respectively. While clathrin coats can more routinely be determined via single-particle approaches due to their more regular structure (Fotin et al., 2006), other vesicular coats display significant structural variability, which requires cryo-ET and subtomogram averaging for visualizing interactions within the coat assembly. Recent work was successful in describing *in vitro* the intricate coat architecture of part of the COPI and retromer coat at subnanometer resolution (Dodonova et al., 2017; Kovtun et al., 2018) and the COPII coat at resolutions beyond 5 Å (Hutchings et al., 2018). *In situ* experiments visualizing COPI and retromer coated vesicles also confirmed the *in vitro* observations (Bykov et al., 2017; Kovtun et al., 2018). These results underscore the potential and versatility offered by cryo-ET and subtomogram averaging to elucidate important protein-protein interactions of relevant large macromolecular assemblies.

6. Conclusions and outlook

This chapter provides a summary of several state-of-the-art applications of cryo-ET and subtomogram averaging for pleomorphic viruses. While high-resolution is currently mostly obtained for optimal specimens, such as retrovirus particles, several developments allow for the optimistic prediction that similar results can be obtained for more challenging samples under even more complex conditions in the future. These developments might need to include a clever combination of hybrid approaches using features from both SPA cryo-EM and cryo-ET/subtomogram averaging, allowing to overcome certain limitations imposed by either method. Some samples, however, will not be applicable to such hybrid approaches and hence continued efforts to improve existing workflows need to be undertaken, ranging from the specimen preparation to data acquisition and image processing.

In particular further developments in sample preparation are possible. On the grid purification, using antibody-affinity based grids, would allow capturing of virions that can be derived only in low-copy numbers from cell cultures or even clinical samples (Stass et al., 2018). Potential improvements in image processing (such as more robust CTF determination algorithms for low SNR projections), need to be combined with a faster and more stable data acquisition to make high-resolution cryo-ET a high-throughput method. As described, progress in this direction has been recently provided by the introduction of novel tilt schemes, exploiting more stable and eucentric microscope stages (Chreifi et al., 2019).

One of the most important limitations is the maximum sample thickness that can be imaged by cryo-ET, which usually cannot exceed 500 nanometers. Along with the development of cryo-focused ion beam milling (cryo-FIBSEM; see Villa et al., 2013), cryo-ET and subtomogram averaging will become even more relevant for studying virus-related events *in situ*, e.g., virion morphogenesis, virus-host interactions or viral capsid uncoating, to name just a few.

Structures of the matrix or (nucleo)capsid structures in many pleomorphic viruses remain to be elucidated, including bunyaviruses (such as hantaviruses or arenaviruses), rhabdoviruses (e.g., Rabies virus), or orthomyxoviruses (e.g., influenza). Continued developments for cryo-ET and subtomogram averaging will be able to provide important answers to applications even beyond virology.

Acknowledgments

We would like to thank Rob Dick and Carrie Bernecky for helpful comments on the text. This work was funded by a grant from the Austrian Science Fund (FWF, Grant P31445). The authors also acknowledge financial support from IST Austria.

References

Accola, M.A., Höglund, S., Göttlinger, H.G., 1998. A putative α-helical structure which overlaps the capsid-p2 boundary in the human immunodeficiency virus type 1 gag precursor is crucial for viral particle assembly. J. Virol. 72, 2072–2078.

Bartesaghi, A., Lecumberry, F., Sapiro, G., Subramaniam, S., 2012. Protein secondary structure determination by constrained single-particle cryo-electron tomography. Structure 20, 2003–2013.

Bartonova, V., Igonet, S., Sticht, J., Glass, B., Habermann, A., Vaney, M.C., Sehr, P., Lewis, J., Rey, F.A., Krausslich, H.G., 2008. Residues in the HIV-1 capsid assembly inhibitor binding site are essential for maintaining the assembly-competent quaternary structure of the capsid protein. J. Biol. Chem. 283, 32024–32033.

Battisti, A.J., Meng, G., Winkler, D.C., McGinnes, L.W., Plevka, P., Steven, A.C., Morrison, T.G., Rossmann, M.G., 2012. Structure and assembly of a paramyxovirus matrix protein. Proc. Natl. Acad. Sci. 109, 13996–14000.

Beck, M., Baumeister, W., 2016. Cryo-Electron tomography: can it reveal the molecular sociology of cells in atomic detail? Trends Cell Biol. 26, 825–837.

Bharat, T.A.M., Riches, J.D., Kolesnikova, L., Welsch, S., Krähling, V., Davey, N., Parsy, M.-L., Becker, S., Briggs, J.A.G., 2011. Cryo-electron tomography of Marburg virus particles and their morphogenesis within infected cells. PLoS Biol. 9, e1001196.

Bharat, T.A.M., Davey, N.E., Ulbrich, P., Riches, J.D., de Marco, A., Rumlova, M., Sachse, C., Ruml, T., Briggs, J.A.G., 2012a. Structure of the immature retroviral capsid at 8 Å resolution by cryo-electron microscopy. Nature 487, 385–389.

Bharat, T.A.M., Noda, T., Riches, J.D., Kraehling, V., Kolesnikova, L., Becker, S., Kawaoka, Y., Briggs, J.A.G., 2012b. Structural dissection of Ebola virus and its assembly determinants using cryo-electron tomography. Proc. Natl. Acad. Sci. 109, 4275–4280.

Bharat, T.A.M., Castillo Menendez, L.R., Hagen, W.J.H., Lux, V., Igonet, S., Schorb, M., Schur, F.K.M., Krausslich, H.-G., Briggs, J.A.G., 2014. Cryo-electron microscopy of tubular arrays of HIV-1 Gag resolves structures essential for immature virus assembly. Proc. Natl. Acad. Sci. 111, 8233–8238.

Bharat, T.A.M., Russo, C.J., Löwe, J., Passmore, L.A., Scheres, S.H.W., 2015. Advances in single-particle electron cryomicroscopy structure determination applied to sub-tomogram averaging. Structure 23, 1743–1753.

Briggs, J.A., Krausslich, H.G., 2011. The molecular architecture of HIV. J. Mol. Biol. 410, 491–500.

Briggs, J.A.G., Wilk, T., Welker, R., Kräusslich, H., Fuller, S.D., 2003. Structural organization of authentic, mature HIV-1 virions and cores. EMBO J. 22, 1707–1715.

Briggs, J.A., Riches, J.D., Glass, B., Bartonova, V., Zanetti, G., Krausslich, H.G., 2009. Structure and assembly of immature HIV. Proc. Natl. Acad. Sci. U.S.A. 106, 11090–11095.

Brilot, A.F., Chen, J.Z., Cheng, A., Pan, J., Harrison, S.C., Potter, C.S., Carragher, B., Henderson, R., Grigorieff, N., 2012. Beam-induced motion of vitrified specimen on holey carbon film. J. Struct. Biol. 177, 630–637.

Bush, D.L., Vogt, V.M., 2014. In vitro assembly of retroviruses. Annu. Rev. Virol. 1, 561–580.

Bush, D.L., Monroe, E.B., Bedwell, G.J., Prevelige Jr., P.E., Phillips, J.M., Vogt, V.M., 2014. Higher-order structure of the Rous sarcoma virus SP assembly domain. J. Virol. 88, 5617–5629.

Bykov, Y.S., Schaffer, M., Dodonova, S.O., Albert, S., Plitzko, J.M., Baumeister, W., Engel, B.D., Briggs, J.A., 2017. The structure of the COPI coat determined within the cell. Elife 6, e32493.

Campos-Olivas, R., Newman, J.L., Summers, M.F., 2000. Solution structure and dynamics of the Rous sarcoma virus capsid protein and comparison with capsid proteins of other retroviruses. J. Mol. Biol. 296, 633–649.

Carlson, L.A., Briggs, J.A., Glass, B., Riches, J.D., Simon, M.N., Johnson, M.C., Muller, B., Grunewald, K., Krausslich, H.G., 2008. Three-dimensional analysis of budding sites and released virus suggests a revised model for HIV-1 morphogenesis. Cell Host Microbe 4, 592–599.

Caspar, D.L.D., Klug, A., 1962. Physical principles in the construction of regular viruses. Cold Spring Harb. Symp. Quant. Biol. 27, 1–24.

Castano-Diez, D., Kudryashev, M., Arheit, M., Stahlberg, H., 2012. Dynamo: a flexible, user-friendly development tool for subtomogram averaging of cryo-EM data in high-performance computing environments. J. Struct. Biol. 178, 139–151.

Cheng, Y., Grigorieff, N., Penczek, P.A., Walz, T., 2015. A primer to single-particle cryo-electron microscopy. Cell 161, 438–449.

Cheslock, S.R., Poon, D.T.K., Fu, W., Rhodes, T.D., Henderson, L.E., Nagashima, K., McGrath, C.F., Hu, W.-S., 2003. Charged assembly helix motif in Murine Leukemia virus capsid: an important region for virus assembly and particle size determination. J. Virol. 77, 7058–7066.

Chreifi, G., Chen, S., Metskas, L.A., Kaplan, M., Jensen, G.J., 2019. Rapid tilt-series acquisition for electron cryotomography. J. Struct. Biol. 205, 163–169.

Cornilescu, C.C., Bouamr, F., Yao, X., Carter, C., Tjandra, N., 2001. Structural analysis of the N-terminal domain of the human T-cell leukemia virus capsid protein. J. Mol. Biol. 306, 783–797.

Craigie, R., Bushman, F.D., 2012. HIV DNA integration. Cold Spring Harb. Perspect. Med. 2, 1–18.
Cyrklaff, M., Risco, C., Fernández, J.J., Jiménez, M.V., Estéban, M., Baumeister, W., Carrascosa, J.L., 2005. Cryo-electron tomography of vaccinia virus. Proc. Natl. Acad. Sci. U.S.A. 102, 2772–2777.
Dai, X., Li, Z., Lai, M., Shu, S., Du, Y., Zhou, Z.H., Sun, R., 2017. In situ structures of the genome and genome-delivery apparatus in a single-stranded RNA virus. Nature 541, 112–116.
Danev, R., Nagayama, K., 2010. Phase plates for transmission electron microscopy. Methods Enzymol. 481, 343–369.
Datta, S.A.K., Curtis, J.E., Ratcliff, W., Clark, P.K., Crist, R.M., Lebowitz, J., Krueger, S., Rein, A., 2007. Conformation of the HIV-1 gag protein in solution. J. Mol. Biol. 365, 812–824.
Datta, S.A.K., Temeselew, L.G., Crist, R.M., Soheilian, F., Kamata, A., Mirro, J., Harvin, D., Nagashima, K., Cachau, R.E., Rein, A., 2011a. On the role of the SP1 domain in HIV-1 particle assembly: a molecular switch? J. Virol. 85, 4111–4121.
Datta, S.A.K., Zuo, X., Clark, P.K., Campbell, S.J., Wang, Y.-X., Rein, A., 2011b. Solution properties of murine leukemia virus gag protein: differences from HIV-1 gag. J. Virol. 85, 12733–12741.
Daum, B., Quax, T.E.F., Sachse, M., Mills, D.J., Reimann, J., Yildiz, Ö., Häder, S., Saveanu, C., Forterre, P., Albers, S.-V., Kühlbrandt, W., Prangishvili, D., 2014. Self-assembly of the general membrane-remodeling protein PVAP into sevenfold virus-associated pyramids. Proc. Natl. Acad. Sci. U.S.A. 111, 3829–3834.
de Marco, A., Mueller, B., Glass, B., Riches, J.D., Kraeusslich, H.-G., Briggs, J.A.G., 2010b. Structural analysis of HIV-1 maturation using cryo-electron tomography. PLoS Pathog. 6, 1–9.
de Marco, A., Davey, N.E., Ulbrich, P., Phillips, J.M., Lux, V., Riches, J.D., Fuzik, T., Ruml, T., Krausslich, H.G., Vogt, V.M., Briggs, J.A., 2010a. Conserved and variable features of Gag structure and arrangement in immature retrovirus particles. J. Virol. 84, 11729–11736.
Dodonova, S.O., Aderhold, P., Kopp, J., Ganeva, I., Röhling, S., Hagen, W.J.H., Sinning, I., Wieland, F., Briggs, J.A.G., 2017. 9Å structure of the COPI coat reveals that the Arf1 GTPase occupies two contrasting molecular environments. Elife 6, 1–29.
Effantin, G., Estrozi, L.F., Aschman, N., Renesto, P., Stanke, N., Lindemann, D., Schoehn, G., Weissenhorn, W., 2016. Cryo-electron microscopy structure of the native prototype foamy virus glycoprotein and virus architecture. PLoS Pathog. 12, e1005721.
Egelman, E.H., 2007. The iterative helical real space reconstruction method: surmounting the problems posed by real polymers. J. Struct. Biol. 157, 83–94.
Emanuel, J., Marzi, A., Feldmann, H., 2018. Filoviruses: ecology, molecular biology, and evolution. Adv. Virus Res. 100, 189–221.
Ertel, K.J., Benefield, D., Castaño-Diez, D., Pennington, J.G., Horswill, M., den Boon, J.A., Otegui, M.S., Ahlquist, P., 2017. Cryo-electron tomography reveals novel features of a viral RNA replication compartment. Elife 6, 1–24.
Fernandez, J.J., Li, S., Crowther, R.A., 2006. CTF determination and correction in electron cryotomography. Ultramicroscopy 106, 587–596.
Fontana, J., Jurado, K.A., Cheng, N., Ly, N.L., Fuchs, J.R., Gorelick, R.J., Engelman, A.N., Steven, A.C., 2015. Distribution and redistribution of HIV-1 nucleocapsid protein in immature, mature, and integrase-inhibited Virions: a role for integrase in maturation. J. Virol. 89, 9765–9780.
Förster, F., Medalia, O., Zauberman, N., Baumeister, W., Fass, D., 2005. Retrovirus envelope protein complex structure in situ studied by cryo-electron tomography. Proc. Natl. Acad. Sci. U.S.A. 102, 4729–4734.

Fotin, A., Kirchhausen, T., Grigorieff, N., Harrison, S.C., Walz, T., Cheng, Y., 2006. Structure determination of clathrin coats to subnanometer resolution by single particle cryo-electron microscopy. J. Struct. Biol. 156, 453–460.

Freed, E.O., 2015. HIV-1 assembly, release and maturation. Nat. Rev. Microbiol. 13, 484–496.

Fromm, S.A., Bharat, T.A.M., Jakobi, A.J., Hagen, W.J., Sachse, C., 2015. Seeing tobacco mosaic virus through direct electron detectors. J. Struct. Biol. 189, 87–97.

Fuller, S.D., Wilk, T., Gowen, B.E., Kräusslich, H.-G., Vogt, V.M., 1997. Cryo-electron microscopy reveals ordered domains in the immature HIV-1 particle. Curr. Biol. 7, 729–738.

Galaz-Montoya, J.G., Flanagan, J., Schmid, M.F., Ludtke, S.J., 2015. Single particle tomography in EMAN2. J. Struct. Biol. 190, 279–290.

Gamble, T.R., Yoo, S., Vajdos, F.F., von Schwedler, U.K., Worthylake, D.K., Wang, H., McCutcheon, J.P., Sundquist, W.I., Hill, C.P., 1997. Structure of the carboxyl-terminal dimerization domain of the HIV-1 capsid protein. Science 278, 849–853 (80-.).

Ganser-Pornillos, B.K., Cheng, A., Yeager, M., 2007. Structure of full-length HIV-1 CA: a model for the mature capsid lattice. Cell 131, 70–79.

Ge, P., Tsao, J., Schein, S., Green, T.J., Luo, M., Zhou, Z.H., 2010. CryoEM model of the bullet-shaped vesicular stomatitis virus. Science 327, 689.

Goh, B.C., Perilla, J.R., England, M.R., Heyrana, K.J., Craven, R.C., Schulten, K., 2015. Atomic modeling of an immature retroviral lattice using molecular dynamics and mutagenesis. Structure.

Grant, T., Grigorieff, N., 2015. Measuring the optimal exposure for single particle cryo-EM using a 2.6 Å reconstruction of rotavirus VP6. Elife 4, e06980.

Grigorieff, N., Harrison, S.C., 2011. Near-atomic resolution reconstructions of icosahedral viruses from electron cryo-microscopy. Curr. Opin. Struct. Biol. 21, 265–273.

Gross, I., Hohenberg, H., Wilk, T., Wiegers, K., Grättinger, M., Müller, B., Fuller, S., Kräusslich, H.G., 2000. A conformational switch controlling HIV-1 morphogenesis. EMBO J. 19, 103–113.

Grünewald, K., Desai, P., Winkler, D.C., Heymann, J.B., Belnap, D.M., Baumeister, W., Steven, A.C., 2003. Three-dimensional structure of herpes simplex virus from cryo-electron tomography. Science 302, 1396–1398 (80-.).

Hadravová, R., de Marco, A., Ulbrich, P., Štokrová, J., Doležal, M., Pichová, I., Ruml, T., Briggs, J.A.G., Rumlová, M., 2012. In vitro assembly of virus-like particles of a gammaretrovirus, the murine leukemia virus XMRV. J. Virol. 86, 1297–1306.

Hagen, W.J.H., Wan, W., Briggs, J.A.G., 2017. Implementation of a cryo-electron tomography tilt-scheme optimized for high resolution subtomogram averaging. J. Struct. Biol. 197, 191–198.

Halldorsson, S., Li, S., Li, M., Harlos, K., Bowden, T.A., Huiskonen, J.T., 2018. Shielding and activation of a viral membrane fusion protein. Nat. Commun. 9, 349.

Hampton, C.M., Strauss, J.D., Ke, Z., Dillard, R.S., Hammonds, J.E., Alonas, E., Desai, T.M., Marin, M., Storms, R.E., Leon, F., Melikyan, G.B., Santangelo, P.J., Spearman, P.W., Wright, E.R., 2017. Correlated fluorescence microscopy and cryo-electron tomography of virus-infected or transfected mammalian cells. Nat. Protoc. 12, 150–167.

Harrison, S.C., Olson, A.J., Schutt, C.E., Winkler, F.K., Bricogne, G., 1978. Tomato bushy stunt virus at 2.9 Å resolution. Nature 276, 368–373.

Hill, C.P., Worthylake, D., Bancroft, D.P., Christensen, A.M., Sundquist, W.I., 1996. Crystal structures of the trimeric human immunodeficiency virus type 1 matrix protein: implications for membrane association and assembly. Proc. Natl. Acad. Sci. U.S.A. 93, 3099–3104.

Himes, B.A., Zhang, P., 2018. emClarity: software for high-resolution cryo-electron tomography and subtomogram averaging. Nat. Methods 15, 955–961.
Hrabe, T., Chen, Y., Pfeffer, S., Kuhn Cuellar, L., Mangold, A.-V., Förster, F., 2012. PyTom: a python-based toolbox for localization of macromolecules in cryo-electron tomograms and subtomogram analysis. J. Struct. Biol. 178, 177–188.
Hutchings, J., Zanetti, G., 2018. Fine details in complex environments: the power of cryo-electron tomography. Biochem. Soc. Trans. 46, 807–816.
Hutchings, J., Stancheva, V., Miller, E.A., Zanetti, G., 2018. Subtomogram averaging of COPII assemblies reveals how coat organization dictates membrane shape. Nat. Commun. 9, 4154.
Ke, Z., Dillard, R., Chirkova, T., Leon, F., Stobart, C., Hampton, C., Strauss, J., Rajan, D., Rostad, C., Taylor, J., Yi, H., Shah, R., Jin, M., Hartert, T., Peebles, R., Graham, B., Moore, M., Anderson, L., Wright, E., Ke, Z., Dillard, R.S., Chirkova, T., Leon, F., Stobart, C.C., Hampton, C.M., Strauss, J.D., Rajan, D., Rostad, C.A., Taylor, J.V., Yi, H., Shah, R., Jin, M., Hartert, T.V., Peebles, R.S., Graham, B.S., Moore, M.L., Anderson, L.J., Wright, E.R., 2018. The morphology and assembly of respiratory syncytial virus revealed by cryo-electron tomography. Viruses 10, 446.
Keller, P.W., Adamson, C.S., Heymann, J.B., Freed, E.O., Steven, A.C., 2011. HIV-1 maturation inhibitor bevirimat stabilizes the immature Gag lattice. J. Virol. 85, 1420–1428.
Khorasanizadeh, S., Campos-Olivas, R., Clark, C., Summers, M., 1999. Sequence-specific 1H, 13C and 15N chemical shift assignment and secondary structure of the HTLV-I capsid protein. J. Biomol. NMR 14, 199–200.
Kingston, R.L., Fitzon-Ostendorp, T., Eisenmesser, E.Z., Schatz, G.W., Vogt, V.M., Post, C.B., Rossmann, M.G., 2000. Structure and self-association of the Rous sarcoma virus capsid protein. Structure 8, 617–628.
Kovtun, O., Leneva, N., Bykov, Y.S., Ariotti, N., Teasdale, R.D., Schaffer, M., Engel, B.D., Owen, D.J., Briggs, J.A.G., Collins, B.M., 2018. Structure of the membrane-assembled retromer coat determined by cryo-electron tomography. Nature 561, 561–564.
Kräusslich, H.G., Fäcke, M., Heuser, A.M., Konvalinka, J., Zentgraf, H., 1995. The spacer peptide between human immunodeficiency virus capsid and nucleocapsid proteins is essential for ordered assembly and viral infectivity. J. Virol. 69, 3407–3419.
Kühlbrandt, W., 2014. Biochemistry. The resolution revolution. Science 343, 1443–1444.
Kunz, M., Frangakis, A.S., 2017. Three-dimensional CTF correction improves the resolution of electron tomograms. J. Struct. Biol. 197, 114–122.
Lee, J.H., Ozorowski, G., Ward, A.B., 2016. Cryo-EM structure of a native, fully glycosylated, cleaved HIV-1 envelope trimer. Science 351, 1043–1048.
Li, S., Hill, C.P., Sundquist, W.I., Finch, J.T., 2000. Image reconstructions of helical assemblies of the HIV-1 CA protein. Nature 407.
Li, X., Mooney, P., Zheng, S., Booth, C.R., Braunfeld, M.B., Gubbens, S., Agard, D.A., Cheng, Y., 2013. Electron counting and beam-induced motion correction enable near-atomic-resolution single-particle cryo-EM. Nat. Methods 10, 584–590.
Liljeroos, L., Huiskonen, J.T., Ora, A., Susi, P., Butcher, S.J., 2011. Electron cryotomography of measles virus reveals how matrix protein coats the ribonucleocapsid within intact virions. Proc. Natl. Acad. Sci. 108, 18085–18090. LP.
Liu, J., Bartesaghi, A., Borgnia, M.J., Sapiro, G., Subramaniam, S., 2008. Molecular architecture of native HIV-1 gp120 trimers. Nature 455, 109–113.
Liu, H., Jin, L., Koh, S.B.S., Atanasov, I., Schein, S., Wu, L., Zhou, Z.H., 2010. Atomic structure of human adenovirus by cryo-EM reveals interactions among protein networks. Science 329, 1038–1043.
Lucic, V., Forster, F., Baumeister, W., 2005. Structural studies by electron tomography: from cells to molecules. Annu. Rev. Biochem. 74, 833–865.

Macek, P., Chmelík, J., Křížová, I., Kadeřávek, P., Padrta, P., Žídek, L., Wildová, M., Hadravová, R., Chaloupková, R., Pichová, I., Ruml, T., Rumlová, M., Sklenář, V., 2009. NMR structure of the N-terminal domain of capsid protein from the Mason–Pfizer monkey virus. J. Mol. Biol. 392, 100–114.
Mak, J., de Marco, A., 2018. Recent advances in retroviruses via cryo-electron microscopy. Retrovirology 15, 23.
Mastronarde, D.N., 2005. Automated electron microscope tomography using robust prediction of specimen movements. J. Struct. Biol. 152, 36–51.
Mattei, S., Glass, B., Hagen, W.J.H., Kräusslich, H.G., Briggs, J.A.G., 2016a. The structure and flexibility of conical HIV-1 capsids determined within intact virions. Science 354, 1434–1437 (80-.).
Mattei, S., Schur, F.K.M., Briggs, J.A., 2016b. Retrovirus maturation—an extraordinary structural transformation. Curr. Opin. Virol. 18, 27–35.
Mattei, S., Tan, A., Glass, B., Müller, B., Kräusslich, H.-G., Briggs, J.A.G., 2018. High-resolution structures of HIV-1 Gag cleavage mutants determine structural switch for virus maturation. Proc. Natl. Acad. Sci. 115, E9401–E9410.
Messaoudi, I., Amarasinghe, G.K., Basler, C.F., 2015. Filovirus pathogenesis and immune evasion: insights from Ebola virus and Marburg virus. Nat. Rev. Microbiol. 13, 663–676.
Mortuza, G.B., Haire, L.F., Stevens, A., Smerdon, S.J., Stoye, J.P., Taylor, I.A., 2004. High-resolution structure of a retroviral capsid hexameric amino-terminal domain. Nature 431, 481–485.
Nermut, M.V., Hockley, D.J., Jowett, J.B.M., Jones, I.M., Garreau, M., Thomas, D., 1994. Fullerene-like organization of HIV gag-protein shell in virus-like particles produced by recombinant Baculovirus. Virology 198, 288–296.
Nicastro, D., Schwartz, C., Pierson, J., Gaudette, R., Porter, M.E., McIntosh, J.R., 2006. The molecular architecture of axonemes revealed by cryoelectron tomography. Science 313, 944–948.
Nickell, S., Forster, F., Linaroudis, A., Net, W.D., Beck, F., Hegerl, R., Baumeister, W., Plitzko, J.M., 2005. TOM software toolbox: acquisition and analysis for electron tomography. J. Struct. Biol. 149, 227–234.
Nickell, S., Kofler, C., Leis, A.P., Baumeister, W., 2006. A visual approach to proteomics. Nat. Rev. Mol. Cell Biol. 7, 225–230.
Noda, T., Hagiwara, K., Sagara, H., Kawaoka, Y., 2010. Characterization of the Ebola virus nucleoprotein-RNA complex. J. Gen. Virol. 91, 1478–1483.
Phillips, J.M., Murray, P.S., Murray, D., Vogt, V.M., 2008. A molecular switch required for retrovirus assembly participates in the hexagonal immature lattice. EMBO J. 27, 1411–1420.
Pornillos, O., Ganser-Pornillos, B.K., Kelly, B.N., Hua, Y., Whitby, F.G., Stout, C.D., Sundquist, W.I., Hill, C.P., Yeager, M., 2009. X-ray structures of the hexameric building block of the HIV capsid. Cell 137, 1282–1292.
Pornillos, O., Ganser-Pornillos, B.K., Yeager, M., 2011. Atomic-level modelling of the HIV capsid. Nature 469, 424–427.
Qu, K., Glass, B., Doležal, M., Schur, F.K.M., Murciano, B., Rein, A., Rumlová, M., Ruml, T., Kräusslich, H.-G., Briggs, J.A.G., 2018. Structure and architecture of immature and mature murine leukemia virus capsids. Proc. Natl. Acad. Sci. U.S.A. 115, E11751–E11760.
Qualley, D.F., Cooper, S.E., Ross, J.L., Olson, E.D., Cantara, W.A., Musier-Forsyth, K., 2019. Solution conformation of bovine leukemia virus gag suggests an elongated structure. J. Mol. Biol. 431, 1203–1216.
Rhee, S.S., Hunter, E., 1990. A single amino acid substitution within the matrix protein of a type D retrovirus converts its morphogenesis to that of a type C retrovirus. Cell 63, 77–86.

Schur, F.K.M., 2019. Toward high-resolution in situ structural biology with cryo-electron tomography and subtomogram averaging. Curr. Opin. Struct. Biol. 58, 1–9.
Schur, F.K.M., Hagen, W.J.H., De Marco, A., Briggs, J.A.G., 2013. Determination of protein structure at 8.5Å resolution using cryo-electron tomography and sub-tomogram averaging. J. Struct. Biol. 184, 394–400.
Schur, F.K.M., Dick, R.A., Hagen, W.J.H., Vogt, V.M., Briggs, J.A.G., 2015a. The structure of immature virus-like Rous sarcoma virus gag particles reveals a structural role for the p10 domain in assembly. J. Virol. 89, 10294–10302.
Schur, F.K.M., Hagen, W.J.H., Rumlová, M., Ruml, T., Müller, B., Kraüsslich, H.G., Briggs, J.A.G., 2015b. Structure of the immature HIV-1 capsid in intact virus particles at 8.8 Å resolution. Nature 517, 505–508.
Schur, F.K.M., Obr, M., Hagen, W.J.H., Wan, W., Jakobi, A.J., Kirkpatrick, J.M., Sachse, C., Kräusslich, H.G., Briggs, J.A.G., 2016. An atomic model of HIV-1 capsid-SP1 reveals structures regulating assembly and maturation. Science 353, 506–508 (80-.).
Seitz, S., Urban, S., Antoni, C., Böttcher, B., 2007. Cryo-electron microscopy of hepatitis B virions reveals variability in envelope capsid interactions. EMBO J. 26, 4160.
Stass, R., Ilca, S.L., Huiskonen, J.T., 2018. Beyond structures of highly symmetric purified viral capsids by cryo-EM. Curr. Opin. Struct. Biol. 52, 25–31.
Subramaniam, S., Bartesaghi, A., Liu, J., Bennett, A.E., Sougrat, R., 2007. Electron tomography of viruses. Curr. Opin. Struct. Biol. 17, 596–602.
Sugita, Y., Matsunami, H., Kawaoka, Y., Noda, T., Wolf, M., 2018. Cryo-EM structure of the Ebola virus nucleoprotein–RNA complex at 3.6 Å resolution. Nature 563, 137–140.
Suloway, C., Shi, J., Cheng, A., Pulokas, J., Carragher, B., Potter, C.S., Zheng, S.Q., Agard, D.A., Jensen, G.J., 2009. Fully automated, sequential tilt-series acquisition with Leginon. J. Struct. Biol. 167, 11–18.
Sundquist, W.I., Krausslich, H.G., 2012. HIV-1 assembly, budding, and maturation. Cold Spring Harb. Perspect. Med. 2, a006924.
Tang, C., Ndassa, Y., Summers, M.F., 2002. Structure of the N-terminal 283-residue fragment of the immature HIV-1 Gag polyprotein. Nat. Struct. Biol. 9, 537–543.
Turoňová, B., Schur, F.K.M., Wan, W., Briggs, J.A.G., 2017. Efficient 3D-CTF correction for cryo-electron tomography using NovaCTF improves subtomogram averaging resolution to 3.4 Å. J. Struct. Biol. 199, 187–195.
Villa, E., Schaffer, M., Plitzko, J.M., Baumeister, W., 2013. Opening windows into the cell: focused-ion-beam milling for cryo-electron tomography. Curr. Opin. Struct. Biol. 23, 771–777.
Voortman, L.M., Stallinga, S., Schoenmakers, R.H., van Vliet, L.J., Rieger, B., 2011. A fast algorithm for computing and correcting the CTF for tilted, thick specimens in TEM. Ultramicroscopy 111, 1029–1036.
Wan, W., Briggs, J.A.G., 2016. Chapter thirteen—cryo-electron tomography and subtomogram averaging. In: Crowther, R.A. (Ed.), Methods In Enzymology. Academic Press, pp. 329–367.
Wan, W., Kolesnikova, L., Clarke, M., Koehler, A., Noda, T., Becker, S., Briggs, J.A.G., 2017. Structure and assembly of the Ebola virus nucleocapsid. Nature 551, 394–397.
White, T.A., Bartesaghi, A., Borgnia, M.J., Meyerson, J.R., de la Cruz, M.J.V., Bess, J.W., Nandwani, R., Hoxie, J.A., Lifson, J.D., Milne, J.L.S., Subramaniam, S., 2010. Molecular architectures of trimeric SIV and HIV-1 envelope glycoproteins on intact viruses: strain-dependent variation in quaternary structure. PLoS Pathog. 6, e1001249.
Wilen, C.B., Tilton, J.C., Doms, R.W., 2012. HIV: cell binding and entry. Cold Spring Harb. Perspect. Med. 2, 1–13.
Wilk, T., Fuller, S.D., 1999. Towards the structure of the human immunodeficiency virus: divide and conquer? Curr. Opin. Struct. Biol. 9, 231–243.

Worthylake, D.K., Wang, H., Yoo, S., Sundquist, W.I., Hill, C.P., 1999. Structures of the HIV-1 capsid protein dimerization domain at 2.6 A resolution. Acta Crystallogr. Sect. D 55, 85–92.

Wright, E.R., Schooler, J.B., Ding, H.J., Kieffer, C., Fillmore, C., Sundquist, W.I., Jensen, G.J., 2007. Electron cryotomography of immature HIV-1 virions reveals the structure of the CA and SP1 Gag shells. EMBO J. 26, 2218–2226.

Xiong, Q., Morphew, M.K., Schwartz, C.L., Hoenger, A.H., Mastronarde, D.N., 2009. CTF determination and correction for low dose tomographic tilt series. J. Struct. Biol. 168, 378–387.

Yeager, M., Wilson-Kubalek, E.M., Weiner, S.G., Brown, P.O., Rein, A., 1998. Supramolecular organization of immature and mature murine leukemia virus revealed by electron cryo-microscopy: implications for retroviral assembly mechanisms. Proc. Natl. Acad. Sci. 95, 7299–7304.

Zanetti, G., Briggs, J.A.G., Grunewald, K., Sattentau, Q.J., Fuller, S.D., 2006. Cryo-electron tomographic structure of an immunodeficiency virus envelope complex in situ. PLoS Pathog. 2, e83.

Zanetti, G., Riches, J.D., Fuller, S.D., Briggs, J.A., 2009. Contrast transfer function correction applied to cryo-electron tomography and sub-tomogram averaging. J. Struct. Biol. 168, 305–312.

Zeltins, A., 2013. Construction and characterization of virus-like particles: a review. Mol. Biotechnol. 53, 92–107.

Zhang, Z., Ma, J., Zhang, X., Su, C., Yao, Q.-C., Wang, X., 2016. Equine infectious anemia virus gag assembly and export are directed by matrix protein through trans-Golgi networks and cellular vesicles. J. Virol. 90, 1824–1838.

Zhao, G., Perilla, J.R., Yufenyuy, E.L., Meng, X., Chen, B., Ning, J., Ahn, J., Gronenborn, A.M., Schulten, K., Aiken, C., Zhang, P., 2013. Mature HIV-1 capsid structure by cryo-electron microscopy and all-atom molecular dynamics. Nature 497, 643–646.

Zhu, P., Liu, J., Bess, J., Chertova, E., Lifson, J.D., Grisé, H., Ofek, G.A., Taylor, K.A., Roux, K.H., 2006. Distribution and three-dimensional structure of AIDS virus envelope spikes. Nature 441, 847–852.

CHAPTER SIX

The application of atomic force microscopy for viruses and protein shells: Imaging and spectroscopy

Pedro J. de Pablo*
Departamento de Física de la Materia Condensada and Instituto de Física de la Materia Condensada (IFIMAC), Universidad Autónoma de Madrid, Madrid, Spain
*Corresponding author: e-mail address: p.j.depablo@uam.es

Contents

Abstract

Atomic force microscopy (AFM) probes surface-adsorbed samples at the nanoscale by using a sharp stylus of nanometric size located at the end of a micro-cantilever. This technique can also work in a liquid environment and offers unique possibilities to study individual protein assemblies, such as viruses, under conditions that resemble their natural liquid milieu. Here, I show how AFM can be used to explore the topography of viruses and protein cages, including that of structures lacking a well-defined symmetry. AFM is not limited for imaging and allows the manipulation of individual viruses with force spectroscopy approaches, such as single indentation and mechanical fatigue assays. These pushing experiments deform the protein cages to obtain their mechanical information and can be used to monitor the structural changes induced by maturation or the exposure to different biochemical environments, such as pH variation. We discuss how studying capsid rupture and self-healing events offers insight into virus uncoating pathways. On the other hand, pulling tests can provide information about the virus-host interaction established between the viral fibers and the cell membrane.

Advances in Virus Research, Volume 105
ISSN 0065-3527
https://doi.org/10.1016/bs.aivir.2019.07.006

Abbreviations

AFM atomic force microscopy
F_l lateral force
F_n normal force
FZ force vs. distance
HAdV human adenovirus
HOPG highly oriented pyrolitic graphite
JM jumping mode
nN nanoNewton
pN picoNewton

1. Introduction

A characteristic structural element in viruses is the protein capsid, which combines multiple functions, including packaging of the viral genome and recognition of the target cell. These protein cages or shells can be defined as a structure built out of protein subunits enclosing a cavity at the nanometer scale. Although viruses illustrate the definition of a protein cage, non-viral structures, such as bacterial microcompartments (Cheng et al., 2008), vault particles (Querol-Audí et al., 2009), clathrin cages (Fotin et al., 2004) and artificial virus-like structures (Lai et al., 2014; Wimmer et al., 2009; Wörsdörfer et al., 2011) are other examples. Virus protein capsids are built up of repeating protein subunits (capsomers) that pack the viral genome (Flint et al., 2004). Viral particles are endorsed with meta-stable properties that permit fulfilling each task of the viral cycle in a sequential fashion (Mateu, 2013). These capacities have induced using viral capsids as protein containers of artificial cargoes (drugs, polymers, enzymes, minerals) (Douglas and Young, 1998) with applications in materials and biomedical sciences.

Protein cages have to protect their cargo against a variety of physico-chemical stringent environments, including molecular crowding (Minton, 2001), osmotic shock (Cordova et al., 2003), thermal and chemical stress (Agirre et al., 2011) and desiccation processes (Carrasco et al., 2009; Martín-González et al., 2018). To improve the design of protein cages it is important to know about their stability and the evolution of structural changes. To this end, electron microscopy (EM) and X-ray crystallography are used to determine detailed high-resolution structures of protein cages (Baker et al., 1999). However, these methodologies require the averaging of a great number of particles, ranging from millions (in the case of crystallography) to thousands of structures for the model reconstruction (in the case of cryo-EM). Therefore, any structural difference between individual particles is lost when averaging the whole specimen population. This complicates structural studies of viruses that lack a well-defined symmetry such as the influenza virus (Calder et al., 2010). In addition, these approaches require environments (vacuum, vitreous ice, etc.) that are far away from the physiological conditions in which protein shells normally operate, and preclude the characterization of their dynamic properties in real time. The advent of single molecule technologies working in liquid environment has demonstrated that the mechanics of biological molecular aggregates are essential to their function (Hinterdorfer and Van Oijen, 2009). The exploration of these properties would complement EM and crystallography to gain insight into the interplay between structure and function of protein cages.

Atomic force microscopy (AFM) allows the structural determination of individual protein particles at nanometric resolution in liquid milieu. In addition, mechanical properties can be measured by manipulating the particles with the AFM probe. Environmental conditions, such as pH, can be varied within the AFM liquid cell to investigate the evolution of the protein shell structure and its mechanical properties. In addition, the nano-dissection abilities of AFM allow the local disruption of protein shells to learn about their assembly/disassembly (Castellanos et al., 2012). AFM can even monitor the highly dynamic entry and budding of individual viruses in the host.

In this review, I provide a general overview of how AFM methods can be applied to viruses. It starts with a description of the most successful modes for imaging protein shells with AFM and examples of virus particles AFM topographies. Subsequently the nano-indentation methodology is introduced, which allows probing the stiffness and rupture force of individual protein cages. In this part I describe how the variation of the structure of viruses with mutations (Minute Virus of Mice—MVM) and accessory proteins (lambda and P22 phages) change their mechanical properties. Also, I briefly refer here to the self-healing abilities of protein cages. I then focus on using mechanical fatigue to understand human adenovirus and rotavirus, and subsequently discuss how the nano-dissection abilities of AFM can be used to induce the genome ejection by uncorking individual particles of T7 bacteriophage. In this frame, I discuss the combination of AFM with fluorescence microscopy to monitor the mechanical unpacking of dsDNA from individual human adenovirus particles. In the last part, I refer to the AFM abilities to monitor the entry and budding of individual virus particles.

2. Imaging viruses with AFM

AFM requires the immobilization of the specimen to be study on a solid surface (substrate). Viral cages are normally physisorbed on the substrate, using polar, non-polar, and van der Waals forces (Muller et al., 1997). Physisorption traps protein cages on the surface without creating chemical bonds that might alter their structure. Each type of protein shell has individual features such as local charge densities and/or hydrophobic patches (Armanious et al., 2016), which can be employed for adsorption, via electrostatic and/or hydrophobic interactions, on different substrates, such as glass, mica and HOPG (Highly Oriented Pyrolytic Graphite) (see details in Moreno-Madrid et al., 2017). One could say that each type of virus has a preferred substrate, since each kind of protein cage exposes different residues in the external layer, thus requiring an adsorption methodology.

Fig. 1 shows Herpex Simplex, P22 bacteriophage and Human Adenovirus (HAdV), particles, adsorbed on silanized glass, HOPG and mica, respectively (Llauro et al., 2016c; Ortega-Esteban et al., 2013; Snijder et al., 2017). These are typical cases of non-enveloped viruses with icosahedral shape. Thus, after adsorption on the surface they can present two-, five- and threefold symmetry axes (inset, Fig. 1A–C). In spite of their undefined morphology conferred by the external lipid membrane, AFM can also image enveloped particles, such as HIV (Ramalho et al., 2016) (Fig. 1D) and influenza virus (Li et al., 2011) (Fig. 1E). The visualization of the soft viral appendages protruding from the virus body is difficult because they bend under the AFM tip (Ares et al., 2014). However, the rotavirus spikes are short enough to keep their integrity for a few images and can be imaged (Fig. 1F) (Jiménez-Zaragoza et al., 2018). Fig. 1G shows the cylindrical structure of tobacco mosaic virus (TMV) (Calo et al., 2016). AFM can also scan huge viruses, such as mimivirus (Fig. 1H) (Kuznetsov et al., 2010).

In AFM the tip mounted at the end of a flexible microcantilever scans the sample in X, Y and Z directions by using piezo actuators. While X and Y scanners move the tip over a square region, the interaction with the surface topography will lead to a bending of the cantilever. The cantilever deflects almost perpendicularly to the surface applying a normal force (F_n), and will also slightly twist due to a lateral dragging force parallel to the surface (F_l) (de Pablo, 2018). Both F_n and F_l are monitored by focusing a laser beam at the end of the cantilever, whose reflection is registered by a four-quadrant photodiode. Thus, each pixel of the image located at a position of planar

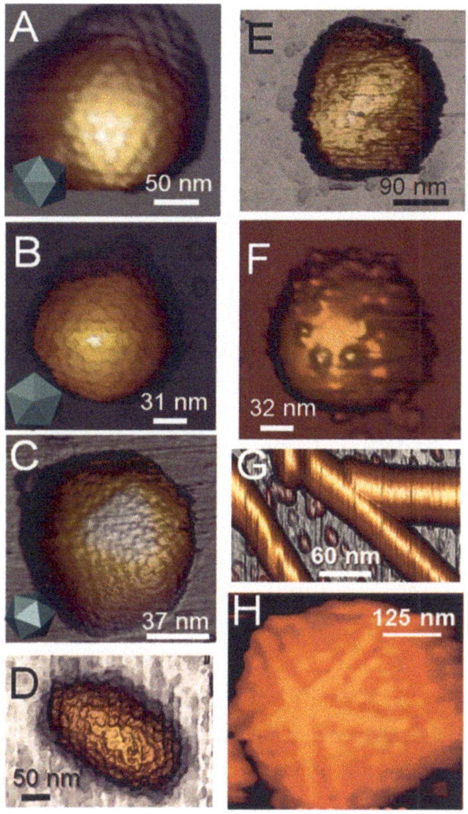

Fig. 1 AFM images of single viruses. (A) HAdV particle adsorbed on a triangular facet (Ortega-Esteban et al., 2013). (B) Herpex Simplex virus core showing a twofold orientation (Snijder et al., 2017). (C) P22 bacteriophage resting on its fivefold symmetry axis (Llauro et al., 2016b). (D) Enveloped HIV virus (Ramalho et al., 2016). (E) Enveloped influenza virus (Li et al., 2011). (F) TMV virus particles (Calo et al., 2016). (G) Viral fibers of T4 bacteriophage (Ares et al., 2014). (H) The star-fish shape feature of a mimivirus (Kuznetsov et al., 2010).

coordinates (X, Y) will be associated with certain bending values of the cantilever F_n and F_l. If the virus particle is too soft or not strongly attached, it will be damaged or displaced by the bending forces. To avoid this effect as much as possible, a feedback loop is engaged to F_n to move the *Z*-piezo position to keep F_n constant. In this working method, termed as contact mode, the AFM topography map will have X, Y and Z coordinates, where Z is the displacement applied to the Z-piezo to keep F_n constant. However, the torsional bending force F_l of the cantilever is ~40 times higher than the flexural bending force F_n (Carpick et al., 1997). These dragging forces are not problematic with specimens that are held by close packing, such as the purple membrane (Butt et al., 1991) and solid surfaces, such as mica (Ohnesorge and Binnig, 1993). However, isolated samples, such as DNA molecules, are especially prone to damage in contact mode (Moreno-Herrero et al., 2004). Individual protein cages are also susceptible to undesired modifications by lateral forces. Their size of tens of nanometers offers a large topographical aspect ratio that is difficult to follow for the feedback loop. A typical approach for overcoming this limitation is the use of fixation agents, such as glutaraldehyde. In such conditions AFM provides images whose resolution is comparable to that of some EM images (Xiao et al., 2009). Nevertheless, glutaraldehyde hampers the characterization of intact native viruses, such as disassembly or mechanical properties (Roos et al., 2010) because it structurally reinforces the specimens (Carrasco et al., 2011; Vinckier et al., 1995). Other approaches include the development of imaging modes that avoid dragging forces as much as possible. In jumping mode (JM), also called pulse force mode (de Pablo et al., 1998; Miyatani et al., 1997), the tip moves laterally when it is not in mechanical contact with the sample, thereby avoiding almost completely dragging forces. JM performs consecutive approach-release cycles at every pixel of the sample. In each cycle, known as a force vs. distance (FZ) curve, the *Z*-piezo moves the tip to the sample until establishing mechanical contact and reaching a certain feedback force. After a few milliseconds, the Z-piezo retracts the tip for about 100 nm releasing the surface. Subsequently the scanner moves laterally to the next pixel, and the process repeats (de Pablo et al., 1998; Ortega-Esteban et al., 2012). Especially in liquid milieu, where adhesion forces between tip and sample are majorly absent, JM has proven very successful (de Pablo, 2013) and sometimes it can be used to obtain mechanical properties such as the stiffness during imaging (Schaap et al., 2006). Variants of JM are now also increasingly implemented by multiple AFM manufacturers (peak force mode, QI^{tm} mode). Nevertheless, the most common imaging mode for biological samples is still amplitude modulation mode. Here, the AFM cantilever is oscillated at small amplitude (~10 nm) and when the tip comes close to the surface this amplitude will be modulated (reduced) by the tip-sample interaction. By using the oscillation amplitude as input for the feedback loop, the interaction forces with the sample are kept constant, while dragging forces are minimized because the tip spends most of its time away from the sample. A limitation of such AFM dynamic modes is that it is difficult to quantify the applied force in real time (Legleiter et al., 2006).

AFM has been widely applied to study the morphology of viruses at high resolution in liquid milieu (Kuznetsov and McPherson, 2011). However, the AFM abilities are well beyond the topographical characterization of samples (Muller et al., 2002). Because the AFM cantilever is a force transducer it can be used to manipulate individual atoms (Sugimoto et al., 2005) and biomolecules (Rief et al., 1997; Schaap et al., 2006). The number of mechanical studies of protein cages has seen a steep increase during the last decade. The virus mechanics have been found to depend on the packed genome (Carrasco et al., 2006, 2008; Hernando-Perez et al., 2012; Roos et al., 2009; Snijder et al., 2013; Zeng et al., 2017), maturation state (Carrasco et al., 2011; Hernando-Pérez et al., 2014b; Ortega-Esteban et al., 2013; Roos et al., 2012), artificial cargo (Llauro et al., 2016b) and structural modification of the cage by protein engineering (Llauro et al., 2016c) and site-directed mutagenesis (Carrillo et al., 2017). The internal pressure of viruses, caused by tightly packed nucleic acid inside some bacteriophages (Purohit et al., 2003), helps during the first stages of genome translocation into the host (Gonzalez-Huici et al., 2004). This pressure can be measured (Smith et al., 2001) and modulated by counter-ions on individual particles (Hernando-Perez et al., 2012). Protein shells of non-viral origin have been also investigated, such as vault particles (Llauro et al., 2014) and encapsulin (Snijder et al., 2016).

3. Single indentation assay

In this spectroscopic technique single FZ experiments are performed on top of a selected protein cage (Fig. 2A). By knowing the spring constant of the cantilever, the bending can be converted into a force, while the indentation of the protein cage is given by the motion of the *Z*-piezo minus the bending of the cantilever. To find the top of the protein cage the particle is first imaged at high magnification (~50 nm × 50 nm) after which an FZ curve is performed on the central region of the particle. It is important to minimize the time between acquiring the image and performing the FZ curve so that thermal drift does not affect the positioning of the AFM tip on the cage too much. Alternatively, an array of force curves can be performed on the particle from which a height image can be reconstructed, after which the central curves can be selected (Li et al., 2011). For studying the protein shell mechanics, the FZ curves have to be performed slow enough to allow the water to leave the capsid during indentation: speeds between

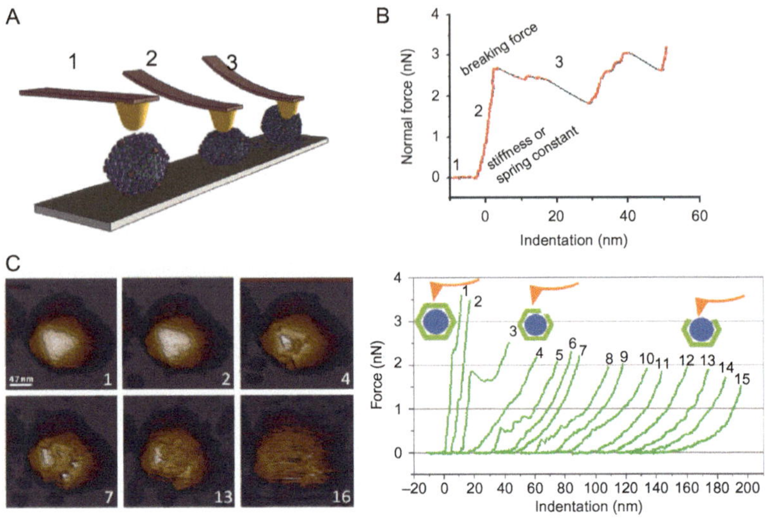

Fig. 2 Nano-indentation assay. (A) Cartoon showing the three main stages during a nano-indentation experiment on a protein cage: before contact (1), during indentation (2) and after breaking (3). (B) Evolution of F_n as function of the z-piezo displacement. The forward curve exhibits the three stages commented in (A). (C) AFM topographies (left) of a single HAdV wild type particle that is sequentially indented 15 times (indentation curves, right). The number of each AFM topography indicates the result of the correspondent indentation number. *Panel (B): Adapted from Moreno-Madrid, F., Martin-Gonzalez, N., Llauro, A., Ortega-Esteban, A., Hernando-Perez, M., Douglas, T., Schaap, I.A., De Pablo, P.J., 2017. Atomic force microscopy of virus shells. Biochem. Soc. Trans. 45, 499–511. Panel (C): Adapted from Ortega-Esteban, A., Condezo, G.N., Perez-Berna, A.J., Chillon, M., Flint, S.J., Reguera, D., San Martin, C., De Pablo, P.J., 2015. Mechanics of viral chromatin reveals the pressurization of human adenovirus. ACS Nano 9, 10826–10833.*

50 nm/s and 1 μm/s are typically used (Li et al., 2011; Snijder et al., 2012). After the contact between tip and particle is established, the FZ curves normally show an approximate linear behavior, which corresponds to the elastic regime of the protein cage (Fig. 2B, label 2).

When the indentation depth is kept small it is possible to perform repetitive FZ curves, and the particle elastically deforms in a reversible way. However, when the indentation surpasses the elastics limits, often at around 20% of the particle's height, the particle breaks, which is visible as a sequence of steps in the FZ curve that resemble the penetration of the tip apex trough the cage (Fig. 2A and B, label 3). The break also depends on the nature of the genomic cargo (Pérez-Berná et al., 2012). Thin shell theory relates the protein cage stiffness k with the Young's modulus E as $k \approx E\frac{h^2}{R}$, where h is the thickness of the shell and R its radius (Landau and Lifshizt, 1986). k is obtained from the linear force vs. indentation curves under the assumption of a spherical shell subjected to two opposing point forces and can thus be used to estimate the effective Young's modulus of viruses (Falvo et al., 1997; Ivanovska et al., 2004). Finite Element Analysis is more realistic that point forces (Klug et al., 2006) because it includes the contact boundary conditions considering a flat substrate and a rounded AFM tip.

However, when the viral genome contributes to the viral stiffness, the thin shell approximation is not valid and the response will show a non-linear Hertzian behavior (Ortega-Esteban et al., 2015b). Performing consecutive FZ curves on the same virus allows to test the stiffness of the viral genome. Fig. 2C shows a human adenovirus (HAdV) particle that has been indented 15 times (Ortega-Esteban et al., 2015b). In the third indentation (Fig. 2C, right graph) the shell breaks (Fig. 2C, frame #4) and facilitates the access of the AFM tip into the virus core, which consists of dsDNA and condensing proteins (San Martín, 2012). Fitting the resulting exponential curves obtained on the core with the Hertz model (Dimitriadis et al., 2002) provides the Young's modulus of the core. In this case Young's moduli of 1.2 and 0.3 MPa were found for the immature and mature cores, respectively. While the higher Young's modulus corresponds to a highly condensed genome before maturation, the lower Young's modulus reveals a less condensed genome for the mature cores. This decondensation has been speculated to be important for the transport of the genome through the nuclear pore (Ortega-Esteban et al., 2015b).

By performing FZ curves that exceed the elastic limits of the protein cage, the rupture events between the protein subunits can also be studied (Ivanovska et al., 2011). For instance, in vault particles the stepped shape of the indentation data after the first rupture event were associated with the individual proteins unzipping while the particle was being disrupted during the nano-indentation experiment (Llauro et al., 2014). Most protein cages show a similar fracture behavior (Hernando-Pérez et al., 2014b), which implies little plastic deformation before breaking like a brittle egg shell. However, in some cases a plastic behavior has been observed (Llauro et al., 2015), i.e., a virus particle deforming like clay before breaking.

Mechanical properties like stiffness, Young's modulus and rupture forces of viral particles have been directly associated with capsid stability. However, during transmission, viruses are confronted with both chemical and mechanical attacks, such as pH variation (Snijder et al., 2013) and osmotic shocks (Cordova et al., 2003). Thus, a correct interpretation of virus mechanics in terms of stability requires an adequate linkage between the ability of a protein shell to resist chemical and mechanical stress. For instance, mature T7 bacteriophage capsids shows higher stability to urea and GuHCl than immature T7 capsids. This increased chemical resistance is accompanied by a lower fragility (Hernando-Pérez et al., 2014b).

Mechanical properties of virus capsids can also be modified by genetic engineering or the specific binding of intrinsic or extrinsic decorating proteins. In the case of the minute virus of mice (MVM), specific mutations of amino acid side chains have been shown to influence the capsid stiffness (spring constant) at each symmetry axis (Carrillo et al., 2017). No point mutation tested substantially reduced stiffness, whereas most mutations increased stiffness (Fig. 3A). In this work, it is found that this stiffening was associated with reduced virus yields during cell infection. It is proposed that a low stiffness at simmetry3/symmetry2 capsid regions may have been biologically selected because it facilitates capsid assembly. In addition to mutations, it has been found that decoration of capsids with accessory either autologous or heterologous proteins also induce changes in the mechanical stability of the protein shell. The maturation of lambda phage implies the addition of the gpD protein to the hydrophobic patches at the external surface of the expanded shell (Medina et al., 2012). AFM experiments (Hernando-Pérez et al., 2014a) revealed that decorated particles exhibit increased spring constant (stiffness) and breaking force (Fig. 3B). This mechanical reinforcement might be required for the virus to resist the stringent environment conditions of the gut, whose molecular crowding is higher than inside *Escherichia coli* (Hernando-Pérez et al., 2014a). Heterologous decorating proteins can be used to artificially alter the stability of protein shells. The mature expanded capsids (EX) of the P22 bacteriophage (Llauro et al., 2016c) can be heated to make them lose their pentons, inducing 12 vacancies in the shell that result in a "whiffle ball" (WB) (Uchida and Douglas, 2013) (Fig. 2A). WB structures are softer than EX, break before the intact P22 structures (Fig. 3C) (Llauro et al., 2015) and are chemically less resistant to sodium dodecyl sulfate. Interestingly, the stability of both EX and WB can be recovered by adding Dec proteins, a decoration protein of the bacteriophage L (Tang et al., 2006). Binding of the Dec protein to WB particles recovered both their mechanical stability (Fig. 2D) and chemical resistance (Llauro et al., 2016c).

Sometimes rupture events have been found to be reversible. Such self-healing ability has been demonstrated on monolayers of capsid protein of human immunodeficiency virus (HIV) (Valbuena and Mateu, 2015)

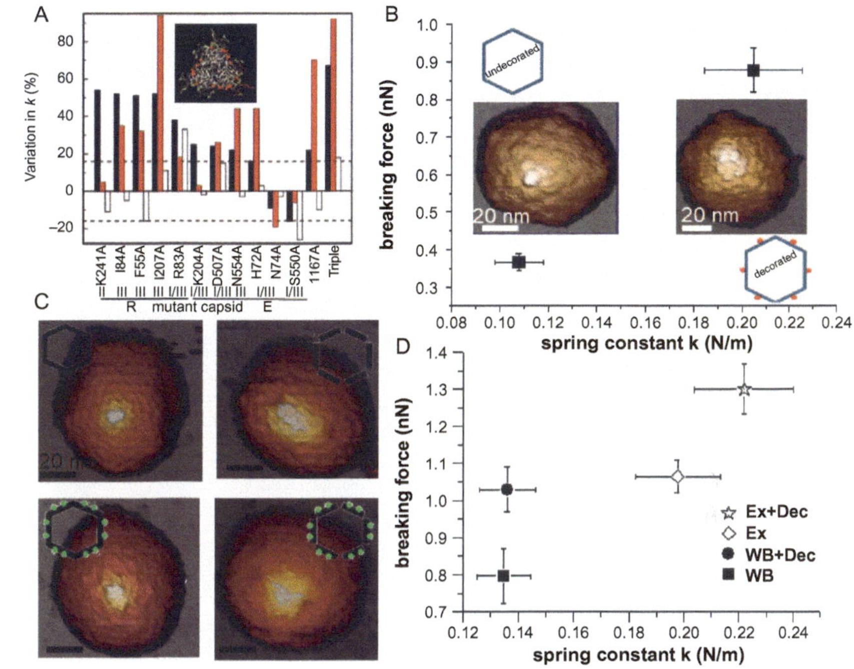

Fig. 3 Modifying the virus mechanical properties by altering the protein shell structure with mutants and decoration proteins. (A) Mutations of the minute virus of mice to remove hydrogen bonds (I), ionic interactions (II) and van der Waals interaction (III) are introduced. For each mutant capsid, bars indicate the % increase or decrease in elastic constant k value relative to that of the control wt capsid, at S2 (black bars), S3 (red bars), or S5 (white bars) regions. R and E refer to rigid and elastic mutants, respectively. Inset: X-ray data showing the positions of the mutated residues colored in red (rigid) or green (elastic). (B) Mechanical properties of lambda phage decorated (right) and undecorated (left) particles (Hernando-Pérez et al., 2014a). (C) P22 structures: expanded (top left), "whiffle ball" (top right), decorated expanded (down left) and decorated "whiffle ball" (down right). (D) Evolution of the breaking force and spring constant for the structures depicted in (C) (Llauro et al., 2016c). *Panel (A): Adapted from Carrillo, P.J.P., Medrano, M., Valbuena, A., Rodríguez-Huete, A., Castellanos, M., Pérez, R., Mateu, M.G., 2017. Amino acid side chains buried along intersubunit interfaces in a viral capsid preserve low mechanical stiffness associated with virus infectivity. ACS Nano 11, 2194–2208.*

(Fig. 4A). Three dimensional structures such as microtubule protein shells and vault particles have demonstrated self-healing capabilities (Llauro et al., 2014; Schaap et al., 2006) (Fig. 3B). Bacteriophage T7 capsid has also shown this recovery ability in very rare occasions (Fig. 4C) (de Pablo et al., 2018). Also, reversible stepwise rupture events (Voros et al., 2017) in the T7 bacteriophage have been attributed to the reversible buckling of individual capsomers during the nanoindentation process.

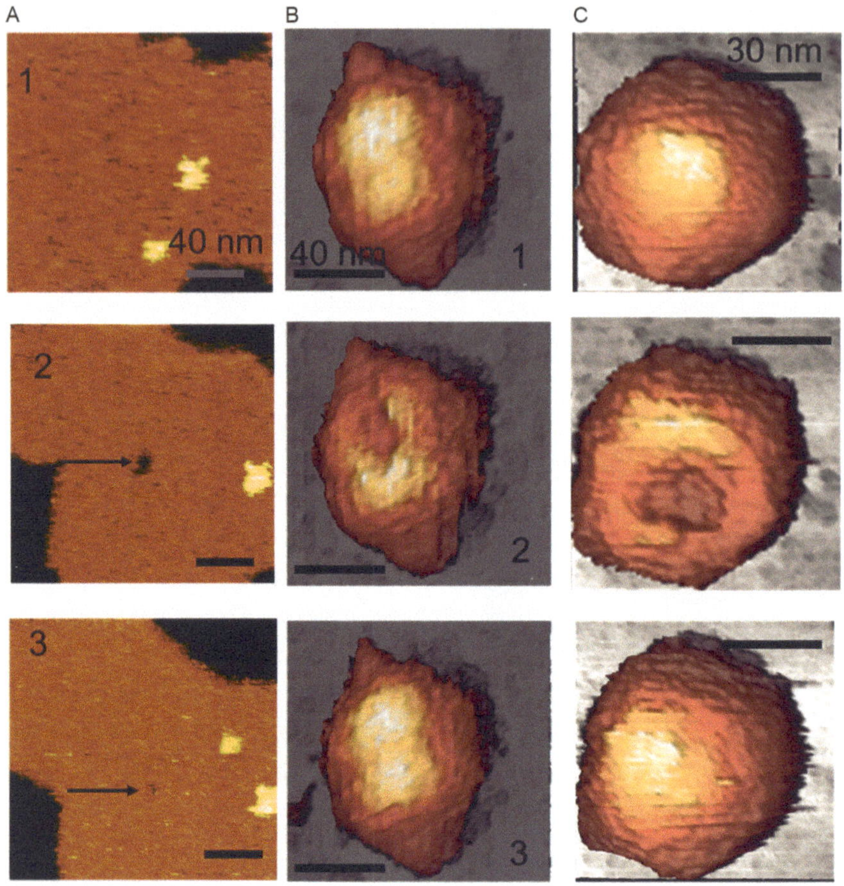

Fig. 4 Self-healing of protein shells. (A) Frame #1 shows a monolayer of HIV protein. Frame #2 presents (arrow) a hole induced by a single indentation assay. Finally, image #3 shows that the hole has disappeared (arrow) (Valbuena and Mateu, 2015). (B) Frame #1 represents the topography of an intact vault particle. In frame #2 the indentation produced a conspicuous break at the middle of the structure. Image #3 shows the same particle a few minutes later, the crack has disappeared (Llauro et al., 2014). (C) Up: Intact T7 bacteriophage. Middle: The AFM tip performed an evident crack on the virus capsid. Bottom: The crack has disappeared after a few minutes (de Pablo et al., 2018).

4. Mechanical fatigue

The rupture or breaking force describes the maximal force that a protein cage can sustain before it collapses, exhibiting large and expeditious structure alterations. Therefore, it is challenging to convey the disassembly pathways, because in the life cycle of many virus shells, disassembly takes place by losing individual capsomers in a progressive way (Greber et al., 1993). A protein cage must also resist a continuous barrage of collisions in crowded environments (Zhou et al., 2008). Following the equipartition theorem, the energy transferred in a molecular collision equals $\sim\frac{3}{2}k_BT$, which is far below the energy supplied by single AFM indentation experiments (100s of k_BT) (Hernando-Pérez et al., 2014a). Jumping mode performs a load cycle at each pixel of the virus during imaging, reaching about 100 pN (Ortega-Esteban et al., 2012) and transfer $\sim$10 k_BT to the particle at every cycle (Hernando-Pérez et al., 2014a). This energy is about 10 times the value provided by the impact of a macromolecule. Consequently, this imaging process probes the resistance of the cage to mechanical inputs of moderate energies. Such molecular fatigue experiments have demonstrated to reproduce the natural pathway of adenovirus uncoating much more accurately than did single nano-indentation experiments (Fig. 2C) at higher forces (Ortega-Esteban et al., 2013). Adenovirus uncoating in the cell happens in a stepwise manner: first, the receptor binding at the plasma membrane induces loss of the fibers (Nakano et al., 2000). Second, in the early endosome, the particle releases a few vertex capsomers (pentons) and peripheral core components (Puntener et al., 2011). Third, after escaping the endosome, a semi-disrupted particle travels to the nuclear pore, where the final disassembly step takes place and the genome gains entry into the nucleus (Greber et al., 1997). This sequential virus uncoating is determined by the maturation process, during which a viral protease cleaves several capsid and core proteins, making the virions metastable and primed for sequential uncoating (Pérez-Berná et al., 2012). Therefore, it is necessary to perform comparative studies of mature and immature particles to establish the differences during the disassembly process. For example, the mutant termed ts1 does not undergo the proteolytic cleavages and is stalled at the immature state and cannot infect cells (Weber, 1976). Mechanical fatigue exerted by jumping mode induces pentons release in a gradual fashion (Fig. 5A) (Ortega-Esteban et al., 2013). Plotting the normalized population of remaining pentons along time for both wt and ts1 mutant particles (Fig. 5B) revealed that wt loses pentons faster than ts1. This is an indication of the higher stability of the mutant, which cannot lose pentons fast enough to escape the early endosome (Ortega-Esteban et al., 2013). In the G33 mutant, all proteins but protein VI are cleaved (Denning et al., 2019), resulting in an immature particle that does not infect the host cell. The comparative study carried out with molecular fatigue on wt and G33 mutant particles revealed again that the immature virus loses less pentons than wt (Fig. 5C) over time. This effect is attributed to the increase of internal

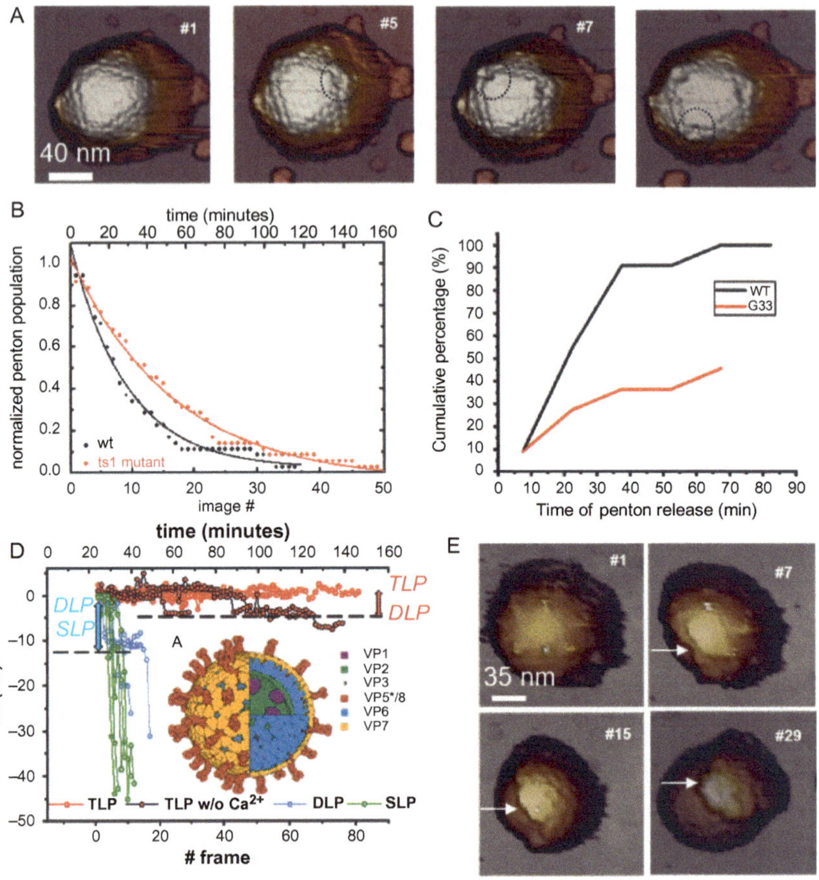

Fig. 5 Mechanical fatigue of viruses. (A) Evolution of a single adenovirus particle under mechanical fatigue, showing the stepwise penton removal (dotted circles). (B) Remaining penton population along time for wt (black) and ts1 mutant (red) (Ortega-Esteban et al., 2013). (C) Cumulative percentage of removed pentons in adenovirus for wt (black) and G33 mutant (red) (Denning et al., 2019). (D) Height loss of TLP (red), DLP (blue) and SLP (green) rotavirus particles during mechanical fatigue experiments. (E) Topographical evolution of a TLP rotavirus, initially showing the spikes (frame #1) and finally the VP2 layer (frame #29). Arrows point to the border of individual RV layers (Jiménez-Zaragoza et al., 2018).

pressure that occurs during maturation due to the dsDNA untightening (Ortega-Esteban et al., 2015b). The genome is less condensed in the wt because the interactions between the dsDNA and the proteins is weaker than in the mutant case. It has been proposed that the DNA-DNA electrostatic repulsion generates an outward pressure than destabilizes the pentons. Fatigue has been also used to study the stability of capsomers in multilayered viruses. The rotavirus (RV) particle is a 100 nm non-enveloped triple-layered particle (TLP) built of three concentric protein shells (Fig. 5D, inset) (Settembre et al., 2011). The inner layer is a protein shell formed by 60 dimers of the VP2 protein that encloses the dsRNA and the transcription machinery. This thin single-layered particle (SLP) is surrounded by a thick $T=13$ layer formed by 260 trimers of VP6 (Settembre et al., 2011) in the double-layered particle (DLP) (Fig. 5D, inset). This particle, which does not disassemble during the infection, constitutes the transcriptional machinery that initiates the main steps of the viral replication cycle in the host cell cytoplasm (Lawton et al., 1997). Since the DLP cannot recognize, bind to and enter the host target cell it is not infectious. This activity is included in the outer shell of the TLP, composed of proteins VP7 and VP4. There are 260 VP7 calcium-stabilized glycoprotein trimers embracing and capping each VP6 trimer of the DLP through the VP7 N-terminal arm (Settembre et al., 2011). Depressions of the VP7 layer can accommodate a maximum of 60 spikes of protein VP4, which take care of the recognition of the host cell. The application of mechanical fatigue to RV particles can remove each protein layer in a sequential fashion, revealing the interplay between these layers (Jiménez-Zaragoza et al., 2018). Fig. 5E presents the topographical evolution of a rotavirus particle as a function of the frame number at a force of 100 pN. The spikes of the TLP drop in a few frames (Fig. 5E, #1 and #7). While in the presence of calcium ions the VP7 protein layer endures more than 70 frames (Fig. 5D, red solid circles), calcium chelation allows removing VP7 from VP6 neatly (Fig. 5D, red open circles). Thus, VP7 is endorsed with a protection role that is curved in the absence of calcium ions, which happens during the early stages of infection. Fatigue experiments performed on DLP (Fig. 5D, blue) revealed that the VP6-VP2 layers interaction is very weak because the VP2 layer is exposed after a few frames. This labile interaction might be related to the conformational changes undergone by VP6 during transcription, which would be prevented by a rigid and stable VP6 layer. The VP2 layer is rapidly demolished during fatigue experiments (Fig. 5D, green). Fatigue has also been applied to induce disassembly of viral protein monolayers in two-dimensional geometries (Valbuena and Mateu, 2017).

Mechanical fatigue experiments have been demonstrated to constitute an efficient methodology for probing the stability of the genetic cargo (Ortega-Esteban et al., 2013). Specifically, in birnaviruses, the height evolution during mechanical fatigue experiments has been attributed to the stabilization role of ribonucleotide dimers (Mertens et al., 2015).

5. Genome release

Beyond offering protection to the load, a protein shell must deliver the cargo at the right moment and location. Tailed bacteriophages release their genome trough the tail into the bacterial host, leaving the protein shell outside. A number of non-enveloped eukaryotic viruses in turn undergo endocytosis and the protein shell disassembles to release the nucleic acid into the cell cytoplasm. Combining AFM with other methods, such as mass spectrometry, has shown to be a useful approach for investigating the interaction of the virus shell with its genome (Snijder et al., 2013). In addition, the ability of AFM to manipulate individual protein shells can be used to induce the mechanical unpacking of genome from single virus particles. In the case of bacteriophages, this process could be thought as uncorking the virus particle. It has been shown that the T7 bacteriophage (Fig. 6A) ejects the dsDNA after being subjected to mechanical stress (Fig. 6B) (Kellermayer et al., 2018). AFM imaging indicated that the DNA leaves the particle through the connector after the tail has been altered (Fig. 6B, circle). In this case, the attachment of the DNA molecule to the surface is a requisite for imaging by AFM. If the DNA molecule does not adsorb on the surface, AFM cannot be used. Presently, we discuss the combination of AFM and Total Internal Reflection Fluorescence Microscopy (TIRFM) for studying the dsDNA unpacking of HAdV. HAdV particles disassemble in a stepwise manner while they travel through the cell. Finally, when hAdV particles reach the nuclear pores, the dsDNA is released and enters the nucleus. HAdV packs 35 kbps of dsDNA and 25 MDa of histone-like proteins that form the core. These proteins are cleaved during maturation by the viral protease. AFM uncoating experiments (Ortega-Esteban et al., 2013) suggested that the immature core appeared as a condensed blob, whereas the core of the mature virus appeared dispersed in multiple pieces. These results indicated differences in the core organization that could have consequences for the diffusion of the dsDNA after disassembly of the virus capsid. In addition, dsDNA chains do not necessarily attach to the sample substrate after diffusing out of the broken cage, and it is thus difficult to monitor this process with AFM. However, fluorescence microscopy allows monitoring non-immobilized biomolecules. By combining both techniques, AFM can be used to fracture single capsids. A DNA-specific intercalating fluorescent dye (YOYO-1) allows tracking dsDNA exposure after the capsid has been crack-opened (Fig. 6C). By using TIRFM only the fluorescence signal close

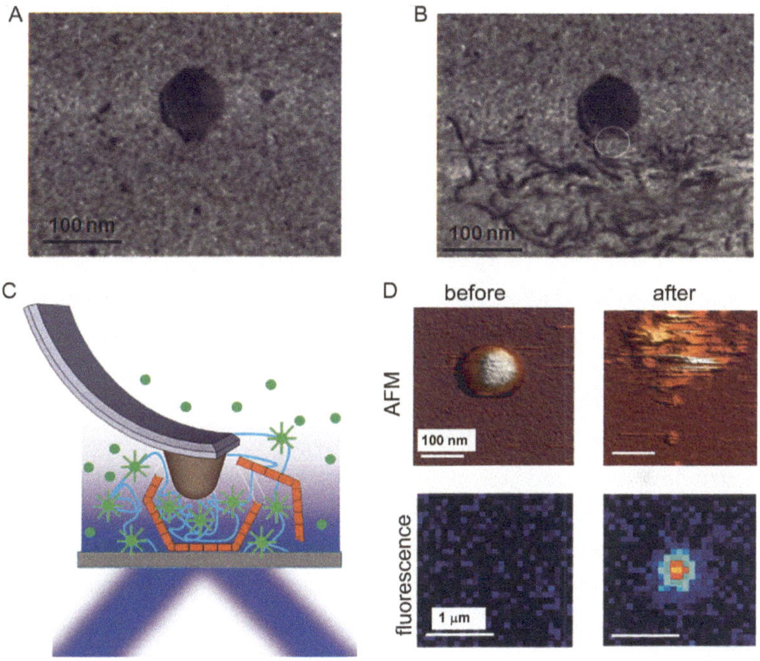

Fig. 6 Mechanical unpacking of the viral genome. (A) Intact bacteriophage T7 phage. (B) The same particle showing the DNA that has been mechanically unpacked. (C) Cartoon of the AFM/fluorescence system. (D) AFM and fluorescence frames of a HAdV capsid before and after releasing DNA. *Panel (B): Adapted from Kellermayer, M.S.Z., Vörös, Z., Csík, G., Herényi, L., 2018. Forced phage uncorking: viral DNA ejection triggered by a mechanically sensitive switch. Nanoscale 10, 1898–1904. Panel (D): Adapted from Ortega-Esteban, A., Bodensiek, K., San Martin, C., Suomalainen, M., Greber, U.F., De Pablo, P.J., Schaap, I.A., 2015. Fluorescence tracking of genome release during mechanical unpacking of single viruses. ACS Nano 9, 10571–10579.*

to the surface is detected (~100 nm), which is comparable to the virus size and where the experiment is confined. Background fluorescence is largely avoided and also the tip apex and cantilever remain mostly out of the evanescent excitation field (Gaiduk et al., 2005). Fig. 5D shows simultaneous AFM and fluorescence images of a single HAdV virus before and after nano-indentation: fluorescence emission is visible after the particle was disrupted. Fluorescence blobs are larger for mature particles, suggesting that the structural change of the core during maturation allows the decondensation of the DNA and make it more accessible to the fluorophore (Ortega-Esteban et al., 2015a). This genome expansion may be an essential feature to allow diffusion through the nuclear pore.

6. pH variation and protein shell mechanics

Many viruses encounter variations in ionic conditions in the host cell that trigger structural changes that in turn lead to disassembly of the capsid and the release of the viral genome (Fontana et al., 2012; Suomalainen and Greber, 2013; Wetz and Kucinski, 1991). AFM force spectroscopy can be employed to investigate such structural transitions (Llauro et al., 2015). To this end, it is necessary to mimic in the AFM liquid cell the changes in cellular environment that the virus encounters during infection. In multiple experiments, viruses have been studied at different pH values. To mimic the acidification in the endosomal pathway that influenza viruses experience on their way through the cell, Li et al. measured the stiffness of influenza virus particles subjected to a pH that was lowered from 7.4 to 5.0 (Li et al., 2014). Fig. 7A shows that influenza virus particles soften in a two-step process that correspond to the different pH values within the endosomes when it evolves from the early to the late stage. The first softening step (pH 7.4–6.0), which is reversible, was attributed to a softening of the glycoprotein spike layer. The second softening step (pH 6.0–5.5), which is irreversible, was associated to the dissociation of the M1 protein layer from the inside of the viral lipid envelope. The M1 protein coats the inside of the lipid envelope with a semi-continuous protein layer (Harris et al., 2006). As a consequence, the stiffness is about 10 times less than that of other viral protein capsids. Because the influenza virus particle can sustain much higher deformations, the force to break open the viruses is still comparable to what is found for protein capsids of non-enveloped viruses (Li et al., 2011; Schaap et al., 2012).

The stiffness of other viruses that have been tested at different pH values is weaker at high pH than at low pH values. Snijder et al. showed that the force to break the capsid of a triatoma virus particle was multifold reduced when the pH was increased from 6.8 to 9.0, which was attributed to the loss of interaction between genome and capsid, which destabilized the latter (Snijder et al., 2012). Cowpea chlorotic mottle virus was found to swell and to soften when the pH was increased from 4.8 to 7.5 (Wilts et al., 2015). The inversed response at pH shifts of the triatoma virus and Cowpea chlorotic mottle virus as compared to that of the influenza virus may indicate a need to endure acidic environments. Both the triatoma virus and Cowpea chlorotic mottle virus use insects as transmission vectors and a high stability at low pH may be important to persist in the intestines of the insects. Ionic

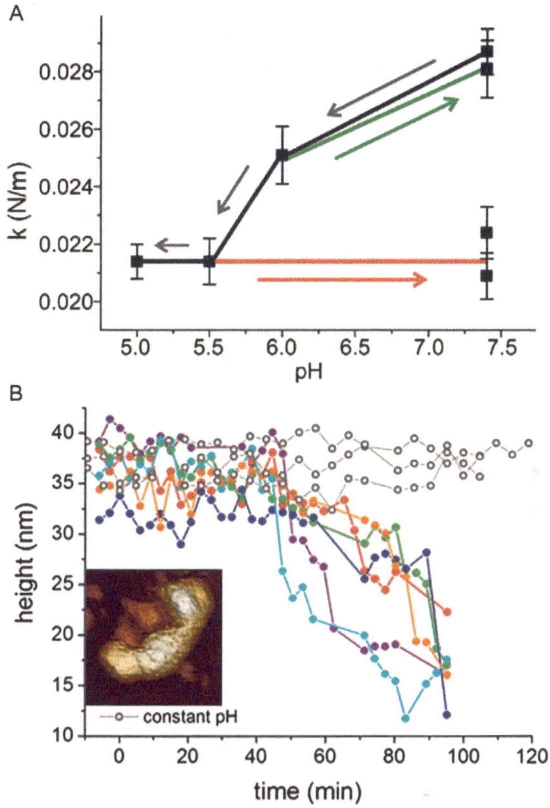

Fig. 7 pH variation and protein shell mechanics. (A) Influenza viruses soften in a two-step process that correspond to the different pH values. The green arrow indicated a reversible change, the red arrow an irreversible change in stiffness (Li et al., 2014). (B) Temporal evolution of some vault particles topographies when lowering the pH from 7.5 to 5.2 (colored curves) and at a constant pH of 7.5 (gray curves). The inset shows a vault particle at pH 5.2 (Llauro et al., 2016a).

conditions are also expected to affect other protein cages than viruses. For instance, it has been suggested that the acidification of the environment has a strong influence on vault particle dynamics by opening the particle into two halves (Goldsmith et al., 2007). AFM has been used to monitor the structural changes of individual vault particles while the pH was varied (Llauro et al., 2016a). These experiments revealed that lowering the pH promoted a global destabilization of the particle governed by the weakening of inter-monomeric contacts. This destabilization resulted in the collapse of vault height when the pH was lowered from 7.5 to 5.2 (colored curves of Fig. 7B).

7. Monitoring virus transmission: Binding and budding

So far, I have reviewed how the mechanics of viruses and other protein cages can be related with their stability and functionality. However, viruses are dynamic entities that will have to cross the cellular membrane (or penetrate the bacterial cell wall in case of bacteriophages) to deliver their genome inside the host cells. Once the genome has been replicated and expressed, new functional virus particles have to cross the cell membrane to leave the hosts. In eukaryotic viruses this process implies virus transport in and out of the host cells. Many viruses use the endocytotic pathway to enter the cell (Flint et al., 2004). Before engulfment by the cellular membrane, viruses must be trapped on the plasma membrane via receptors that interact with the virus particle in a specific fashion. The interaction between HAdV fibers and CAR proteins at the plasma membrane constitutes an illustrative example (San Martín, 2012). AFM allows measuring the force of these interactions (Alsteens et al., 2017). In these experiments, single virus particles were covalently attached to the AFM tip (Fig. 8A, inset). The combination with confocal microscopy enabled the co-localization of the cells in culturing conditions with the tip. Thus, the AFM tip can scan the relevant area of the cell to obtain the adhesive force between the virus particle attached to the tip and the receptors in the membrane. The recording of this

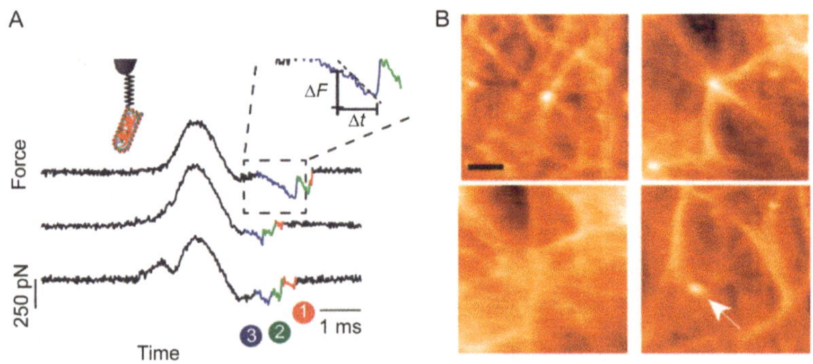

Fig. 8 Binding and budding of viruses. (A) The temporal evolution of the adhesion force between the fibers of a rabies virus and the receptors of the cell membrane. (B) Time evolution of the budding process of a single HIV virus with 10 min between frames (scale bar is 1.5 μm). The white arrow points to a virus. *Panel (A): Adapted from Alsteens, D., Newton, R., Schubert, R., Martinez-Martin, D., Delguste, M., Roska, B., Muller, D.J., 2017. Nanomechanical mapping of first binding steps of a virus to animal cells. Nat. Nanotechnol. 12, 177–183. Panel (B): Adapted from Gladnikoff, M., Rousso, I., 2008. Directly monitoring individual retrovirus budding events using atomic force microscopy. Biophys. J. 94, 320–326.*

adhesion as function of time revealed binding events between the virus particle and the receptor (Fig. 8A). The colored steps correspond to the rupture forces of 1, 2 and 3 virus fibers, respectively. Such force spectroscopy experiments allowed to extract parameters such as k_{off} and the free energy of the fiber-receptor interaction. Also, the binding of influenza virus particles to cells have similarly been characterized by a combination of AFM and optical trapping force spectroscopy (Sieben et al., 2012).

Also virus budding, the process by which enveloped viruses acquire their external membrane from the host cell membrane when they leave the cell, have been studied with AFM (Kuznetsov et al., 2002). In principle, AFM can also image the temporal evolution of this process by recording sequential topographies of the cell surface. Fig. 8B presents the evolution of the structural changes of the cell membrane attributed to the budding process of a single HIV virus particle (Gladnikoff et al., 2009). In this case the budding process appears to be associated with a remodeling of the actin cortex under the cell membrane. The upper left panel shows the initial stage with a conspicuous blob (virus). The upper right panel presents the maximum bulge induced by the virus just before leaving the cell (bottom left). Finally, a new virus particle starts the budding process again (bottom right).

8. Conclusions

AFM enables unique possibilities for studying individual virus particles and other protein cages, complementing classical EM and X-ray diffraction studies. First, the high signal to noise ratio of AFM imaging allows obtaining nanometer resolution on single particles. This resolution enables the identification of virus elements that are not symmetrically ordered (Cartagena et al., 2013). Also, it opens the way to study dynamics and mechanics of single virus particles. Second, because AFM can operate under buffered conditions, dynamic events, such as the response to ionic changes and budding processes, can be studied in real time. Third, structural virology studies are typically focused on establishing structure-function relationships (Baker et al., 1999). Because AFM can measure mechanical properties of viral particles in real time, the above relationship can be extended to structure-function-mechanics. Current developments such as high-speed AFM (Ando et al., 2003) and combined approaches such as AFM with single molecule fluorescence (Ortega-Esteban et al., 2015a) will greatly enhance our ability to study the structural dynamics of protein cages at shorter timescales, which will help us better understand their functionality.

Acknowledgments

I acknowledge my students, former students, collaborators and projects, FIS2017-89549-R. Fundación BBVA and "María de Maeztu" Program for Units of Excellence in R&D (MDM-2014-0377).

References

Agirre, J., Aloria, K., Arizmendi, J.M., Iloro, I., Elortza, F., Sánchez-Eugenia, R., Marti, G.A., Neumann, E., Rey, F.A., Guérin, D.M.A., 2011. Capsid protein identification and analysis of mature Triatoma virus (TrV) virions and naturally occurring empty particles. Virology 409, 91–101.

Alsteens, D., Newton, R., Schubert, R., Martinez-Martin, D., Delguste, M., Roska, B., Muller, D.J., 2017. Nanomechanical mapping of first binding steps of a virus to animal cells. Nat. Nanotechnol. 12, 177–183.

Ando, T., Kodera, N., Naito, Y., Kinoshita, T., Furuta, K., Toyoshima, Y.Y., 2003. A high-speed atomic force microscope for studying biological macromolecules in action. Chemphyschem 4, 1196–1202.

Ares, P., Garcia-Doval, C., Llauro, A., Gomez-Herrero, J., Van Raaij, M.J., De Pablo, P.J., 2014. Interplay between the mechanics of bacteriophage fibers and the strength of virus-host links. Phys. Rev. E Stat. Nonlin. Soft Matter Phys. 89, 052710.

Armanious, A., Aeppli, M., Jacak, R., Refardt, D., Sigstam, T., Kohn, T., Sander, M., 2016. Viruses at solid-water interfaces: a systematic assessment of interactions driving adsorption. Environ. Sci. Technol. 50, 732–743.

Baker, T.S., Olson, N.H., Fuller, S.D., 1999. Adding the third dimension to virus life cycles: three-dimensional reconstruction of icosahedral viruses from cryo-electron micrographs. Microbiol. Mol. Biol. Rev. 63, 862–922.

Butt, H.J., Prater, C.B., Hansma, P.K., 1991. Imaging purple membranes dry and in water with the atomic force microscope. J. Vac. Sci. Technol. B 9, 1193–1196.

Calder, L.J., Wasilewski, S., Berriman, J.A., Rosenthal, P.B., 2010. Structural organization of a filamentous influenza A virus. Proc. Natl. Acad. Sci. U. S. A. 107, 10685–10690.

Calo, A., Eleta-Lopez, A., Stoliar, P., De Sancho, D., Santos, S., Verdaguer, A., Bittner, A.M., 2016. Multifrequency force microscopy of helical protein assembly on a virus. Sci. Rep. 6, 21899.

Carpick, R.W., Ogletree, D.F., Salmeron, M., 1997. Lateral stiffness: a new nanomechanical measurement for the determination of shear strengths with friction force microscopy. Appl. Phys. Lett. 70, 1548–1550.

Carrasco, C., Carreira, A., Schaap, I.A.T., Serena, P.A., Gomez-Herrero, J., Mateu, M.G., De Pablo, P.J., 2006. DNA-mediated anisotropic mechanical reinforcement of a virus. Proc. Natl. Acad. Sci. U. S. A. 103, 13706–13711.

Carrasco, C., Castellanos, M., De Pablo, P.J., Mateu, M.G., 2008. Manipulation of the mechanical properties of a virus by protein engineering. Proc. Natl. Acad. Sci. U. S. A. 105, 4150–4155.

Carrasco, C., Douas, M., Miranda, R., Castellanos, M., Serena, P., Carrascosa, J., Mateu, M., Marques, M., De Pablo, P., 2009. The capillarity of nanometric water menisci confined inside closed-geometry viral cages. Proc. Natl. Acad. Sci. U. S. A. 106, 5475–5480.

Carrasco, C., Luque, A., Hernando-Perez, M., Miranda, R., Carrascosa, J.L., Serena, P.A., De Ridder, M., Raman, A., Gomez-Herrero, J., Schaap, I.A.T., Reguera, D., De Pablo, P.J., 2011. Built-in mechanical stress in viral shells. Biophys. J. 100, 1100–1108.

Carrillo, P.J.P., Medrano, M., Valbuena, A., Rodríguez-Huete, A., Castellanos, M., Pérez, R., Mateu, M.G., 2017. Amino acid side chains buried along intersubunit interfaces in a viral capsid preserve low mechanical stiffness associated with virus infectivity. ACS Nano 11, 2194–2208.

Cartagena, A., Hernando-Perez, M., Carrascosa, J.L., De Pablo, P.J., Raman, A., 2013. Mapping in vitro local material properties of intact and disrupted virions at high resolution using multi-harmonic atomic force microscopy. Nanoscale 5, 4729–4736.

Castellanos, M., Pérez, R., Carrillo, P.J., De Pablo, P.J., Mateu, M.G., 2012. Mechanical disassembly of single virus particles reveals kinetic intermediates predicted by theory. Biophys. J. 102, 2615–2624.

Cheng, S., Liu, Y., Crowley, C.S., Yeates, T.O., Bobik, T.A., 2008. Bacterial microcompartments: their properties and paradoxes. Bioessays 30, 1084–1095.

Cordova, A., Deserno, M., Gelbart, W.M., Ben-Shaul, A., 2003. Osmotic shock and the strength of viral capsids. Biophys. J. 85, 70–74.

De Pablo, P.J., 2013. Atomic force microscopy of viruses. Subcell. Biochem. 68, 247–271.

De Pablo, P.J., 2018. Atomic force microscopy of virus shells. Semin. Cell Dev. Biol. 73, 199–208.

De Pablo, P.J., Colchero, J., Gomez-Herrero, J., Baro, A.M., 1998. Jumping mode scanning force microscopy. Appl. Phys. Lett. 73, 3300–3302.

De Pablo, P.J., Hernando-Pérez, M., Carrasco, C., Carrascosa, J.L., 2018. Direct visualization of single virus restoration after damage in real time. J. Biol. Phys. 44, 225–235.

Denning, D., Bennett, S., Mullen, T., Moyer, C., Vorselen, D., Wuite, G.J.L., Nemerow, G., Roos, W.H., 2019. Maturation of adenovirus primes the protein nano-shell for successful endosomal escape. Nanoscale 11, 4015–4024.

Dimitriadis, E.K., Horkay, F., Maresca, J., Kachar, B., Chadwick, R.S., 2002. Determination of elastic moduli of thin layers of soft material using the atomic force microscope. Biophys. J. 82, 2798–2810.

Douglas, T., Young, M., 1998. Host-guest encapsulation of materials by assembled virus protein cages. Nature 393, 152–155.

Falvo, M.R., Washburn, S., Superfine, R., Finch, M., Brooks, F.P., Chi, V., Taylor, R.M., 1997. Manipulation of individual viruses: friction and mechanical properties. Biophys. J. 72, 1396–1403.

Flint, S.J., Enquist, L.W., Racaniello, V.R., Skalka, A.M., 2004. Principles of Virology. ASM Press, Washington, DC.

Fontana, J., Cardone, G., Heymann, J.B., Winkler, D.C., Steven, A.C., 2012. Structural changes in influenza virus at low pH characterized by cryo-electron tomography. J. Virol. 86, 2919–2929.

Fotin, A., Cheng, Y., Sliz, P., Grigorieff, N., Harrison, S.C., Kirchhausen, T., Walz, T., 2004. Molecular model for a complete clathrin lattice from electron cryomicroscopy. Nature 432, 573.

Gaiduk, A., Kuhnemuth, R., Antonik, M., Seidel, C.A., 2005. Optical characteristics of atomic force microscopy tips for single-molecule fluorescence applications. Chemphyschem 6, 976–983.

Gladnikoff, M., Shimoni, E., Gov, N.S., Rousso, I., 2009. Retroviral assembly and budding occur through an actin-driven mechanism. Biophys. J. 97, 2419–2428.

Goldsmith, L.E., Yu, M., Rome, L.H., Monbouquette, H.G., 2007. Vault nanocapsule dissociation into halves triggered at low pH. Biochemistry 46, 2865–2875.

Gonzalez-Huici, V., Salas, M., Hermoso, J.M., 2004. The push-pull mechanism of bacteriophage O29 DNA injection. Mol. Microbiol. 52, 529–540.

Greber, U.F., Willetts, M., Webster, P., Helenius, A., 1993. Stepwise dismantling of adenovirus-2 during entry into cells. Cell 75, 477–486.

Greber, U.F., Suomalainen, M., Stidwill, R.P., Boucke, K., Ebersold, M.W., Helenius, A., 1997. The role of the nuclear pore complex in adenovirus DNA entry. EMBO J. 16, 5998–6007.

Harris, A., Cardone, G., Winkler, D.C., Heymann, J.B., Brecher, M., White, J.M., Steven, A.C., 2006. Influenza virus pleiomorphy characterized by cryoelectron tomography. Proc. Natl. Acad. Sci. U. S. A. 103, 19123–19127.

Hernando-Perez, M., Miranda, R., Aznar, M., Carrascosa, J.L., Schaap, I.A.T., Reguera, D., De Pablo, P.J., 2012. Direct measurement of phage phi29 stiffness provides evidence of internal pressure. Small 8, 2366–2370.

Hernando-Pérez, M., Lambert, S., Nakatani-Webster, E., Catalano, C., De Pablo, P., 2014a. Cementing proteins provide extra mechanical stabilization to viral cages. Nat. Commun. 5, 4520.

Hernando-Pérez, M., Pascual, E., Aznar, M., Ionel, A., Caston, J.R., Luque, A., Carrascosa, J.L., Reguera, D., De Pablo, P.J., 2014b. The interplay between mechanics and stability of viral cages. Nanoscale 6, 2702–2709.

Hinterdorfer, P., Van Oijen, A., 2009. Handbook of Single-Molecule Biophysics. Springer, Dordrecht; New York.

Ivanovska, I.L., De Pablo, P.J., Ibarra, B., Sgalari, G., Mackintosh, F.C., Carrascosa, J.L., Schmidt, C.F., Wuite, G.J.L., 2004. Bacteriophage capsids: tough nanoshells with complex elastic properties. Proc. Natl. Acad. Sci. U. S. A. 101, 7600–7605.

Ivanovska, I.L., Miranda, R., Carrascosa, J.L., Wuite, G.J.L., Schmidt, C.F., 2011. Discrete fracture patterns of virus shells reveal mechanical building blocks. Proc. Natl. Acad. Sci. U. S. A. 108, 12611–12616.

Jiménez-Zaragoza, M., Yubero, M.P., Martín-Forero, E., Castón, J.R., Reguera, D., Luque, D., De Pablo, P.J., Rodríguez, J.M., 2018. Biophysical properties of single rotavirus particles account for the functions of protein shells in a multilayered virus. eLife 7 e37295.

Kellermayer, M.S.Z., Vörös, Z., Csík, G., Herényi, L., 2018. Forced phage uncorking: viral DNA ejection triggered by a mechanically sensitive switch. Nanoscale 10, 1898–1904.

Klug, W.S., Bruinsma, R.F., Michel, J.P., Knobler, C.M., Ivanovska, I.L., Schmidt, C.F., Wuite, G.J.L., 2006. Failure of viral shells. Phys. Rev. Lett. 97, 228101.

Kuznetsov, Y.G., McPherson, A., 2011. Atomic force microscopy in imaging of viruses and virus-infected cells. Microbiol. Mol. Biol. Rev. 75, 268–285.

Kuznetsov, Y.G., Datta, S., Kothari, N.H., Greenwood, A., Fan, H., Mcpherson, A., 2002. Atomic force microscopy investigation of fibroblasts infected with wild-type and mutant murine leukemia virus (MuLV). Biophys. J. 83, 3665–3674.

Kuznetsov, Y.G., Xiao, C.A., Sun, S.Y., Raoult, D., Rossmann, M., McPherson, A., 2010. Atomic force microscopy investigation of the giant mimivirus. Virology 404, 127–137.

Lai, Y.-T., Reading, E., Hura, G.L., Tsai, K.-L., Laganowsky, A., Asturias, F.J., Tainer, J.A., Robinson, C.V., Yeates, T.O., 2014. Structure of a designed protein cage that self-assembles into a highly porous cube. Nat. Chem. 6, 1065–1071.

Landau, L.D., Lifshizt, E., 1986. Theory of Elasticity. Pergamon, London.

Lawton, J.A., Estes, M.K., Prasad, B.V., 1997. Three-dimensional visualization of mRNA release from actively transcribing rotavirus particles. Nat. Struct. Biol. 4, 118–121.

Legleiter, J., Park, M., Cusick, B., Kowalewski, T., 2006. Scanning probe acceleration microscopy (SPAM) in fluids: mapping mechanical properties of surfaces at the nanoscale. Proc. Natl. Acad. Sci. U. S. A. 103, 4813–4818.

Li, S., Eghiaian, F., Sieben, C., Herrmann, A., Schaap, I.A., 2011. Bending and puncturing the influenza lipid envelope. Biophys. J. 100, 637–645.

Li, S., Sieben, C., Ludwig, K., Höfer, C.T., Chiantia, S., Herrmann, A., Eghiaian, F., Schaap, I.A.T., 2014. pH-controlled two-step uncoating of influenza virus. Biophys. J. 106, 1447–1456.

Llauro, A., Guerra, P., Irigoyen, N., Rodriguez, J.F., Verdaguer, N., De Pablo, P.J., 2014. Mechanical stability and reversible fracture of vault particles. Biophys. J. 106, 687–695.

Llauro, A., Coppari, E., Imperatori, F., Bizzarri, A.R., Caston, J.R., Santi, L., Cannistraro, S., De Pablo, P.J., 2015. Calcium ions modulate the mechanics of tomato bushy stunt virus. Biophys. J. 109, 390–397.

Llauro, A., Schwarz, B., Koliyatt, R., De Pablo, P.J., Douglas, T., 2016a. Tuning viral capsid nanoparticle stability with symmetrical morphogenesis. ACS Nano 10, 8465–8473.
Llauro, A., Guerra, P., Kant, R., Bothner, B., Verdaguer, N., De Pablo, P.J., 2016b. Decrease in pH destabilizes individual vault nanocages by weakening the inter-protein lateral interaction. Sci. Rep. 6, 34143.
Llauro, A., Luque, D., Edwards, E., Trus, B.L., Avera, J., Reguera, D., Douglas, T., Pablo, P.J., Caston, J.R., 2016c. Cargo-shell and cargo-cargo couplings govern the mechanics of artificially loaded virus-derived cages. Nanoscale 8, 9328–9336.
Martín-González, N., Darvas, S.M.G., Durana, A., Marti, G.A., Guérin, D.M., De Pablo, P.J., 2018. Exploring the role of genome and structural ions in preventing viral capsid collapse during dehydration. J. Phys. Condens. Matter 30, 104001.
Mateu, M.G.E., 2013. Structure and Physics of Viruses. Springer.
Medina, E., Nakatani, E., Kruse, S., Catalano, C.E., 2012. Thermodynamic characterization of viral procapsid expansion into a functional capsid shell. J. Mol. Biol. 418, 167–180.
Mertens, J., Casado, S., Mata, C.P., Hernando-Pérez, M., De Pablo, P.J., Carrascosa, J.L., Castón, J.R., 2015. A protein with simultaneous capsid scaffolding and dsRNA-binding activities enhances the birnavirus capsid mechanical stability. Sci. Rep. 5, 13486.
Minton, A.P., 2001. The influence of macromolecular crowding and macromolecular confinement on biochemical reactions in physiological media. J. Biol. Chem. 276, 10577–10580.
Miyatani, T., Horii, M., Rosa, A., Fujihira, M., Marti, O., 1997. Mapping of electrical double-layer force between tip and sample surfaces in water with pulsed-force-mode atomic force microscopy. Appl. Phys. Lett. 71, 2632–2634.
Moreno-Herrero, F., Colchero, J., Gomez-Herrero, J., Baro, A.M., 2004. Atomic force microscopy contact, tapping, and jumping modes for imaging biological samples in liquids. Phys. Rev. E 69, 031915.
Moreno-Madrid, F., Martin-Gonzalez, N., Llauro, A., Ortega-Esteban, A., Hernando-Perez, M., Douglas, T., Schaap, I.A., De Pablo, P.J., 2017. Atomic force microscopy of virus shells. Biochem. Soc. Trans. 45, 499–511.
Muller, D.J., Amrein, M., Engel, A., 1997. Adsorption of biological molecules to a solid support for scanning probe microscopy. J. Struct. Biol. 119, 172–188.
Muller, D.J., Janovjak, H., Lehto, T., Kuerschner, L., Anderson, K., 2002. Observing structure, function and assembly of single proteins by AFM. Prog. Biophys. Mol. Biol. 79, 1–43.
Nakano, M.Y., Boucke, K., Suomalainen, M., Stidwill, R.P., Greber, U.F., 2000. The first step of adenovirus type 2 disassembly occurs at the cell surface, independently of endocytosis and escape to the cytosol. J. Virol. 74, 7085–7095.
Ohnesorge, F., Binnig, G., 1993. True atomic-resolution by atomic force microscopy through repulsive and attractive forces. Science 260, 1451–1456.
Ortega-Esteban, A., Horcas, I., Hernando-Perez, M., Ares, P., Perez-Berna, A.J., San Martin, C., Carrascosa, J.L., De Pablo, P.J., Gomez-Herrero, J., 2012. Minimizing tip-sample forces in jumping mode atomic force microscopy in liquid. Ultramicroscopy 114, 56–61.
Ortega-Esteban, A., Pérez-Berná, A., Menéndez-Conejero, R., Flint, S., San Martín, C., De Pablo, P., 2013. Monitoring dynamics of human adenovirus disassembly induced by mechanical fatigue. Sci. Rep. 3, 1434.
Ortega-Esteban, A., Bodensiek, K., San Martin, C., Suomalainen, M., Greber, U.F., De Pablo, P.J., Schaap, I.A., 2015a. Fluorescence tracking of genome release during mechanical unpacking of single viruses. ACS Nano 9, 10571–10579.
Ortega-Esteban, A., Condezo, G.N., Perez-Berna, A.J., Chillon, M., Flint, S.J., Reguera, D., San Martin, C., De Pablo, P.J., 2015b. Mechanics of viral chromatin reveals the pressurization of human adenovirus. ACS Nano 9, 10826–10833.

Pérez-Berná, A.J., Ortega-Esteban, A., Menéndez-Conejero, R., Winkler, D.C., Menéndez, M., Steven, A.C., Flint, S.J., De Pablo, P.J., San Martín, C., 2012. The role of capsid maturation on adenovirus priming for sequential uncoating. J. Biol. Chem. 287, 31582–31595. https://doi.org/10.1074/jbc.M112.389957.

Puntener, D., Engelke, M.F., Ruzsics, Z., Strunze, S., Wilhelm, C., Greber, U.F., 2011. Stepwise loss of fluorescent core protein V from human adenovirus during entry into cells. J. Virol. 85, 481–496.

Purohit, P.K., Kondev, J., Phillips, R., 2003. Mechanics of DNA packaging in viruses. Proc. Natl. Acad. Sci. U. S. A. 100, 3173–3178.

Querol-Audí, J., Casañas, A., Usón, I., Luque, D., Castón, J.R., Fita, I., Verdaguer, N., 2009. The mechanism of vault opening from the high resolution structure of the N-terminal repeats of MVP. EMBO J. 28, 3450.

Ramalho, R., Rankovic, S., Zhou, J., Aiken, C., Rousso, I., 2016. Analysis of the mechanical properties of wild type and hyperstable mutants of the HIV-1 capsid. Retrovirology 13, 17.

Rief, M., Gautel, M., Oesterhelt, F., Fernandez, J.M., Gaub, H.E., 1997. Reversible unfolding of individual titin immunoglobulin domains by AFM. Science 276, 1109–1112.

Roos, W.H., Radtke, K., Kniesmeijer, E., Geertsema, H., Sodeik, B., Wuite, G.J.L., 2009. Scaffold expulsion and genome packaging trigger stabilization of herpes simplex virus capsids. Proc. Natl. Acad. Sci. U. S. A. 106, 9673–9678.

Roos, W.H., Bruinsma, R., Wuite, G.J.L., 2010. Physical virology. Nat. Phys. 6, 733–743.

Roos, W.H., Gertsman, I., May, E.R., Brooks, C.L., Johnson, J.E., Wuite, G.J.L., 2012. Mechanics of bacteriophage maturation. Proc. Natl. Acad. Sci. U. S. A. 109, 2342–2347.

San Martín, C., 2012. Latest insights on adenovirus structure and assembly. Viruses 4, 847–877.

Schaap, I.A.T., Carrasco, C., De Pablo, P.J., Mackintosh, F.C., Schmidt, C.F., 2006. Elastic response, buckling, and instability of microtubules under radial indentation. Biophys. J. 91, 1521–1531.

Schaap, I.A.T., Eghiaian, F., Des Georges, A., Veigel, C., 2012. Effect of envelope proteins on the mechanical properties of influenza virus. J. Biol. Chem. 287, 41078–41088.

Settembre, E.C., Chen, J.Z., Dormitzer, P.R., Grigorieff, N., Harrison, S.C., 2011. Atomic model of an infectious rotavirus particle. EMBO J. 30, 408–416.

Sieben, C., Kappel, C., Zhu, R., Wozniak, A., Rankl, C., Hinterdorfer, P., Grubmüller, H., Herrmann, A., 2012. Influenza virus binds its host cell using multiple dynamic interactions. Proc. Natl. Acad. Sci. U. S. A. 109, 13626–13631.

Smith, D.E., Tans, S.J., Smith, S.B., Grimes, S., Anderson, D.L., Bustamante, C., 2001. The bacteriophage phi 29 portal motor can package DNA against a large internal force. Nature 413, 748–752.

Snijder, J., Ivanovska, I.L., Baclayon, M., Roos, W.H., Wuite, G.J., 2012. Probing the impact of loading rate on the mechanical properties of viral nanoparticles. Micron 43, 1343–1350.

Snijder, J., Uetrecht, C., Rose, R.J., Sanchez-Eugenia, R., Marti, G.A., Agirre, J., Guerin, D.M., Wuite, G.J., Heck, A.J., Roos, W.H., 2013. Probing the biophysical interplay between a viral genome and its capsid. Nat. Chem. 5, 502–509.

Snijder, J., Kononova, O., Barbu, I.M., Uetrecht, C., Rurup, W.F., Burnley, R.J., Koay, M.S., Cornelissen, J.J., Roos, W.H., Barsegov, V., Wuite, G.J., Heck, A.J., 2016. Assembly and mechanical properties of the cargo-free and cargo-loaded bacterial nanocompartment encapsulin. Biomacromolecules 17, 2522–2529.

Snijder, J., Radtke, K., Anderson, F., Scholtes, L., Corradini, E., Baines, J., Heck, A.J.R., Wuite, G.J.L., Sodeik, B., Roos, W.H., 2017. Vertex-specific proteins pUL17 and pUL25 mechanically reinforce herpes simplex virus capsids. J. Virol. 91, e00123–17.

Sugimoto, Y., Abe, M., Hirayama, S., Oyabu, N., Custance, O., Morita, S., 2005. Atom inlays performed at room temperature using atomic force microscopy. Nat. Mater. 4, 156–U36.

Suomalainen, M., Greber, U.F., 2013. Uncoating of non-enveloped viruses. Curr. Opin. Virol. 3, 27–33.

Tang, L., Gilcrease, E.B., Casjens, S.R., Johnson, J.E., 2006. Highly discriminatory binding of capsid-cementing proteins in bacteriophage L. Structure 14, 837–845.

Uchida, M., Douglas, T., 2013. Biophysical chemistry: unravelling capsid transformations. Nat. Chem. 5, 444–445.

Valbuena, A., Mateu, M.G., 2015. Quantification and modification of the equilibrium dynamics and mechanics of a viral capsid lattice self-assembled as a protein nanocoating. Nanoscale 7, 14953–14964.

Valbuena, A., Mateu, M.G., 2017. Kinetics of surface-driven self-assembly and fatigue-induced disassembly of a virus-based nanocoating. Biophys. J. 112, 663–673.

Vinckier, A., Heyvaert, I., Dhoore, A., McKittrick, T., Van Haesendonck, C., Engelborghs, Y., Hellemans, L., 1995. Immobilizing and imaging microtubules by atomic-force microscopy. Ultramicroscopy 57, 337–343.

Voros, Z., Csik, G., Herenyi, L., Kellermayer, M.S.Z., 2017. Stepwise reversible nanomechanical buckling in a viral capsid. Nanoscale 9, 1136–1143.

Weber, J., 1976. Genetic analysis of adenovirus type 2. III. Temperature sensitivity of processing viral proteins. J. Virol. 17, 462–471.

Wetz, K., Kucinski, T., 1991. Influence of different ionic and pH environments on structural alterations of poliovirus and their possible relation to virus uncoating. J. Gen. Virol. 72, 2541–2544.

Wilts, B.D., Schaap, I.A., Schmidt, C.F., 2015. Swelling and softening of the cowpea chlorotic mottle virus in response to pH shifts. Biophys. J. 108, 2541–2549.

Wimmer, E., Mueller, S., Tumpey, T.M., Taubenberger, J.K., 2009. Synthetic viruses: a new opportunity to understand and prevent viral disease. Nat. Biotechnol. 27, 1163.

Wörsdörfer, B., Woycechowsky, K.J., Hilvert, D., 2011. Directed evolution of a protein container. Science 331, 589.

Xiao, C., Kuznetsov, Y.G., Sun, S.Y., Hafenstein, S.L., Kostyuchenko, V.A., Chipman, P.R., Suzan-Monti, M., Raoult, D., McPherson, A., Rossmann, M.G., 2009. Structural studies of the giant mimivirus. PLoS Biol. 7, 958–966.

Zeng, C., Moller-Tank, S., Asokan, A., Dragnea, B., 2017. Probing the link among genomic cargo, contact mechanics, and nanoindentation in recombinant adeno-associated virus 2. J. Phys. Chem. B 121, 1843–1853.

Zhou, H.X., Rivas, G., Minton, A.P., 2008. Macromolecular crowding and confinement: biochemical, biophysical, and potential physiological consequences. Annu. Rev. Biophys. 37, 375–397.

CHAPTER SEVEN

Structural mass spectrometry goes viral

Jasmin Dülfer[a], Alan Kadek[a,b], Janine-Denise Kopicki[a], Boris Krichel[a], Charlotte Uetrecht[a,b,*]

[a]Heinrich Pette Institute, Leibniz Institute for Experimental Virology, Hamburg, Germany
[b]European XFEL GmbH, Schenefeld, Germany
*Corresponding author: e-mail address: charlotte.uetrecht@xfel.eu

Contents

Abstract

Over the last 20 years, mass spectrometry (MS), with its ability to analyze small sample amounts with high speed and sensitivity, has more and more entered the field of structural virology, aiming to investigate the structure and dynamics of viral proteins as close to their native environment as possible. The use of non-perturbing labels in hydrogen-deuterium exchange MS allows for the analysis of interactions between viral proteins and host cell factors as well as their dynamic responses to the environment. Cross-linking MS, on the other hand, can analyze interactions in viral protein complexes and identify virus-host interactions in cells. Native MS allows transferring viral proteins, complexes and capsids into the gas phase and has broken boundaries to overcome size limitations, so that now even the analysis of intact virions is possible. Different MS approaches not only inform about size, stability, interactions and dynamics of virus assemblies, but also bridge the gap to other biophysical techniques, providing valuable constraints for integrative structural modeling of viral complex assemblies that are often inaccessible by single technique approaches. In this review, recent advances are highlighted, clearly showing that structural MS approaches in virology are moving towards systems biology and ever more experiments are performed on cellular level.

Advances in Virus Research, Volume 105
ISSN 0065-3527
https://doi.org/10.1016/bs.aivir.2019.07.003

Abbreviations

Virus names

ASV	avian sarcoma virus
BMV	brome mosaic virus
CCMV	cowpea chlorotic mottle virus
DENV	dengue virus
HCV	hepatitis C virus
HIV	human immunodeficiency virus
PIV	parainfluenza virus
PLRV	potato leafroll virus
RYMV	rice yellow mottle virus
TMV	tobacco mosaic virus

General abbreviations

3D	three dimensional
AFM	atomic force microscopy
CCS	collision cross section
CDMS	charge detection mass spectrometry
CID	collision induced dissociation
EM	electron microscopy
EMD	electrophoretic mobility diameter
Env	envelope protein of HIV
ESI	electrospray ionization
ET	electron tomography
ETD	electron transfer dissociation
EX1/2	exchange regime 1/2 in HDX
FPOP	fast photochemical oxidation of proteins
FT-ICR	Fourier-transform ion cyclotron resonance
GEMMA	gas-phase electrophoretic mobility molecular analysis
HA	hemagglutinin
HBGA	histo-blood group antigen
HDX	hydrogen-deuterium exchange
HPLC	high-pressure liquid chromatography
IM	ion mobility
IRMPD	infrared multiphoton dissociation
LC-MS	liquid chromatography mass spectrometry
m/z	mass to charge
MALDI	matrix-assisted laser desorption-ionization
MD	molecular dynamics
MS	mass spectrometry
MS/MS	tandem MS
MX	macromolecular crystallography
NEMS	nano(electro)mechanical MS
NMR	nuclear magnetic resonance
PTM	post-translational modification
SAXS	small angle X-ray scattering

SID	surface induced dissociation
ss	single-stranded
ToF	time-of-flight
UPLC	ultra-high-pressure liquid chromatography
UVPD	ultraviolet photodissociation
VLP	virus-like particle
XFEL	X-ray free-electron laser
XL-MS	cross-linking mass spectrometry

1. Introduction

The ultimate goal in structural virology is to understand how the structure of a virus influences its interactions with a host organism and thus governs its lifecycle. Therefore, virologists nowadays employ a broad range of structural biology methods to uncover the organization of viral particles and to understand their function at the molecular level. Macromolecular crystallography techniques (MX) as well as nuclear magnetic resonance (NMR) spectroscopy routinely provide three dimensional (3D) structures with resolution down to single atoms, while electron microscopy-based (EM) methods are fast approaching this boundary with cryogenic sample preparation and recent developments in detector technology (Merk et al., 2016). However, all these techniques have their own challenges and limitations. These can range from the often frustrating efforts to produce sufficiently large, diffracting crystals, through sample size limitations, up to often very significant demands on sample amount and homogeneity or limited ability to assign subunit positions or identify post-translational modifications (PTMs). Most importantly, however, there is very rarely a single technique to answer all questions at hand. Therefore, structural biology is moving more and more in the direction of so-called integrative approaches, where high-resolution structural data from different methods are combined with lower resolution information. These data then ultimately feed into computational modeling in order to obtain the most detailed functional understanding possible of the biological system (Marklund and Benesch, 2019; Politis and Schmidt, 2018; Ward et al., 2013).

One of the experimental techniques, which provide invaluable input for the integrative structural efforts, is mass spectrometry (MS). Originally belonging purely to the domains of physics and analytical chemistry, MS has changed significantly since the revolutionary developments of electrospray ionization (ESI) and matrix-assisted laser desorption-ionization (MALDI) in the late 1980s (Karas et al., 1987; Tanaka et al., 1988; Yamashita and Fenn, 1984). These two so-called "soft-ionization techniques" have enabled the analyses of biomolecules without their unwanted fragmentation and allowed MS to expand hugely into the field of biochemistry. There, it has ever since been used in proteomics as a method of choice for identifying proteins, quantifying them and for analyzing their PTMs (Luo and Muesing, 2014; Yates et al., 2009). Moreover, and more importantly from the point of view of higher order macromolecular structure, so-called structural MS has also established itself especially in the last two decades as a powerful method to study proteins. This was enabled by two major lines of development. On one hand, pioneering works by the groups of Carol Robinson, Albert Heck and Joseph Loo have exploited the softness of the ESI process to transfer whole non-covalently bound protein assemblies into the gas phase inside a mass spectrometer without disrupting their stoichiometry and interactions in a process termed native MS (Leney and Heck, 2017; Loo, 1995; Rostom and Robinson, 1999; Van Berkel et al., 2000). Additionally, as was shown later, the conformations of such protein complex ions often survive largely unperturbed, which opened up possibilities for further gas phase structural studies and "gas-phase structural biology" (Ruotolo et al., 2005; Seo et al., 2016; Uetrecht et al., 2008b). On the other hand, many techniques were developed which probe protein conformations, interactions and dynamics directly in solution. Usually, such methods use some form of chemical labeling—be it through isotopic exchange (Katta and Chait, 1991), chemical cross-linking (Rossi et al., 1995) or radical chemistry (Maleknia et al., 2001)—and capitalize from the use of MS as a detection method.

Together, all the various structural MS approaches offer a broad portfolio of experimental techniques, which can probe a wide range of protein structural features, while at the same time retrieving information about protein identity and PTMs. Moreover, structural MS nicely complements the three traditional high-resolution structural biology techniques, albeit with lower overall resolution. However, this is often more than balanced by the ability of MS to work with low amounts of samples, in low concentrations, and with high throughput. Importantly, structural MS can often work within complex mixtures, with several pioneering methods probing protein structures directly within cells (Chavez et al., 2018; Chorev et al., 2018; Espino et al., 2015; Gan et al., 2017; Garcia-Moreno et al., 2019; Nguyen et al., 2018). Additionally, structural MS is in theory unlimited in terms of molecular weight, offering information on systems ranging from small isolated protein domains all the way up to huge multi-megadalton assemblies. Perhaps most importantly, structural MS can also provide insight into protein conformational dynamics and coexisting transient species in solution, which is frequently missing or can be very challenging to access by static averaged high-resolution structures provided by MX and EM.

In this review, an overview of current MS-based methods is provided as employed for structural virology yielding valuable information on protein structure, especially for integrative modeling efforts. Moreover, as MS and its role in general structural biology has been extensively covered previously (Benesch and Ruotolo, 2011; Konermann et al., 2014b; Liko et al., 2016; Lossl et al., 2016), the focus is specifically at applications in the context of structural virology. Finally, as this field was already reviewed in detail in 2011 (Uetrecht and Heck, 2011), recent developments in the area, reflecting instrumental and methodological advances over the last years, will mainly be discussed. As structural MS encompasses a rather broad spectrum of experimental techniques, which study various aspects of protein structure under different conditions, the covered techniques are grouped and divided based on the overall methodological approach.

2. Investigating interactions and dynamics of viral proteins using covalent labeling techniques

In the following section, we highlight a few examples where MS coupled with labeling techniques shed light on interactions and dynamics of viral proteins that play important roles in different parts of the viral lifecycle. Furthermore, we demonstrate how insights derived from labeling MS experiments can assist drug and antibody development. Starting from tiny labels, which are less likely to alter structure, we move to larger ones, which carry a higher intrinsic propensity for impacting structure but are chemically more stable and therefore better suited for highly heterogeneous samples, where additional sample handling is needed before analysis.

2.1 Hydrogen-deuterium exchange mass spectrometry

Hydrogen-deuterium exchange (HDX)-MS is a sensitive technique to localize regions of conformational dynamics in proteins and identify amino acids engaged in ligand binding under close to native conditions in solution. As opposed to other MS experiments, HDX can be performed in practically any buffer condition and with additives or co-factors that are required for proper protein folding and function—the sole difference is buffer preparation from deuterated water. The method exploits the naturally occurring exchange between solvent-accessible protein backbone hydrogens and deuterium in solution over time, measured as mass increase using MS (Konermann et al., 2011; Trabjerg et al., 2018).

The rate of exchange is influenced by several factors, including solution conditions such as pH and temperature, as well as by the peptides' local environment (Bai et al., 1993). Most importantly, from the viewpoint of protein structure and interactions, the involvement of hydrogens in hydrogen bonding within higher order elements of protein structure (e.g., in α-helices and β-sheets) or in interactions with ligands slows down the exchange rate for respective backbone amide hydrogens. Protection from solvent (e.g., in a hydrophobic core of the protein or at a protein-protein interface) also typically leads to a decrease in exchange in the affected protein region. Additionally, HDX can detect more distant effects induced by an interaction for example through allostery (Konermann et al., 2014a). Finally, conformational dynamics and flexibility of protein regions play a role in influencing the exchange rate too, as the exposure of regions to the surrounding deuterated solvent accelerates the HDX. This can sometimes even be reflected in more complex exchange kinetics (so-called EX1 and EX2 regimes, discussed in more detail later), which uncover additional information on distinct conformational dynamics (Brown and Wilson, 2017).

Considering the range of monitored changes within proteins, HDX-MS is a versatile tool to gain insights into the dynamics and structural responses of protein backbones to perturbed solution conditions and ligand or protein binding (Chalmers et al., 2011; Engen, 2009). Furthermore, conformational dynamics at different pH and temperature can also be probed, even though this requires slight adjustments in the experimental procedure and data analysis (Bai et al., 1993; Engen, 2009).

2.1.1 HDX-MS workflow

The overall workflow of the most common HDX experiment is depicted in Fig. 1A. The studied protein is labeled in a deuterated buffer under native conditions. After various time intervals, the labeling reaction is quenched by rapidly reducing the pH to 2.5 and the temperature to 0 °C, where the HDX rate and therefore also the rate of D/H back exchange, is minimal. Under these conditions the half-life of the deuterium label (throughout subsequent LC-MS analysis performed in non-deuterated solvents) is sufficiently long (Englander, 2006). Therefore, deuterium incorporation can be localized when the protein is split up into smaller peptides prior to LC-MS analysis.

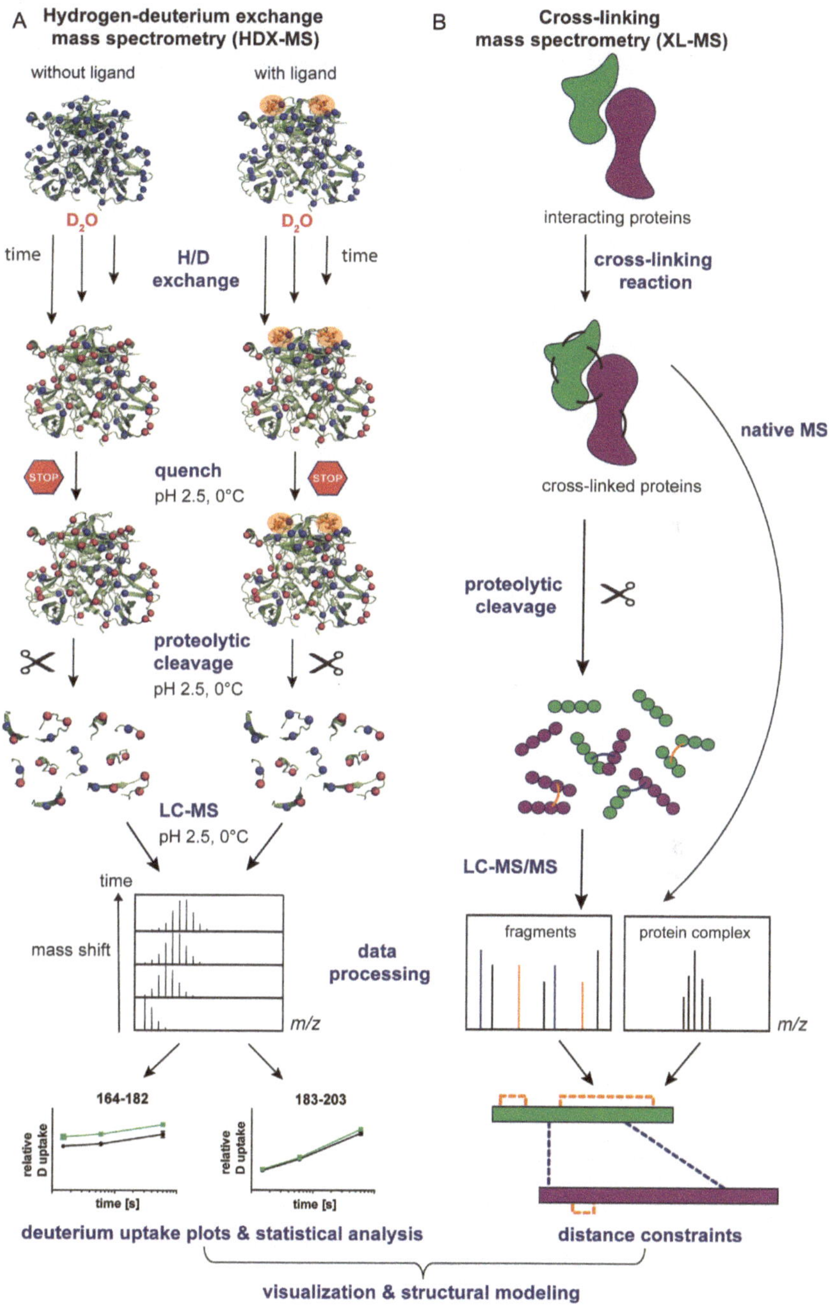

Fig. 1 Comparison of MS-based covalent labeling workflows. (A) HDX-MS workflow for a continuous labeling experiment. After labeling for several time intervals, followed by a quench step, which reduces HDX to a minimum, the protein alone or with an interaction

(Continued)

These peptides can be generated via digestion with an acidic, usually aspartic, protease, either in bulk solution or more conveniently using protease immobilized into a flow-through column (Wang et al., 2002). Due to the pH required to keep deuterium back exchange loss minimal, porcine pepsin is routinely used as the protease of choice for HDX-MS applications. However, a number of alternative aspartic proteases are used, which can yield different peptides and more precise deuteration localization (Cravello et al., 2003; Kadek et al., 2014; Rey et al., 2009; Yang et al., 2015).

Following digestion, peptides are separated in a liquid chromatography (LC) system and directly electrosprayed into a mass spectrometer. There, peptides are first identified by their mass-to-charge ratio (m/z) and tandem MS (MS/MS) fragmentation patterns from non-deuterated samples. Subsequently, their deuterium content is measured in the samples exposed to deuterium and compared between experimental conditions to localize any differences. The experimental procedure is reviewed in more technical detail in Guttman and Lee (2016). Besides the so-called bottom-up approach presented above, where the protein is cleaved by a protease, so-called top-down approaches exist as well. There, the intact protein is sprayed into the mass spectrometer and fragmented in the gas phase (Leurs et al., 2015).

2.1.2 Methodological advances to tackle challenging systems

The advantages of HDX-MS are a low required amount of protein (~20–100 pmol per deuteration time point) and accessibility of large protein complexes, which makes it possible to measure conformational dynamics and ligand binding of large virus-like particles (VLPs) or even intact viruses (Bereszczak et al., 2013; Lim et al., 2017; van de Waterbeemd et al., 2017).

Fig. 1—Cont'd partner is proteolytically digested and resulting peptides are separated and analyzed by LC-MS. From the observed mass shifts, deuterium uptake plots can be generated and peptides showing differences under the analyzed conditions can be identified. (B) General XL-MS workflow. Either a protein alone or a protein complex is cross-linked using suitable agents to capture the present structure and conformation. The cross-linked complex can be analyzed as a whole using native MS. Usually, however, the XL reaction is followed by proteolytic digestion, which in turn is followed by LC-MS/MS analysis of the different peptides obtained. For MS/MS fragmentation, those precursors are automatically selected that are likely to represent a cross-linked peptide. The MS/MS spectra can be used to identify the respective cross-linking pairs using appropriate software. The results are then used as restraints for structural modeling.

The protein digestion step can sometimes be a limitation in HDX experiments, as poor proteolytic digestion may result in low sequence coverage and therefore loss of information. Obtaining sufficient sequence coverage is often hampered by the occurrence of PTMs, such as glycosylation or disulfide bridges, adding to the spectral complexity or hindering efficient protease cleavage. However, this can in most cases be overcome by protocol optimization, e.g., addition of reducing or denaturing agents during the quench step or choice of a different protease (Trabjerg et al., 2018). In particular, the broad specificity of pepsin can be detrimental on this account as a large number of variably glycosylated peptides may be generated. A way to overcome this can be the implementation of a deglycosylation step before HDX (Guttman et al., 2011; Houde et al., 2009), or even after quenching (Jensen et al., 2016).

The use of an ultra-high-pressure liquid chromatography (UPLC) rather than a high-pressure liquid chromatography (HPLC) system increases chromatographic resolution and analysis speed and facilitates the investigation of more complex datasets (Engen and Wales, 2015). A conventional HDX-MS experiment usually provides resolution on the peptide level, even though more sophisticated fragmentation techniques, such as electron transfer dissociation, can be used to bring resolution down to ultimately the single amino acid level (Rand, 2013).

2.1.3 HDX-MS data interpretation

Depending on the research question, several types of experiments can be performed to provide information on different aspects of the protein structure and dynamics. In most cases, deuterium uptake of a protein incubated under a certain condition (e.g., pH, temperature or ligand presence) is compared with deuterium uptake of the protein under standard conditions (so-called differential approach with continuous labeling). In this way, protein regions undergoing changes in dynamics, and therefore changing their hydrogen-bonding pattern and solvent accessibility, can be identified. This can provide various information, such as conformational changes induced by ligand binding and location of binding epitopes, folding or unfolding depending on solution conditions, protein stability, protein-protein interactions and overall flexibility of different parts of the studied protein (Marcsisin and Engen, 2010; Morgan and Engen, 2009) (Fig. 2). Another approach is so-called pulsed labeling, where the protein is first forced to undergo structural changes (e.g., by being exposed to abruptly changed pH, temperature or addition of a perturbant for different times)

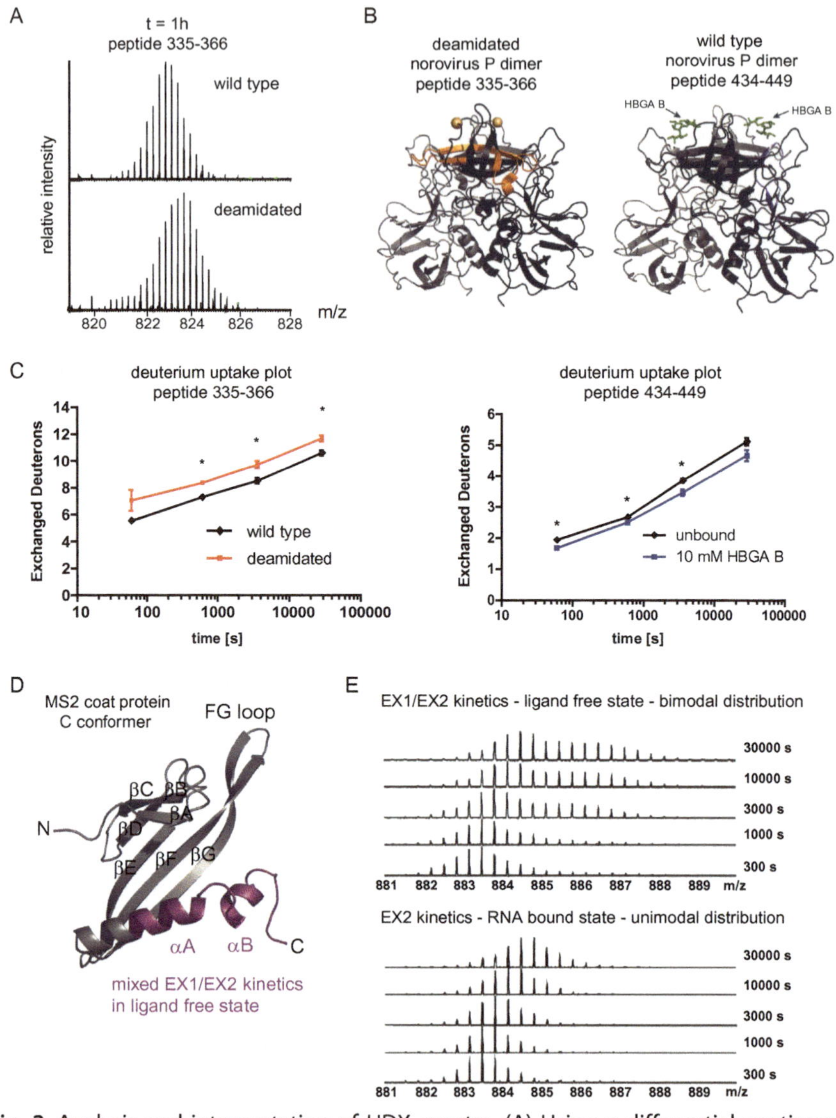

Fig. 2 Analysis and interpretation of HDX spectra. (A) Using a differential continuous labeling approach, regions in a protein that show altered dynamics under different conditions can be identified. Here, a peptide of the human norovirus protruding domain (P) dimer shows higher deuterium uptake (indicated by a larger mass shift in the spectrum) when a specific asparagine is deamidated than in the wildtype state. (B) The peptide from panel (A) is labeled in orange in the crystal structure (PDB: 4X06) and the deamidated amino acid is indicated by a sphere (left). When a ligand (HBGA B) is bound to the wildtype protein, a peptide in the binding site gets protected from HDX (blue, right). (C) Deuterium uptake plots show deprotection for the peptide in the deamidated protein indicating higher flexibility, whereas the peptide engaged in ligand binding in

and is then labeled by a short "pulse" of deuterium exposure. This approach is mainly used to identify and monitor transient folding intermediates (Konermann et al., 2011; Morgan and Engen, 2009).

An advantage of HDX-MS, that offers versatile benefits for virological questions, is that it can detect different protein populations in solution. Two populations of the same protein that differ structurally will incorporate different amounts of deuterium and can be observed as bimodal distributions in the mass spectra (Marcsisin and Engen, 2010). On the other hand, bimodal distributions can also be the result of more complex exchange kinetics of just a single protein species in solution. All proteins naturally fluctuate through transient, locally unfolded states in solution. Deuterium exchange occurs at these moments when a region of protein structure is temporarily loosened up and the hydrogen bonds are momentarily interrupted. If the refolding of such a temporarily more exposed region is slower than the deuterium uptake rate, the respective population will get more deuterated than the folded population, showing as a discreet jump in deuteration in the spectra (EX1 kinetics, Fig. 2E) (Weis et al., 2006). For most proteins, however, the rate of this temporary unfolding and refolding is much faster than the deuterium uptake rate, so that the deuteration occurs gradually and only a single peak series can be detected (EX2 kinetics). EX1 kinetics are often observed for proteins showing large concerted structural transitions in parts of their structure, such as is for example often the case for fusion proteins, and can be a powerful tool to study the chronology of structural intermediates in a complex refolding process (Garcia and Lee, 2016).

---

the wildtype shows protection from HDX. (D) The crystal structure shows an MS2 coat protein monomer (PDB: 2MS2) with different structural elements. Mixed EX1 and EX2 kinetics are detected in the αB-helix and parts of the αA-helix in the ligand free protein (dark purple). (E) HDX data for the peptide exhibiting mixed kinetics in the protein dimer (upper spectrum). When RNA is bound, a switch to pure EX2 kinetics is observed (lower spectrum) pointing towards RNA-induced stabilization. *Panels (A–C): Adapted from Mallagaray, A., Creutznacher, R., Dülfer, J., Mayer, P. H. O., Grimm, L. L., Orduna, J. M., Trabjerg, E., Stehle, T., Rand, K. D., Blaum, B. S., Uetrecht, C., Peters, T. 2019. A post-translational modification of human Norovirus capsid protein attenuates glycan binding. Nat. Commun., 10, 1320 under CC BY 4.0 license. Panel (D) different exchange kinetics are observed in coat proteins of bacteriophage MS2. Spectra republished with permission of Royal Society of Chemistry, from Morton, V. L., Burkitt, W., O'Connor, G., Stonehouse, N. J., Stockley, P. G., Ashcroft, A. E. 2010. RNA-induced conformational changes in a viral coat protein studied by hydrogen/deuterium exchange mass spectrometry. Phys. Chem. Chem. Phys., 12, 13468–75, Copyright 2010, permission conveyed through Copyright Clearance Center, Inc.*

2.1.4 Dynamic interactions of viral proteins with host cell factors

HDX-MS can probe interactions, conformational changes and structural transitions in proteins, whole viral capsids and even intact infectious viral particles (Bereszczak et al., 2013, 2014; Garcia et al., 2015; Lim et al., 2017). Looking at isolated viral proteins *in vitro* (Mallagaray et al., 2019) used a combination of HDX-MS, NMR spectroscopy and X-ray crystallography to study interactions of human norovirus capsid proteins with host cell glycans that play a role in cell attachment, namely histo-blood group antigens (HBGAs, Taube et al., 2018). A specific spontaneous asparagine deamidation (Fig. 2B) was identified in a predominant GII.4 strain, which dramatically decreased HBGA binding affinity. HDX-MS confirmed that the deamidated residue abolished HBGA binding in the protein. Whereas binding led to clear protection in the wildtype protein (Fig. 2C), the deamidated protein took up much more deuterium than the wildtype throughout its binding site-containing domain. The underlying increased flexibility likely causes the large decrease in HBGA binding affinity (Fig. 2A). As deamidation takes place very rapidly at 37 °C, it is likely to occur *in vivo,* potentially influencing infectivity. Here, the study has clearly benefited from combination of structural biology techniques. MS confirmed the presence, exact position and extent of the deamidation by detecting the corresponding mass shift and fragmentation spectra on the peptide level. NMR data revealed an amino acid conversion, which could be assigned to the deamidation only after complete backbone assignment. Additionally, X-ray crystallography proved a loop structure facilitating deamidation.

Moreover, HDX can also probe altered dynamics and allosteric response to ligands that influence capsid assembly. An interesting feature of viruses is their ability to arrange small identical capsid proteins in different conformations to build large capsids of different symmetries. These capsid assembly stages have been studied for a model single-stranded (ss) RNA virus, bacteriophage MS2, by several MS methods before (Knapman et al., 2010). To go a step further, Morton et al. (2010) used HDX-MS to study conformational switches in the MS2 coat protein. They especially focused on potential allosteric effects of RNA interactions on coat protein conformations during assembly. The RNA bound state showed decreased deuterium uptake compared to the free protein in β-strands E and G, confirming the RNA binding sites suggested previously by other techniques. However, HDX provided previously inaccessible insight into the internal dynamics of the protein in solution by revealing evidence of allosteric modulation in the FG-loop. Most importantly, a surprising shift from mixed EX1 to pure EX2 exchange kinetics occurred in a distinct coat protein region between ligand-free and RNA-bound states, respectively. This was explained as a concerted RNA-induced stabilization of α-helices A and B (Fig. 2D and E). As demonstrated by this example, HDX-MS is capable of detecting subtle differences in the dynamics of protein complexes and can provide further details on allosteric and local folding or unfolding events required for or introduced by ligand interactions.

Additionally, HDX is by far not limited to probing protein-protein and protein-RNA interactions. Wijesinghe et al. (2017) focused on interactions of cell membrane lipids with Marburg virus matrix protein VP40. The scope of the study was to probe lipid dependent oligomerization states of the protein. HDX-MS in combination with molecular modeling demonstrated that two basic loops are the main contact areas between VP40 and anionic lipid vesicles. Moreover, several C- and N-terminal peptides were protected from HDX, which form a conserved oligomerization interface for matrix assembly in Ebola viruses. In conclusion, a model of VP40 matrix assembly following membrane association was proposed (Wijesinghe et al., 2017).

2.1.5 Conformational responses to environmental factors

Dynamics of viral capsids are highly influenced by the environment the virus travels through during its lifecycle. Even though high sequence and structural similarity can exist for viruses of a particular strain or serotype, dynamic responses to environmental factors can be diverse. For Dengue virus (DENV), for example, the rise of temperature from 28 °C in the vector to 37 °C in the host triggers structural transitions in the capsid, important for host cell entry. These transitions are particularly prominent in some DENV 2 strains in contrast to DENV 1, 3 and 4 strains, even though viruses of the four DENV serotypes share around 60–75% identity at the amino acid level (Fibriansah et al., 2013; Mukhopadhyay et al., 2005). Lim et al. (2017) used HDX-MS to deduce the structural dynamics of DENV 1 and 2 infectious particles and to explain the observed disparity in temperature-dependent expansion. Due to its icosahedral symmetry, deuterium uptake of envelope proteins (E proteins) in the intact particle represents averages of all E*E*-protein copies on the viral surface. Isolated structural changes in only some of the proteins in the particle would give rise to multimodal peak distributions, whereas equal structural changes in all symmetry units would be seen as unimodal peak distributions in the HDX data. DENV 1 and 2 particles exhibited exclusively unimodal exchange profiles, pointing toward a collective response of the whole virus to temperature increase.

Furthermore, in contrast to DENV 2, DENV 1 only undergoes expansion at 40 °C, a condition that resembles dengue fever.

Garcia et al. (2015) investigated the pH dependent conformational switch leading to influenza hemagglutinin (HA) activation. While crystal structures for the pre- and post-fusion state are available, HDX-MS could provide additional details on the actual sequence of smaller conformational changes during this transition triggered by low pH. The soluble HA ectodomain (Fig. 3A) resembled the conformation of full-length HA in the virion at pH 7.5 as the HDX profiles revealed no differences in all covered protein regions, even though some sequence coverage was lost for HA from the intact virion. The loss of coverage is largely due to complexity in the virion including additional viral proteins, genome and lipids, which create spectral overlap complicating analyses. Precipitation of viral DNA or RNA and implementation of a centrifugation step after quenching can help to reduce the overall complexity of the sample (Garcia et al., 2015). Comparative HDX of soluble HA in pre- (pH 7.5) and post-fusion (pH 4.9) state revealed many sites with significant differences, in agreement with published crystal structures, but could also pinpoint previously unknown areas of local destabilization in post-fusion soluble HA (Fig. 3B–D). HDX at various pH from 7.5 (pre-fusion) down to 5.6 (fusion activation in endosomes) indicated local dynamic changes and intermediate states on the way to the fusion activation pH threshold providing a more complete picture of the fusion mechanism.

2.1.6 HDX analysis of HIV structural proteins for drug and vaccine development

Nowadays, virology moves towards using structural knowledge to develop and improve antiviral therapies and vaccines. Many drugs against human immunodeficiency virus (HIV) target essential non-structural proteins, whereas vaccine development is mainly based on the entry-mediating HIV envelope glycoprotein (Env). The high genetic diversity and especially the high sequence variability in the Env protein gp120 surface subunit still hamper the development of a broadly efficient vaccine (van Schooten and van Gils, 2018). Furthermore, Env is highly glycosylated under native conditions, which presents a major challenge for analysis due to the large glycan heterogeneity. However, MS is in principle able to determine glycosylation sites on the protein and further analyze their structure and composition after removal by glycosidases (Parker et al., 2010).

Davenport et al. (2013) characterized the structural flexibility of gp120 from four HIV-1 isolates, that needs to be accommodated by neutralizing

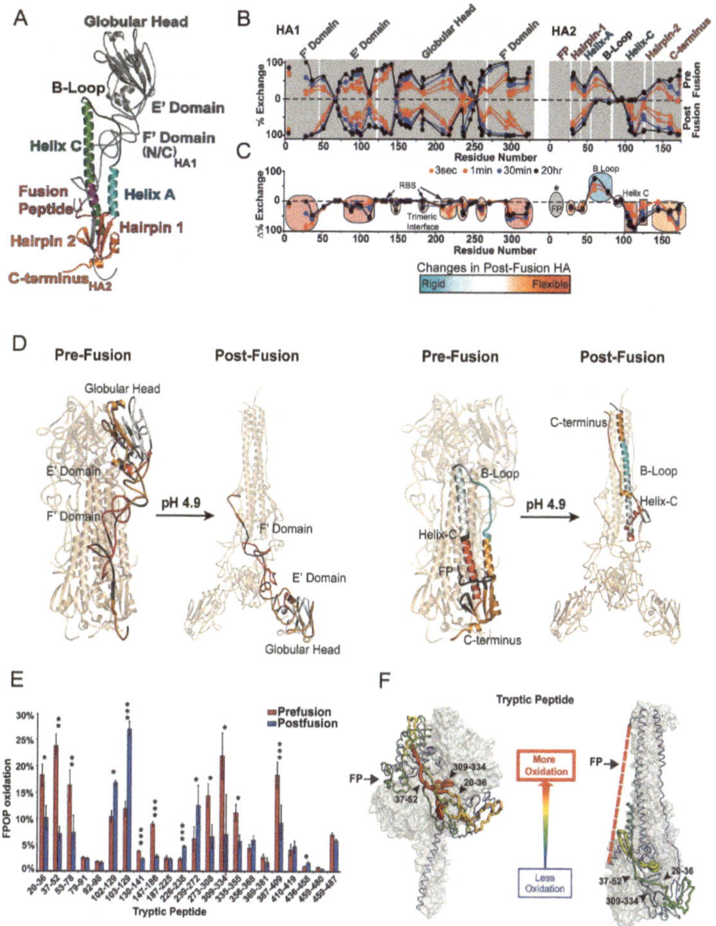

Fig. 3 HDX and hydroxyl radical labeling analysis of viral fusion proteins. (A) The structural organization of pre-fusion influenza HA color-coded on one monomer (PDB: 2HMG). (B) HDX exchange plots of pre-fusion and post-fusion soluble HA after labeling for various time intervals at pH 7.5. Data are plotted on the midpoint of each analyzed peptide sequence. (C) HDX difference map for exchange plots in (B). Positive differences correspond to regions that gain protection, whereas negative differences indicate a loss of protection in post-fusion soluble HA. (D) Areas in post-fusion soluble HA that become highly (red), moderately (orange) or less (cyan) flexible relative to the pre-fusion conformation are mapped onto the pre- and post-fusion trimer (PDB: 2HMG). (E) Comparison of the extent of oxidative modification in tryptic peptides in either the pre-fusion (red bars) or post-fusion (blue bars) conformation of parainfluenza virus 5 fusion protein. (F) Oxidative labeling (FPOP) data mapped onto the pre-fusion crystal structure (left) and post-fusion homology structure (right) of the fusion protein trimer, which was based on the post-fusion crystal structure of hPIV3 (PDB: 1ZTM, 22% identity). Peptides that undergo a drastic decrease in accessibility (reflected in the level of oxidation) between the pre- and post-fusion stated are indicated by arrowheads. *Panels (A–D): Reprinted from Garcia, N. K., Guttman, M., Ebner, J. L., Lee, K. K. 2015. Dynamic changes during acid-induced activation of influenza hemagglutinin. Structure, 23, 665–76, Copyright (2015), with permission from Elsevier. Panels (E and F): Adapted from Poor, T. A., Jones, L. M., Sood, A., Leser, G. P., Plasencia, M. D., Rempel, D. L., Jardetzky, T. S., Woods, R. J., Gross, M. L., Lamb, R. A. 2014. Probing the paramyxovirus fusion (F) protein-refolding event from pre- to postfusion by oxidative footprinting. Proc. Natl. Acad. Sci. U. S. A., 111, E2596–605 with permission from PNAS.*

antibodies, and identified structural features connected with improved antibody recognition. HDX-MS analysis showed that the overall structure of full-length, glycosylated gp120 is similar among diverse isolates, but also revealed clear isolate-specific differences in the conformational dynamics of both unbound and CD4-bound gp120. Moreover, CD4 binding has been shown to stabilize gp120, an effect also detected by reduced deuterium incorporation in all isolates.

In a follow up study, Liang et al. (2016) looked into the dynamics of HIV-1 gp120 with various truncations in its flexible loop. As shown previously by Kwong et al. (2002), conformational switching and structural flexibility are key mechanisms the virus uses to evade immune recognition. Structural stabilization thus could provide better antibody recognition of vaccine candidates. Loop truncation resulted in an increased level of order within the conserved gp120 subunit core, which suggested that the influence of structural dynamics on antibody-antigen interactions should be considered in the characterization and optimization of vaccine candidates (Liang et al., 2016). Similar approaches to identify flexible protein regions have been used before in combination with MX for type III phosphatidylinositol 4 kinase beta (Fowler et al., 2016). Here, HDX could guide crystallization efforts by providing information on intrinsically disordered regions in the protein to create more stable mutants for MX.

2.1.7 *Structural transitions in fusion proteins monitored with oxidative labeling*

Another emerging labeling technique to probe structural dynamics and ligand binding is fast photochemical oxidation of proteins (FPOP). Also in this technique, amino acids are labeled according to their surface accessibility. However, in contrast to HDX, side chains are permanently modified in FPOP by oxidation with hydroxyl radicals. These can be produced by a range of methods, but most common are UV-photolysis of diluted H_2O_2 and direct X-ray photolysis of water in a flow system (Hambly and Gross, 2005; Maleknia et al., 2001). This results in a stable modification and more straightforward sample analysis after labeling as opposed to HDX. However, labeling is highly dependent on the reactivity of side chains and the technique bears a higher chance for over-labeling that can cause changes in the structure. Due to its fast labeling time in the micro- to millisecond timescale (Vahidi and Konermann, 2016), FPOP enables the study of fast kinetics in the native protein structure (Zhang et al., 2018). As with the other labeling techniques, proteins are digested and labels mapped by LC-MS.

The method was used to illustrate the transition from pre- to post-fusion state in the fusion protein of parainfluenza virus (PIV) 5 (Poor et al., 2014). Interaction with the receptor-binding protein HN triggers an extensive refolding event that enables fusion with the host membrane. Using FPOP, the pre-fusion protein becomes more oxidized than the post-fusion protein, indicating that the protein is more flexible and solvent exposed prior to fusion (Fig. 3). This is well in line with the current model for class I fusion proteins, which undergo a series of refolding events until the stable post-fusion state is reached (Bossart et al., 2013). Moreover, FPOP data combined with homology modeling and molecular dynamics (MD) simulations could identify regions accounting for metastability of the pre-fusion state (Poor et al., 2014).

2.2 Cross-linking mass spectrometry

Chemical cross-linking coupled with MS (XL-MS) is a low-resolution method on its own. In combination with other techniques, such as native MS or cryo-EM, complementary data and integrative structural modeling for structural elucidation of large protein assemblies up to atomic resolution can be provided (Bullock et al., 2018; Schmidt and Urlaub, 2017). The strength of this method lies in its ability to identify interactions between protein subunits or proteins and nucleic acids or lipids at the amino acid level (Haberkant et al., 2013; Kaur et al., 2019). The method has in theory no size limitation allowing analysis of large heterogeneous protein assemblies even directly out of the cellular context (Liu et al., 2017). The technique is able to tolerate protein impurities, but higher sample purity unsurprisingly positively affects the quality of the obtained data. Again, as for the other chemical labeling techniques, XL-MS is performed in a bottom-up approach, in which cross-linking of intact proteins is followed by their proteolytic digestion and LC-MS analysis (Fig. 1B). Notably, opposed to small deuterium or medium sized hydroxyl radical labels as in HDX and FPOP, changes in the native structure introduced by the cross-linker are a major concern in XL experiments (Arlt et al., 2017; Rozbesky et al., 2018).

2.2.1 Cross-linking agents

Non-covalently interacting protein assemblies are subjected to a chemical reaction with reactive, bifunctional small molecules, known as cross-linkers. Thereby, a pair of amino acids is intra- or intermolecularly covalently linked, which can in turn be detected by MS (Kaur et al., 2019; Sharon and Sinz, 2015). The functional groups are separated by a spacer arm, which defines the distance being bridged by the cross-linker, and therefore acts as a molecular ruler. A good distance match between the functional groups in the sample and the cross-linker is necessary for successful reaction. This then defines a distance constraint usable for computational modeling. Intermolecular cross-links inform about interaction partners and their binding interface, while intramolecular linkages within a protein inform about its fold. It is also possible that only one end of the cross-linker reacts with the target molecule, while the other is hydrolyzed. The resulting so-called monolinks are not useful for distance measurements but can nevertheless pinpoint solvent-exposed amino acids (Gingras et al., 2007).

Common cross-linkers are for example homobifunctional *N*-hydroxysuccinimide (NHS) esters, which form bonds with lysine residues or the N-terminus due to their primary amine reactivity. Moreover, hydrazines can be used to cross-link acidic residues. However, the reaction is less efficient and the amino acids, e.g., glutamic acid, aspartic acid or the C-terminus, require chemical activation (Holding, 2015; Leitner et al., 2014; Novak and Kruppa, 2008; Piotrowski and Sinz, 2018). Activation of amino acids can also enable direct connections between carboxyl and amine groups—so-called "zero-length" cross-links (Schwarz et al., 2016). Furthermore, a lot of progress has been achieved recently in the field of photoreactive cross-linkers. Here, diazirines are particularly suitable. Upon irradiation, they turn to extremely reactive radical species, which are not limited to specific side chain types (Kaur et al., 2019; Sharon and Sinz, 2015). Finally, hetero-bifunctional linkers can be used to obtain asymmetric cross-links, which are usually composed of NHS esters at one end and maleimide moieties reacting with cysteines or photoreactive groups at the other end (Piotrowski and Sinz, 2018).

Although structural investigations of the interaction between proteins and nucleic acids are of great relevance in virus research due to their biological importance, the development of suitable cross-linkers for this kind of applications remains difficult. Recently, Scalabrin et al. (2018) tried to extend the repertoire of available cross-linkers for nucleic acids by screening a variety of potential bifunctional agents. Despite challenges to find a universal nucleic acid cross-linker and many types of compounds failing to fulfill expectations, several platinum and mustard gas derivatives have shown potential for further optimization (Scalabrin et al., 2018). Nevertheless, formaldehyde as a reactive agent and the use of UV light have long become established for cross-linking proteins and nucleic acids (Chodosh, 2001; Hoffman et al., 2015; Lenz et al., 2007).

2.2.2 Detection of cross-linked proteins with MS

Cross-linked protein complexes can be studied in their entity by means of native MS or even MALDI MS (Chen et al., 2016; Gasilova and Nazabal, 2012; Kool et al., 2011). This procedure is particularly suitable for very labile complexes such as transient states, which under normal conditions would immediately disintegrate in the gas phase (Sharon and Sinz, 2015; Tabb, 2012). Furthermore, it is possible to dissociate and fragment these complexes, and identify resulting peptides in a subsequent experiment to obtain more localized information in a top-down approach (Chen et al., 2016).

More common is however the bottom-up approach as used in HDX-MS. For XL-MS, standard specific proteases, namely trypsin, can be used as pH is of no concern after labeling. On the other hand, to avoid structural alterations, cross-linking happens sub-stoichiometrically resulting in a large number of non-cross-linked peptides that vastly outnumber the cross-linked ones. To facilitate detection of the cross-linked peptides, their enrichment is critical. This can, for example, be achieved biochemically using specific moieties on the cross-linkers or by chromatographic methods (Haupl et al., 2017; O'Reilly and Rappsilber, 2018; Tinnefeld et al., 2017). Despite such enrichment, identifying the cross-linked peptides within large LC-MS datasets is often very challenging. For this purpose, there are different approaches simplifying automated data analysis. Lately, isotope pairs have been used to identify cross-linked peptides; either by digestion in $H_2{}^{18}O$ (Zelter et al., 2010); or by signature fragments from MS-cleavable cross-linkers (Klykov et al., 2018; Sinz et al., 2015; Tang and Bruce, 2010).

2.2.3 Molecular interactions within viral protein complexes

XL-MS is particularly well suited to synergistically support the current methods for structural elucidation. Looking at the latest achievements in structural biology in terms of resolution, especially cryo-EM has come a long way. However, the benefit of XL still holds for integrative studies with lower resolution techniques such as cryo-electron tomography (cryo-ET) and as a method to verify interactions at the peptide and amino acid levels (Schmidt and Urlaub, 2017). In this way, the constraints derived from XL-MS data contributed significantly to the development of a structural model for monomeric and dimeric integrase of avian sarcoma virus (ASV). Integrative structural analysis using X-ray crystallography, small angle X-ray scattering (SAXS) and XL-MS led to a novel pathway for the formation of functional complexes consisting of ASV integrase and host DNA (Bojja et al., 2011). More recently, cross-links between the coat protein and the viral protease of intact turnip yellow virus were found, suggesting that the protease has a regulatory function in coat protein maturation. Moreover, two different structural models for the viral coat protein of turnip yellow virus were obtained—one for the monomer and one for the trimer, which is the essential building block of the capsid in all *Luteoviridae* (Alexander et al., 2017).

XL-MS has been used to study structure of bacteriophage pro-capsids. Kang et al. (2006) predicted secondary structure elements in the coat protein of bacteriophage P22 based on intramolecular linkages. Furthermore, intermolecular connections in the assembled procapsid shell identified by XL-MS were used to build a more detailed structural model based on cryo-EM reconstructions. Another study, performed by Singh et al. (2013), examined the bacteriophage λ major capsid protein in both monomeric and assembled state. The identified chemical cross-links within the assembled λ procapsid shell protein are depicted in Fig. 4A. Combining XL-MS constraints with computational modeling, a pseudo-atomic model for the capsid protein in both states could be generated that was in good agreement with published biochemical data (Fig. 4B).

Cross-links between proteins and nucleic acids can be used to analyze the interaction between viral proteins and viral genomes (Hoffman et al., 2015; Lenz et al., 2007). Hung et al. (2014) explored the interactions of viral RNA and coat proteins of the bamboo mosaic virus using reversible formaldehyde cross-linking. Subsequently, mutation experiments verified that the detected binding sites are significantly involved in RNA accumulation. In the same year, the DNA binding sites in the adenoviral protein V were examined. The adenovirus core is formed by the genome together with protein V bound. Additionally, protein V interacts with protein VI, which is part of the capsid, thereby linking the internal core of the virion with the outer capsid layer. Furthermore, the XL-MS data supported a nucleosome-like structure of the genome (Perez-Vargas et al., 2014). Similarly, the highly ordered RNA genome of human parechovirus (HPeV) interacts with its capsid proteins VP1 and VP3 promoting virus assembly (Shakeel et al., 2017). A large number of RNA-protein interactions within the HPeV1 capsid from XL-MS confirmed these results. Surprisingly, previously undetected interactions between the genome and the VP0 protein, forming parechovirus capsids, were identified as well (Shakeel et al., 2018).

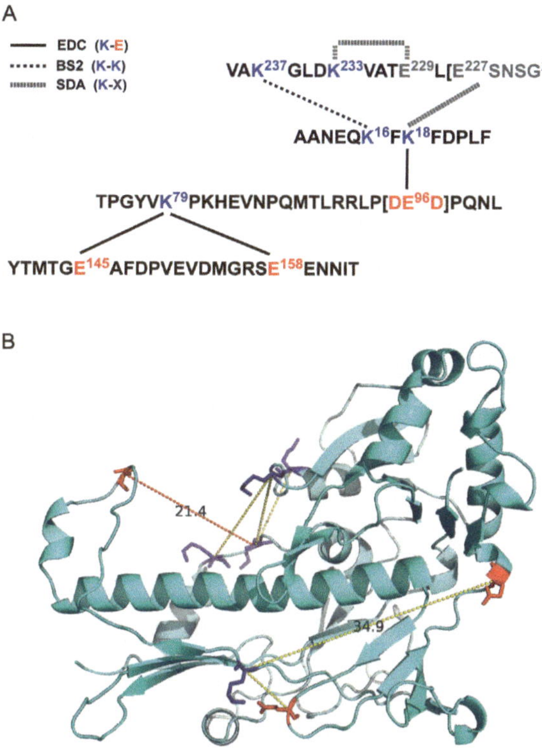

Fig. 4 Cross-links within the assembled bacteriophage λ procapsid shell. (A) Cross-linking experiments were performed on the λ major capsid monomer using three different cross-linkers. (B) Using the XL data, a structural homology model based on the pro-phage CFT073 protein (PDB: 3BQW, 43% identity) could be derived. The linked glutamic acid and lysine residues are marked in red and blue whereas the yellow lines show the measured distances between them. *Reprinted from Singh, P., Nakatani, E., Goodlett, D. R., Catalano, C. E. 2013. A pseudo-atomic model for the capsid shell of bacteriophage lambda using chemical cross-linking/mass spectrometry and molecular modeling. J. Mol. Biol., 425, 3378–3388, Copyright (2013), with permission from Elsevier.*

2.2.4 Characterization of virus-host interplay

In addition to the elucidation of viral structures as such, structural virologists have a considerable interest in discovering how viruses and hosts interact with each other during infection. If identified, critical steps within the viral lifecycle can be potentially exploited as suitable drug targets. Lum and Cristea (2016) have reviewed in detail how proteomics methods, including XL-MS, are useful to characterize virus-host interplay. As exemplified on potato leafroll virus (PLRV), interactions between viral and host proteins during infection were investigated systematically (Chavez et al., 2012; DeBlasio et al., 2016). Using reporter ions, DeBlasio et al. (2016) were eventually able to describe an extensive host-pathogen protein interaction network using XL-MS. Furthermore, the XL-MS data were used to derive a structure for the major capsid protein of PLRV and to propose a model for asymmetric packing within the virion.

In order to minimize artifact-like states and to give a precise picture of the virus-host interaction, an examination directly in a native environment is worth striving for. The following examples illustrate how advanced *in celulo* and *in vivo* XL approaches have become. Sophisticated XL approaches have been used to identify poliovirus proteins that bind to the viral genome during amplification in cell culture in an unbiased screen of host factors. In total, 66 novel interacting proteins were identified (Lenarcic et al., 2013). Furthermore, Phillips et al. (2016) detected novel interactions between the viral genome and host proteins in DENV infected Huh7 cells after UV cross-linking. Using antisense-mediated affinity purification, the complexes were first isolated and afterwards identified by MS. Verification of newly found interactions by RNA immunoprecipitation revealed accurate prediction by XL-MS as only one out of seven interactions missed the validation criteria (Phillips et al., 2016).

3. Protein complex formation and capsid assembly studied by native MS and allied techniques

Native MS employs ESI to interrogate protein structure *in vacuo* (Fenn et al., 1989; Leney and Heck, 2017) and closes a gap by delivering orthogonal data about complex interaction dynamics and heterogeneity. Capable of transferring intact non-covalent protein complexes into the gas phase, native MS has been tackling increasingly big and complex assemblies, ultimately leading to viral capsids and intact viruses (Uetrecht et al., 2008a; Uetrecht and Heck, 2011).

3.1 Native mass spectrometry and allied techniques

The basis of native MS is very soft ESI from volatile buffer surrogates. Nevertheless, other more exotic techniques exist, which have higher tolerance to salts or detergents (Peetz et al., 2019). The main challenge is retaining non-covalent interactions of proteins and their stoichiometry in complexes upon transfer into the gas phase. However, ESI is soft enough that intact viruses have been shown to survive the process and keep their infectivity (Hogan et al., 2006; Siuzdak et al., 1996). The ideal protein ion is free of all solvent and buffer adducts but populated by multiple charges. In ESI, the protein sample is sprayed from an externally charged capillary into a mass spectrometer (Fig. 5A). The physical background regarding desolvation, protein charging and structural integrity of gas phase ions was

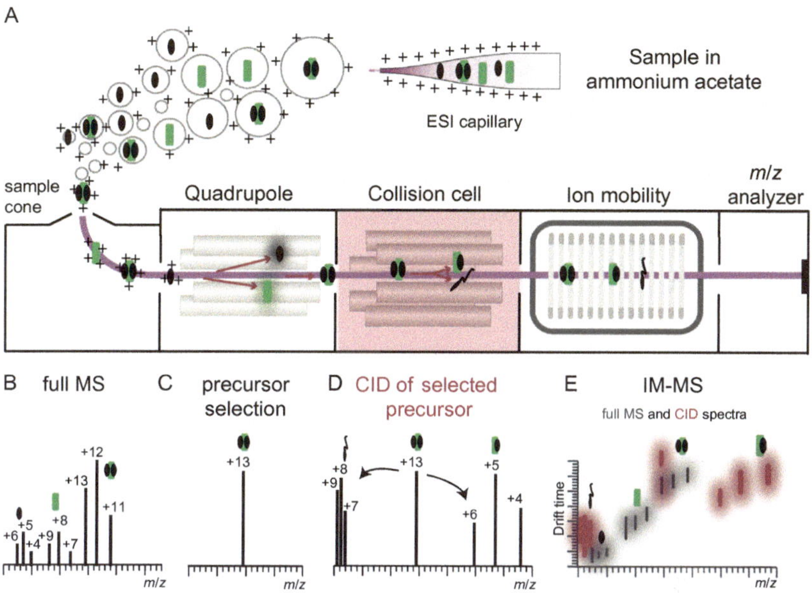

Fig. 5 Exemplary workflow and readout in native MS. (A) A protein complex with two subunits (black and green) is sprayed from ammonium acetate solution, by the application of an electric potential and enters the mass spectrometer as molecular ions. The modules of the mass spectrometer allow for manipulation and analysis *in vacuo* as depicted in schematic spectra. (B) Full ESI-MS spectra reveal a Gaussian distribution of the charge states of each mass of the sample proteins. An additional charge distribution indicates a trimeric complex. (C) In order to confirm subunit stoichiometry of the candidate complex, its molecular ions are filtered for in the quadrupole region of the mass spectrometer. (D) Selected ions are subsequently collided with a neutral gas in the collision cell. The complex dissociates into two subunits of characteristic masses, which confirms 2:1 stoichiometry. The dissociated subunit unfolds and takes a large number of charges with it. It is therefore detected in the lower *m/z* range of the spectrum, whereas the remainder of the complex is detected at higher *m/z*. (E) Additional information about conformation is obtained by ion mobility MS (IM-MS), where drift time allows an approximation of the CCS in the gas phase. Here, IM spectra of full MS (B) and CID (D) are compared, where the smaller subunit has a substantially longer drift time when ejected from the complex, due to its unfolding process during dissociation. IM-MS studies become particularly powerful when a protein complex already has different subunit conformations in solution.

unraveled experimentally and in MD simulations described elsewhere (Konermann et al., 2013; McAllister et al., 2015). The ESI process for native MS presented herein requires buffer exchange into an MS-compatible solution of volatile salt. For that purpose, even though not buffering at neutral pH, ammonium acetate is particularly suitable, due to its adjustable pH (4.6–10) and ionic strength (0 to several M), mimicking the natural environment and providing favorable conditions for the protein until its desolvation (Konermann, 2017). Small amounts of non-volatile salts or small molecules up to a few mM can be added when required, for example bivalent cations or co-factors. Ideally, native MS reveals the charge-state distributions of all mass species, including any intermediates, present in the sample in solution (Fig. 5B). The mass can be deduced from the m/z values of neighboring peaks in a charge-state distribution, where charge states of the same species differ in z by 1. For similar sized analytes, the signal intensity is proportional to the analyte concentration; however, large oligomeric complexes deviate from this direct correlation at higher concentrations, which hampers a reliable quantification (Root et al., 2017). Nevertheless, semi-quantification and hence characterizing binding affinities is possible as has been shown in antibody dimerization (Rose et al., 2011). Determining the mass of a species of interest is in itself not always sufficient to deduce its stoichiometry so that further degrees of separation (e.g., ion mobility mass spectrometry (IM-MS)) or MS/MS fragmentation approaches are necessary.

3.1.1 Dissociation of protein complexes in the gas phase

The charge acquired during the ESI process allows protein manipulation and detection in the gas phase, opening the door to the application of methods that are incompatible with having the protein in solution. For example, to detect the stoichiometry unambiguously, a complex ion can be filtered out or selected in a quadrupole and dissociated subsequently (Fig. 5C and D). For this purpose, collision induced dissociation (CID) is mostly used, in which additional energy is imparted to the analyte ions *via* high-velocity collisions with heavy inert gas atoms. This gradually increases the internal energy of the ion until individual subunits of a non-covalent complex unfold and are eventually ejected from the complex one-by-one (Pagel et al., 2010). Mass information obtained by the detection of the ejected subunits and the remaining subcomplexes unequivocally proves stoichiometry of the original complex. The order of subunit ejection also correlates with peripheral location in the complex and thus provides even more insight into complex architectures. Notably, this dissociation behavior is usually distinct from what is observed in solution, where it is a spontaneous (often reversible) equilibrium reaction rather than an irreversible consequence of physical collisions with gas molecules (Boeri Erba and Petosa, 2015). The described method is especially informative if a specific complex species is singled out according to its *m/z* in the gas phase before CID occurs, so that dissociation products can be traced back to one single complex species.

In principle, any combination of protein fragmentation with (native) MS analysis of intact proteins is called top-down MS. When harsh CID conditions are used, backbone fragmentation of the intact protein occurs on top of subunit dissociation of the protein complex. Additionally, these backbone fragmentations are often induced by other fragmentation techniques such as ultraviolet photodissociation (UVPD) and infrared multiphoton dissociation (IRMPD) (Brodbelt, 2014), impacts with a surface (SID) (Stiving et al., 2019) or electron transfer reactions (ETD) (Lermyte et al., 2018) in the gas phase. Thus, information about the primary structure of a protein, its PTMs as well as its presence in a complex can be obtained. Top down approaches have been used to predict transmembrane regions of Hepatitis C virus (HCV) p7 viroporin sprayed from detergent micelles based on the observed CID fragmentation pattern (Konijnenberg et al., 2015). Furthermore, the existence of PTMs can be derived from fragmentation patterns, e.g., the complex glycosylation profile of influenza HA (Cruz et al., 2018).

3.1.2 Ion mobility mass spectrometry

One approach often coupled with MS to provide a semi-orthogonal separation is ion mobility. IM-MS is able to separate ions according to their "retention" time in a region with elevated pressure of an inert gas (usually nitrogen or helium). The retention or arrival time is proportional to *m/z* and overall shape, which allows distinguishing coexisting protein conformers with the same mass but different structures, as well as protein complex structure and ligand interactions (Eschweiler et al., 2017; Goth and Pagel, 2017). Since the *m/z* is known from MS, the arrival time can be converted into a rotationally averaged collision cross section (CCS) (Uetrecht et al., 2010). This value can be compared with CCS calculated from available structures or from computational models, where it also functions as a constraint (Politis et al., 2014). The CCS allows drawing conclusions from the gas phase structure relating back to the protein structure in solution. To give an example, Uetrecht et al. (2008a) employed native IM-MS for the first time on empty viral capsids revealing that the 3D structure retained in the gas phase is consistent with a hollow sphere.

3.1.3 Gas-phase electrophoretic mobility molecular analysis

Large multiply charged protein complexes can sometimes lead to difficulties in the interpretation of IM-MS derived spectra due to spectral overlaps. This limitation is overcome in gas-phase electrophoretic mobility molecular analysis (GEMMA). Proteins are sprayed from an ammonium acetate solution using ESI like in native MS, but the numerous charges of large macromolecular complexes are reduced to a single charge to simplify spectra. Like in IM-MS, GEMMA measures an electrophoretic mobility diameter (EMD) using an ion mobility drift cell (Kaddis and Loo, 2007). The EMD represents the overall size and shape of the complex, rather than the exact molecular weight. Kaddis et al. (2007) compared the utility of GEMMA for different sized macromolecular complexes up to the size of the 4.6 MDa cowpea chlorotic mottle virus and found that there is a strong correlation between the EMD and the molecular mass of a complex measured by ESI-MS.

3.1.4 Instrument development for higher mass resolution

In recent years, native MS has progressed towards higher resolution mass analyzers, by the introduction of high-mass capable Orbitrap (orbital ion trap) instruments complementing the so far mostly used time-of-flight (ToF) and Fourier-transform ion cyclotron resonance (FT-ICR) devices. However, these platforms are still used in parallel as in general ToFs have a higher dynamic range, speed and sensitivity, while FT-ICRs offer even higher mass resolution than Orbitraps (Fort et al., 2017).

This boost of mass resolving power in a compact instrument had direct influence on MS-based research on viruses. The impact of capsid protein heterogeneity on an adeno-associated virus capsid could for example be determined with an Orbitrap instrument with a precision that ToF was not able to achieve before (Snijder et al., 2014c). However, in a more comprehensive study, the two mass analyzers were compared in order to determine the heterogeneity in genome load of plant viruses cowpea chlorotic mottle virus (CCMV) and brome mosaic virus (BMV) (van de Waterbeemd et al., 2016). Interestingly, the heterogeneity was observed in the ToF instrument, whereas the Orbitrap showed a singular peak distribution. Here, the increased number of species created by different genome sizes resulted in overlapping spectral peaks pushing Orbitrap MS resolving power and exact mass determination to its limits. However, although undesirable, this does not always constitute a critical problem for obtaining useful data. First of all, not every sample of interest requires being fully charge state resolved in order to answer the research questions. This is exemplified in native

MS studies of norovirus assembly as well as in the determination of cargo load in capsids and capsid-like nanocompartments (Brasch et al., 2011; Pogan et al., 2018a; Snijder et al., 2016).

3.1.5 Charge detection mass spectrometry

The major aim of instrument development for conventional native MS is to achieve charge state resolution even for high mass complexes, such as virus capsids. However, larger biomolecules also pick up more charges during the ESI process, which makes it more challenging to achieve peak resolution in the mass spectrum due to very closely packed peaks of different charge states leading to overlaps (Peng et al., 2014). Furthermore, desolvation of such huge molecules tends to be incomplete and, together with inherent sample heterogeneity, adds more complexity to the spectrum. To overcome these challenges and enable the analysis of high mass complexes (approaching 1 GDa), methods that can determine the m/z together with directly measuring the charge of single ions are of advantage (Hogan et al., 2006). Charge detection mass spectrometry (CDMS) was developed already in the 1990s (Fuerstenau and Benner, 1995; Snijder and Heck, 2014). Here, the charge of each ion is directly determined from multiple passages through a flight tube while the m/z ratio is measured from the flight time (Keifer et al., 2017; Snijder and Heck, 2014). This technique was first applied for whole viruses in 2001, where masses of intact rice yellow mottle virus (RYMV) and tobacco mosaic virus (TMV) could be measured (Fuerstenau et al., 2001). Since then, further instrument development and the design of different instrument setups enabled very accurate charge determination even for high-mass samples (Draper et al., 2018; Keifer et al., 2017). In recent years, CDMS has been used to look at the effect of coat protein substitutions on the assembly products of P22 procapsids (Keifer et al., 2016a) and to measure the mass of the infectious P22 particle with accuracy unprecedented at this molecular weight (Keifer et al., 2016b). In addition to ESI in conventional CDMS devices, other laser based ionization techniques have been implemented for intact high-mass particles, up to the size of whole cells. Such setups have been successfully used to measure intact vaccinia viruses (Liang et al., 2018; Peng et al., 2006).

3.1.6 Nanomechanical resonators

Another method to measure the mass of single, huge ions is nano(electro) mechanical MS (NEMS). Here, the analytes are still introduced into the mass spectrometer by ESI, but instead of measuring the m/z ratio of gas-phase ions, changes in the frequency of a resonator are measured when ions adsorb onto it. Already small variations in mass can alter the resonance frequency, so that the mass of single ions can be measured (Gil-Santos et al., 2010; Ruz et al., 2014; Snijder and Heck, 2014). The NEMS detector has in theory an unlimited mass range, so that DNA, protein complexes, bacteria, viruses and even whole cells could in principle be measured (Dominguez-Medina et al., 2018; Malvar et al., 2016). The current limitations are mass resolution, mass accuracy and sensitivity, since only a small portion of the introduced ions hits the detector (Snijder and Heck, 2014). An advancement of this technique has recently been introduced by Dominguez-Medina et al. (2018), where instead of using ESI, analytes were nebulized directly from solution. Combined with some other modifications of the setup, it was possible to measure the mass of empty as well as filled bacteriophage T5 capsids up to 105 MDa with less than one picomole of sample (Dominguez-Medina et al., 2018).

3.2 Investigation of viral complexes with native MS-based methods

As shown before, native MS on its own already provides a variety of different approaches to tackle questions regarding size, mass, stability and shape of viral proteins and capsids. However, combining different MS techniques with each other or with other biophysical methods can be of great value for the analysis of dynamic processes of protein complexes, VLPs and virions. An overview of additional biophysical techniques that can be used to characterize the structure and properties of virus capsids can be found in Kondylis et al. (2019). Starting from viral protein complex interactions, we will move to investigations of assembly and stability of virus capsids and finally show that native MS can even probe the genome release in intact virions.

3.2.1 Viral protein complex formation and stoichiometry

Native MS in combination with IM-MS is capable of studying complex formation and stoichiometry of intrinsically disordered proteins, which are due to their structural heterogeneity and dynamics not easily accessible by X-ray crystallography. Based on charge, peak distributions and CCS, distinct conformational states along the continuum from ordered to disordered structures can be detected by MS (Stuchfield and Barran, 2018). D'Urzo et al. (2015) studied the complex of the intrinsically disordered N_{TAIL}

domain and the phosphoprotein X domain from measles virus, whose interaction is crucial for replication of the virus. A broad charge-state distribution was observed for N_{TAIL} protein alone, consistent with its intrinsically disordered nature. Upon mixing of both proteins, a complex of 1:1 stoichiometry became evident that showed a bimodal charge-state distribution indicating the coexistence of an open and closed conformation (Fig. 6A). CCS measured with IM-MS confirmed this heterogeneity over a broad range of charge states (Fig. 6B).

Furthermore, native MS can help to characterize fragments resulting from either natural proteolytic cleavage or limited proteolysis. In the latter case, Holm et al. (2018) used homology modeling, SAXS and limited proteolysis to investigate the structure of the cap-snatching endonuclease from Crimean-Congo hemorrhagic fever virus. The products of tryptic digestion were analyzed using native MS. This enabled identification of the fragments with an accuracy down to the amino acid level directly from the biologically active complex and thus drawing conclusions about the protein interaction. To give another example, the role of N-terminally cleaved minor capsid protein VI from adenovirus was investigated using an integrative approach of proteomics, native and HDX-MS (Snijder et al., 2014a). The protein could be identified by proteomics in the mature virion and was released during disassembly, which unequivocally proved its location in the capsid.

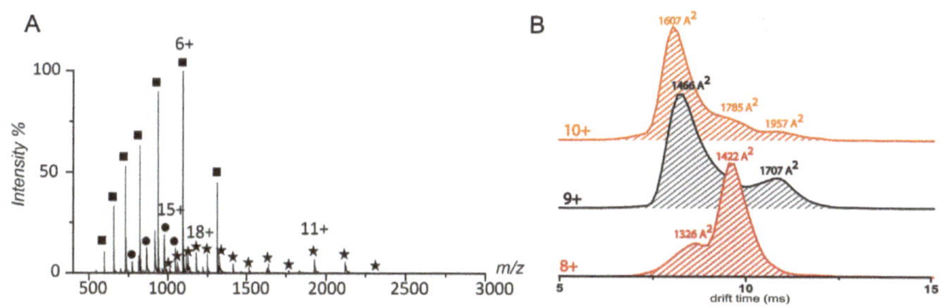

Fig. 6 Native and IM-MS of measles virus N_{TAIL} and phosphoprotein X domain complex. (A) Complex formation (stars) of phosphoprotein X (squares) and N_{TAIL} protein (circles) measured by native MS. (B) The complex displays a significant degree of structural heterogeneity, based on the arrival times and calculated CCS in an IM-MS experiment. *Adapted with permission from Springer Nature: D'Urzo, A., Konijnenberg, A., Rossetti, G., Habchi, J., Li, J., Carloni, P., Sobott, F., Longhi, S., Grandori, R. 2015. Molecular basis for structural heterogeneity of an intrinsically disordered protein bound to a partner by combined ESI-IM-MS and modeling. J. Am. Soc. Mass Spectrom., 26, 472–81, Copyright (2015).*

Further native and HDX-MS based binding studies showed that the protein binds in the cavity of capsid hexon trimers in a concentration and pH dependent manner.

3.2.2 Capsid assembly and stability

Apart from characterization of complexes made of few single proteins, the real power of native MS comes into play, when it comes to characterization of capsids and even fully assembled intact virions. For HBV capsids, $T=3$ and $T=4$ assemblies could be measured with charge state resolution with a truncated capsid protein, missing the RNA binding domain (Uetrecht et al., 2008a). Furthermore, CID confirmed the complex stoichiometry and completeness of the assembly with 180 and 240 subunits, respectively. Additionally, HBV and norovirus VLPs were subjected to IM-MS, to obtain more information about potential assembly intermediates (Uetrecht et al., 2011). Dimers, tetramers, hexamers and also larger assemblies of capsid proteins were found for HBV depending on the conditions. The obtained CCS values were compared with predictions from computational modeling. The CCSs of the hexamer differed depending on the ionic strength with more compact structures under non-assembly (low ionic strength) conditions. The intermediates resembled assembly competent sheet-like structures, rather than globular shapes. The results were later refined by analyzing the assembly core with genetically fused HBV capsid dimers (Holmes et al., 2015). They found that the capsid formed monomers, dimers, trimers and pentamers of dimers as nucleation intermediates. Although the higher stability of fused capsid proteins facilitates detection of transient capsid assembly intermediates, this modification also alters protein structural flexibility leading to more stable capsids and possibly altered assembly pathways compared to the wild type (Holmes et al., 2015).

Using CDMS, Pierson et al. (2014) looked into later stages of HBV capsid assembly. Here, capsids were assembled at high ionic strength and particles with masses corresponding to $T=3$ and $T=4$ particles were detected. Furthermore, high-mass intermediates with masses between those of the $T=3$ and $T=4$ particles were found, which most probably represent incomplete capsid structures also seen in cryo-EM. These results support a model, where after initiation of assembly, the complete capsid is formed by successive addition of dimers to a hole in the incomplete capsid (Pierson et al., 2014). Further studies following HBV capsid assembly in real time also detected kinetically trapped high-mass assembly intermediates mainly depending on ionic strength and concluded that multiple assembly pathways with distinct kinetics could exist (Lutomski et al., 2018). At the end of the assembly reaction, defective and overgrown particles were observed in CDMS spectra, which after some time spontaneously relax back to a perfect icosahedral $T = 4$ geometry. These results indicate that HBV capsid assembly occurs *via* an error-correction mechanism rather than by final incorporation of the last dimeric building block (Lutomski et al., 2017).

Impressively, even intact virions, as exemplified for a picorna-like Triatoma virus, could be investigated by native MS in combination with atomic force microscopy (AFM) (Snijder et al., 2013). Here alkaline-induced uncoating (Fig. 7A) was followed by different complementary techniques, providing insights into assembly states, subunit contacts, capsid intermediates and mechanical properties of the particles. The virion is (at least *in vitro*) unpacked *via* a labile intermediate state followed by disassembly into pentons, whereby genomic RNA and VP4 proteins are released. The remaining pentons reassemble even under alkaline conditions into empty capsids consisting of VP1, VP2 and VP3 only. In neutral conditions, the genome stabilizes capsid pentons, consisting of VP1, VP2, VP3 and VP4, but in alkaline conditions, the genome induces uncoating by relaxing their penton interaction. Furthermore, CID experiments indicated that interpenton connections are mostly stabilized by VP2. Similar approaches were used for norovirus assembly processes, where high-resolution structures are difficult to achieve due to the heterogeneity of assembly intermediates. Assembly and disassembly of norovirus capsids was shown to be reversible and smaller capsids were formed at high ionic strength and alkaline conditions (Shoemaker et al., 2010), which was later shown to also depend on the virus strain (Pogan et al., 2018a, b) (Fig. 7B and C).

A study combining native MS and GEMMA analyzed empty capsids and virions of rhinovirus HRV-A2 (Fig. 8) (Weiss et al., 2015). Using a high mass calibration, the molecular weight of the virion was determined within 1.3% deviation of the theoretical value from the EMD, while native MS was not able to resolve charge states of the virion, preventing exact mass determination. However, GEMMA measurements of the empty capsid substantially deviated from theoretical mass, which is probably due to collapse upon ionization, similar to the EMD reduction observed for CCMV (Kaddis et al., 2007). Consequently, for spherical, filled virions GEMMA can reach accurate mass determination and provides information about the size and shape of the particles.

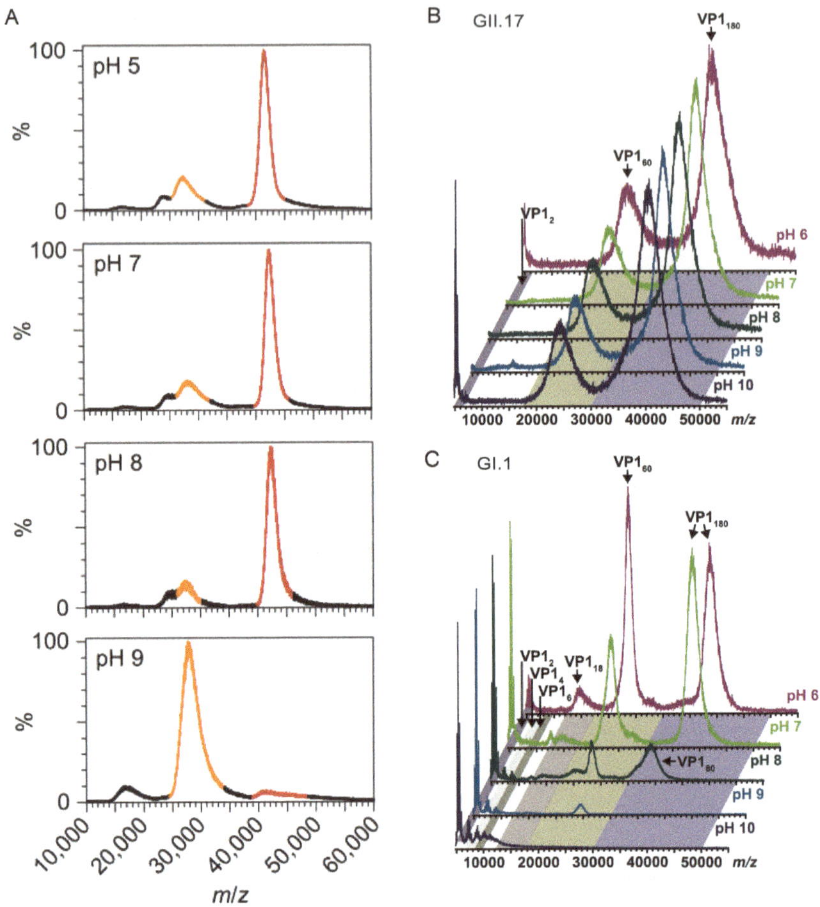

Fig. 7 Comparison of pH dependent capsid stability at alkaline pH by native MS. (A) pH induced genome release and formation of empty capsids (yellow) out of Triatoma virions (red). Genome-packed virions are stable from pH 5 to 8, whereas alkaline conditions trigger genome release as indicated by the signal shift to empty capsids at pH 9. (B and C) Mass spectra of human GII.17 and GI.1 norovirus VLPs, respectively. For the GII.17 strain two stable oligomeric states are observed at all tested pH values. The picture looks different for the GI.1 strain that starts to disassemble into smaller oligomers at higher pH. *Panel (A): Reprinted by permission from Elsevier: Snijder, J., Uetrecht, C., Rose, R., Sanchez-Eugenia, R., Marti, G. A., Agirre, J., Guérin, D., Wuite, G., Heck, A., Roos, W. 2013. Probing the biophysical interplay between a viral genome and its capsid. Nat. Chem., 5, 502, Copyright (2013). Panel (B and C): Adapted from Pogan, R., Schneider, C., Reimer, R., Hansman, G., Uetrecht, C. 2018b. Norovirus-like VP1 particles exhibit isolate dependent stability profiles. J. Phys. Condens. Matter, 30, 064006 under CC BY 3.0 license.*

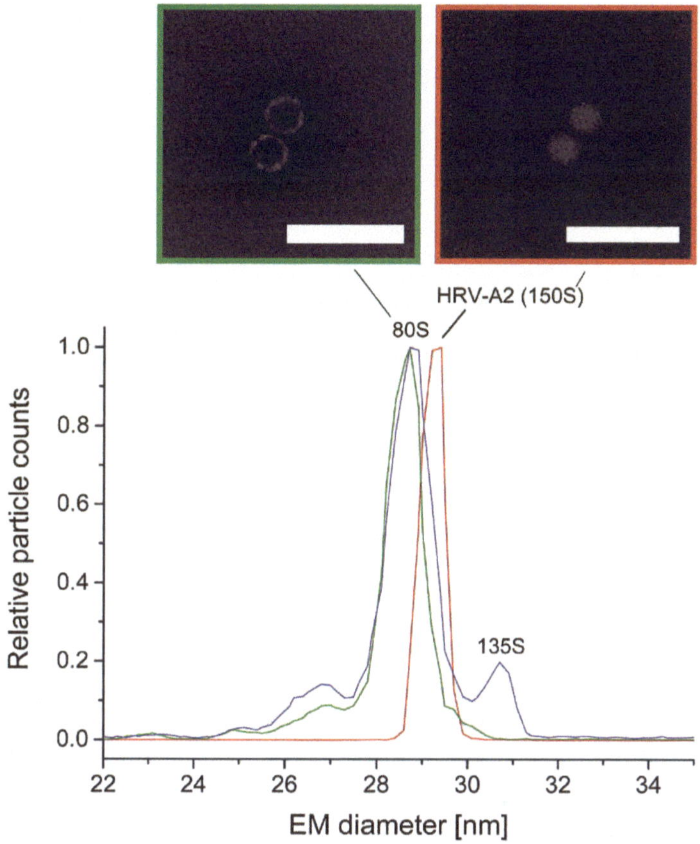

Fig. 8 GEMMA analysis of viral and subviral particles of human rhinovirus serotype 2. The spectrum shows discrimination of intact virions (red) and A- (blue) and B-particles (green) by their EMD. The insets show transmission EM images of corresponding particles (size bar 100 nm). *Reprinted from Weiss, V. U., Bereszcazk, J. Z., Havlik, M., Kallinger, P., Gösler, I., Kumar, M., Blaas, D., Marchetti-Deschmann, M., Heck, A. J., Szymanski, W. W. 2015. Analysis of a common cold virus and its subviral particles by gas-phase electrophoretic mobility molecular analysis and native mass spectrometry. Anal. Chem., 87, 8709–8717 under CC BY 4.0 license.*

4. Integrative structural modeling with MS-derived data

In contrast to MX or EM techniques, structural MS approaches do not observe electron densities within the molecules and thus cannot directly provide high-resolution 3D structures. However, they can uncover a wealth of information about the structure and function of the studied systems, which are not readily accessible by other techniques as exemplified here.

Hence, the most informative approach is usually for the classical structural biology methods and MS-based approaches to work together and complement each other. In the simplest form, a high-resolution molecular structure (if available) is used as a foundation, on top of which MS results are visualized and in the context of which the data are interpreted. In more advanced setups, MS experiments can be used to obtain direct experimental constraints which then, together with pieces of information provided by other techniques, feed into computational modeling (Fig. 9) (Politis and Schmidt, 2018).

This restraint-guided modeling generally involves first a model-generation step, in which any experimental constraints are combined with structural information on the constituents. These can be either high-resolution structures obtained *via* homology modeling, or low resolution

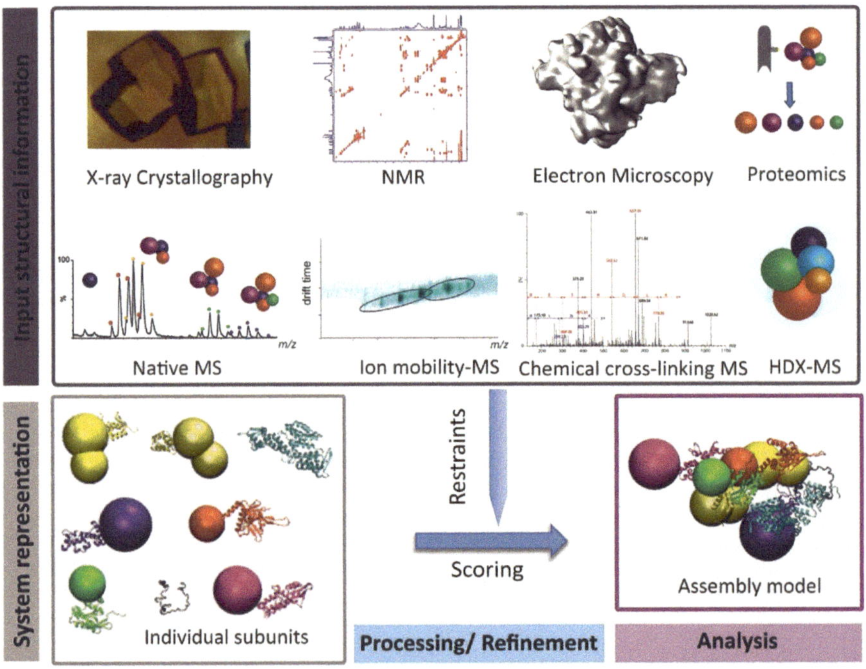

Fig. 9 Integrative structural modeling workflow. Input information from various techniques are collected and encoded into structural restraints. This information is fed into computational modeling generating first models of the complex structure. Using simulation methods, the model, which best fits the experimental constraints according to a scoring function, is evaluated during the processing and refinement step. As a last step, the resulting models are further analyzed to determine the final assembly model. *Reprinted with permission from Science, Politis, A., Borysik, A. J. 2015. Assembling the pieces of macromolecular complexes: hybrid structural biology approaches. Proteomics, 15, 2792–803, Copyright 2015.*

data represented by molecular envelopes or even coarse-grained structure representations. Then, simulation methods (such as MD) are employed to search the conformational space of the assembly, looking for models satisfying the input experimental constraints. Finally, the models are evaluated, scored and optimized in order to judge their credibility and unambiguousness (Politis and Schmidt, 2018). For more thorough discussion of integrative modeling workflows, the interested readers are kindly referred to specialized reviews covering this topic (Politis and Borysik, 2015; Politis and Schmidt, 2018; Schneidman-Duhovny et al., 2014).

The MS-derived experimental constraints utilized for modeling efforts are very diverse in nature owing to the broad range of techniques employed by structural MS. At the whole protein complex level, the information about constituent components, their amino acid sequences and PTMs are available from bottom-up and top-down proteomics. Native MS with IM can on the other hand inform about the general stoichiometry and topology of the intact protein complex (on the level of subunit connectivity), as well as provide a rotationally averaged CCS value reflecting its shape and size. The experimental CCS values can be used for the purpose of comparisons with computationally generated models (Eschweiler et al., 2017). Moreover, a lot of effort has recently focused recently on speeding-up the algorithm implementations for *in silico* CCS calculations on structural models (Marklund et al., 2015; Zanotto et al., 2018). Also, implementations of the CCS calculations have been extended towards utilizing lower resolution molecular envelopes from EM experiments (Degiacomi and Benesch, 2016). Finally, the ability of IM-MS to distinguish several coexisting conformers is recognized as a powerful feature informing MD, while trying to model protein conformational dynamics (Marklund and Benesch, 2019).

In terms of more localized structural information, distances between reactive amino acid side chains provided by chemical XL-MS are probably the most straightforward constraints and have long been used in modeling (Faini et al., 2016). However, even here algorithms are improving, for example by taking into account the flexibility of side chains and cross-linkers. Moreover, distances are measured along protein surfaces instead of the direct path through space when evaluating models (Bullock et al., 2016; Degiacomi et al., 2017). Last but not least, HDX-MS also provides a wealth of structural information about solution accessibility, conformational flexibility and state of hydrogen bonding of protein regions. However, as these data are not point-to-point distances, but rather describe interaction interfaces or larger areas of protein structure, they are somewhat harder to define and implement as direct modeling constraints. Nevertheless, an interface constraint without defined orientation can for example be used in the popular molecular docking platform HADDOCK (Karaca et al., 2010). Additionally, an exciting new approach to select modeled structures solely based on comparing their predicted HDX with experimental data has been pioneered recently by the group of Antoni Borysik (Borysik, 2017; Harris et al., 2019).

Overall, the ultimate goal of computational modeling is to combine different snippets of structural data with constraints from various experimental sources in order to piece together the structure and dynamic motions of the whole assembly. Here, an important challenge lies in properly judging the inherent uncertainties associated with the various sources of data and assigning accordingly their importance during scoring (Politis and Schmidt, 2018; Schneidman-Duhovny et al., 2014). Nevertheless, there are many software packages enabling efficient integrative structural modeling utilizing MS-derived information. To name just a few: HADDOCK (Karaca et al., 2010) with the derived Model Molecular Machines (M3) protocol (Karaca et al., 2017), Integrative Modeling Platform (Russel et al., 2012; Webb et al., 2018) or the MS-based hybrid modeling method (Politis et al., 2014).

Utilizing these advanced tools and MS-derived structural restraints complemented by high-resolution data from other structural methods, it has so far been possible to obtain structural models of several very challenging systems. For example, Politis et al. (2014) combined native MS, ion mobility and cross-linking together with crystal structures of the individual subunits to obtain a model of the until then elusive 19S proteasomal lid. Moreover, MS revealed the assembly pathways of the whole 19S proteasome and defined several key intermediate subcomplexes, ultimately leading to the formulation of a model for the early steps in the 19S proteasome assembly process (Politis et al., 2014). In the same year, a large collaborative effort (Shukla et al., 2014) used negative-staining EM, HDX-MS, cross-linking and molecular docking to study the interaction of β-arrestin with the β2-adrenergic receptor (a member of the pharmaceutically extremely important G-protein coupled receptor family). In this work, MS proposed the binding interface as well as revealed further dynamics in the arrestin molecule, shedding light into the mechanistic details of this very dynamic regulation process.

Finally, a combination of native MS, ion mobility, HDX-MS, cross-linking and cryo-EM was used in two interconnected works by Snijder et al., (2014b, 2017) to derive a structural and functional model of a highly dynamic circadian clock oscillator system based on Kai proteins. MS here initially provided crucial information about the oligomeric state equilibrium of the Kai system, as well as detected allosteric modulation and guided molecular docking (Snijder et al., 2014b). In the latter work, native MS was also used to monitor the assembly process of the Kai complex and the level of its autophosphorylation, which eventually resulted in sufficiently monodisperse samples to obtain cryo-EM structures (Snijder et al., 2017).

5. Future directions

Mass spectrometry has matured into an important technique in the structural biology portfolio and, as exemplified in this review, can also tackle many questions in the realm of virology. Also, it is well placed in the current trend for highly integrative structural approaches. As these studies address ever more complex systems, the integrative efforts will very likely continue and intensify in the future, benefiting from the range of unique information structural MS can provide.

At the same time, MS instrumentation is constantly improving. On one hand, experimental automation and more streamlined data processing help to bring structural MS techniques closer to non-experts. The benefits of this can be nicely showcased by the wide adoption of HDX-MS in vaccine and antibody development within the biopharmaceutical industry (Masson et al., 2017). On the other hand, developments in advanced ion fragmentation methods aim to further increase the amount of structural information obtained from MS analyses of intact protein complexes (Brodie et al., 2018; Mistarz et al., 2018). In this regard, native top-down MS is expected to simultaneously provide information about the identity of protein complex components, their PTMs as well as stoichiometry. Like other MS techniques, native MS in general is also likely to further develop in the direction of systems biology in the future.

Additionally, newly developed technologies will enable completely novel types of experiments. The use of MS to gently deposit proteins in a "soft-landing" process on a surface prior to imaging by AFM or other advanced microscopy techniques holds great promise for single-particle conformation analysis (Longchamp et al., 2017; Mikhailov et al., 2014). In another line of research, X-ray free electron lasers (XFELs), the new extremely bright radiation sources in the structural biology portfolio, are aiming to enable diffractive imaging of single proteins and viral particles without the need for crystallization. The technique is ultimately predicted to be able to provide near-atomic resolution (Neutze et al., 2000; Yoon et al., 2016) and currently work is ongoing to improve the experimentally achieved resolution (Aquila et al., 2015; Kurta et al., 2017; Rose et al., 2018), which is so far limited by desolvation and particle hit rates. In this regard, native MS, with its ability to mass select and manipulate ions in the gas phase, could help alleviate these limitations and enable structural studies of unpurifiable transient species, occurring for example during viral capsid assembly. Therefore, work on a sample delivery system for XFELs based on native MS is under way (Schulz et al., 2013; Uetrecht et al., 2019).

Moreover, in order to provide the most informative results, all structural MS techniques strive to analyze proteins as close to their native environment as possible. In this regard, especially native MS has recently made a significant leap forward. First, novel types of electrospray emitters have shown that the laborious procedures to exchange buffers for MS-compatible volatile solutions prior to analysis may sometimes be avoided. Protein complexes could thus be analyzed directly from biochemically relevant buffers without desalting (Susa et al., 2017) or from crude cell lysates (Gan et al., 2017). Second, largely through the pioneering work by Carol Robinson's group, native MS has also made significant impact in the field of membrane proteins by studying protein complexes and their interactions with lipids in a number of systems. These ranged from micelles, detergents and lipid nanodiscs to the very recent success in the analysis of proteins released directly from their native cellular membranes (Bechara and Robinson, 2015; Chorev et al., 2018). Here, a great opportunity appears for structural virology to extend these techniques to viruses and to learn more about the interactions of viral proteins with host cell membranes and membranes of enveloped viral particles. Finally, as the various labeling structural MS techniques have always been generally well applicable for membrane protein systems, these techniques in turn push for analyses directly within the cellular environment. Both cross-linking and FPOP methods have recently greatly progressed in this area (Chea and Jones, 2018; Espino and Jones, 2019; Piotrowski and Sinz, 2018), with cross-linking offering the possibility to not only study protein complexes and their structure, but also their interactions and interaction networks within the cellular context (Chavez and Bruce, 2019; Garcia-Moreno et al., 2019; Liu et al., 2017). In conclusion, MS-based approaches can ultimately bridge the gap between structural and systems biology in the near future. Moreover, as shown here, already now structural MS has a huge impact on structural virology.

Acknowledgments

The Heinrich-Pette-Institute, Leibniz Institute for Experimental Virology is supported by the Freie und Hansestadt Hamburg and the Bundesministerium für Gesundheit (BMG). C.U. and J.D. acknowledge funding from DFG FOR2327 Virocarb. A.K. gratefully acknowledges a postdoctoral fellowship from the Alexander von Humboldt Foundation. B.K. and C.U. further thank the European Commission for funding through an ERC StG-2017 759661 SPOCk'S MS.

References

Alexander, M.M., Mohr, J.P., Deblasio, S.L., Chavez, J.D., Ziegler-Graff, V., Brault, V., Bruce, J.E., Heck, M.C., 2017. Insights in luteovirid structural biology guided by chemical cross-linking and high resolution mass spectrometry. Virus Res. 241, 42–52.

Aquila, A., Barty, A., Bostedt, C., Boutet, S., Carini, G., Deponte, D., Drell, P., Doniach, S., Downing, K.H., Earnest, T., Elmlund, H., Elser, V., Guhr, M., Hajdu, J., Hastings, J., Hau-Riege, S.P., Huang, Z., Lattman, E.E., Maia, F.R., Marchesini, S., Ourmazd, A., Pellegrini, C., Santra, R., Schlichting, I., Schroer, C., Spence, J.C., Vartanyants, I.A., Wakatsuki, S., Weis, W.I., Williams, G.J., 2015. The linac coherent light source single particle imaging road map. Struct. Dyn. 2, 041701.

Arlt, C., Flegler, V., Ihling, C.H., Schafer, M., Thondorf, I., Sinz, A., 2017. An integrated mass spectrometry based approach to probe the structure of the full-length wild-type tetrameric p53 tumor suppressor. Angew. Chem. Int. Ed. Eng. 56, 275–279.

Bai, Y., Milne, J.S., Mayne, L., Englander, S.W., 1993. Primary structure effects on peptide group hydrogen exchange. Proteins 17, 75–86.

Bechara, C., Robinson, C.V., 2015. Different modes of lipid binding to membrane proteins probed by mass spectrometry. J. Am. Chem. Soc. 137, 5240–5247.

Benesch, J.L., Ruotolo, B.T., 2011. Mass spectrometry: come of age for structural and dynamical biology. Curr. Opin. Struct. Biol. 21, 641–649.

Bereszczak, J.Z., Rose, R.J., Van Duijn, E., Watts, N.R., Wingfield, P.T., Steven, A.C., Heck, A.J., 2013. Epitope-distal effects accompany the binding of two distinct antibodies to hepatitis B virus capsids. J. Am. Chem. Soc. 135, 6504–6512.

Bereszczak, J.Z., Watts, N.R., Wingfield, P.T., Steven, A.C., Heck, A.J., 2014. Assessment of differences in the conformational flexibility of hepatitis B virus core-antigen and e-antigen by hydrogen deuterium exchange-mass spectrometry. Protein Sci. 23, 884–896.

Boeri Erba, E., Petosa, C., 2015. The emerging role of native mass spectrometry in characterizing the structure and dynamics of macromolecular complexes. Protein Sci. 24, 1176–1192.

Bojja, R.S., Andrake, M.D., Weigand, S., Merkel, G., Yarychkivska, O., Henderson, A., Kummerling, M., Skalka, A.M., 2011. Architecture of a full-length retroviral integrase monomer and dimer, revealed by small angle X-ray scattering and chemical cross-linking. J. Biol. Chem. 286, 17047–17059.

Borysik, A.J., 2017. Simulated isotope exchange patterns enable protein structure determination. Angew. Chem. Int. Ed. Eng. 56, 9396–9399.

Bossart, K.N., Fusco, D.L., Broder, C.C., 2013. Paramyxovirus entry. Adv. Exp. Med. Biol. 790, 95–127.

Brasch, M., De La Escosura, A., Ma, Y., Uetrecht, C., Heck, A.J.R., Torres, T., Cornelissen, J.J.L.M., 2011. Encapsulation of phthalocyanine supramolecular stacks into virus-like particles. J. Am. Chem. Soc. 133, 6878–6881.

Brodbelt, J.S., 2014. Photodissociation mass spectrometry: new tools for characterization of biological molecules. Chem. Soc. Rev. 43, 2757–2783.

Brodie, N.I., Huguet, R., Zhang, T., Viner, R., Zabrouskov, V., Pan, J.X., Petrotchenko, E.V., Borchers, C.H., 2018. Top-down hydrogen-deuterium exchange analysis of protein structures using ultraviolet photodissociation. Anal. Chem. 90, 3079–3082.

Brown, K.A., Wilson, D.J., 2017. Bottom-up hydrogen deuterium exchange mass spectrometry: data analysis and interpretation. Analyst 142, 2874–2886.

Bullock, J.M.A., Schwab, J., Thalassinos, K., Topf, M., 2016. The importance of non-accessible crosslinks and solvent accessible surface distance in modeling proteins with restraints from crosslinking mass spectrometry. Mol. Cell. Proteomics 15, 2491–2500.

Bullock, J.M.A., Sen, N., Thalassinos, K., Topf, M., 2018. Modeling protein complexes Using restraints from crosslinking mass spectrometry. Structure 26, 1015–1024 e2.

Chalmers, M.J., Busby, S.A., Pascal, B.D., West, G.M., Griffin, P.R., 2011. Differential hydrogen/deuterium exchange mass spectrometry analysis of protein-ligand interactions. Expert Rev. Proteomics 8, 43–59.

Chavez, J.D., Bruce, J.E., 2019. Chemical cross-linking with mass spectrometry: a tool for systems structural biology. Curr. Opin. Chem. Biol. 48, 8–18.

Chavez, J.D., Cilia, M., Weisbrod, C.R., Ju, H.J., Eng, J.K., Gray, S.M., Bruce, J.E., 2012. Cross-linking measurements of the potato leafroll virus reveal protein interaction topologies required for virion stability, aphid transmission, and virus-plant interactions. J. Proteome Res. 11, 2968–2981.

Chavez, J.D., Lee, C.F., Caudal, A., Keller, A., Tian, R., Bruce, J.E., 2018. Chemical crosslinking mass spectrometry analysis of protein conformations and supercomplexes in heart tissue. Cell Syst. 6, 136–141 e5.

Chea, E.E., Jones, L.M., 2018. Analyzing the structure of macromolecules in their native cellular environment using hydroxyl radical footprinting. Analyst 143, 798–807.

Chen, F., Gulbakan, B., Weidmann, S., Fagerer, S.R., Ibanez, A.J., Zenobi, R., 2016. Applying mass spectrometry to study non-covalent biomolecule complexes. Mass Spectrom. Rev. 35, 48–70.

Chodosh, L.A., 2001. UV crosslinking of proteins to nucleic acids. Curr. Protoc. Mol. Biol. 36 (1), 12.5.1–12.5.8.

Chorev, D.S., Baker, L.A., Wu, D., Beilsten-Edmands, V., Rouse, S.L., Zeev-Ben-Mordehai, T., Jiko, C., Samsudin, F., Gerle, C., Khalid, S., Stewart, A.G., Matthews, S.J., Grunewald, K., Robinson, C.V., 2018. Protein assemblies ejected directly from native membranes yield complexes for mass spectrometry. Science 362, 829–834.

Cravello, L., Lascoux, D., Forest, E., 2003. Use of different proteases working in acidic conditions to improve sequence coverage and resolution in hydrogen/deuterium exchange of large proteins. Rapid Commun. Mass Spectrom. 17, 2387–2393.

Cruz, E., Cain, J., Crossett, B., Kayser, V., 2018. Site-specific glycosylation profile of influenza A (H1N1) hemagglutinin through tandem mass spectrometry. Hum. Vaccin. Immunother. 14, 508–517.

D'Urzo, A., Konijnenberg, A., Rossetti, G., Habchi, J., Li, J., Carloni, P., Sobott, F., Longhi, S., Grandori, R., 2015. Molecular basis for structural heterogeneity of an intrinsically disordered protein bound to a partner by combined ESI-IM-MS and modeling. J. Am. Soc. Mass Spectrom. 26, 472–481.

Davenport, T.M., Guttman, M., Guo, W., Cleveland, B., Kahn, M., Hu, S.L., Lee, K.K., 2013. Isolate-specific differences in the conformational dynamics and antigenicity of HIV-1 gp120. J. Virol. 87, 10855–10873.

Deblasio, S.L., Chavez, J.D., Alexander, M.M., Ramsey, J., Eng, J.K., Mahoney, J., Gray, S.M., Bruce, J.E., Cilia, M., 2016. Visualization of host-polerovirus interaction topologies using protein interaction reporter technology. J. Virol. 90, 1973–1987.

Degiacomi, M.T., Benesch, J.L., 2016. EM intersectionIM: software for relating ion mobility mass spectrometry and electron microscopy data. Analyst 141, 70–75.

Degiacomi, M.T., Schmidt, C., Baldwin, A.J., Benesch, J.L.P., 2017. Accommodating protein dynamics in the modeling of chemical crosslinks. Structure 25, 1751–1757 e5.

Dominguez-Medina, S., Fostner, S., Defoort, M., Sansa, M., Stark, A.K., Halim, M.A., Vernhes, E., Gely, M., Jourdan, G., Alava, T., Boulanger, P., Masselon, C., Hentz, S., 2018. Neutral mass spectrometry of virus capsids above 100 megadaltons with nanomechanical resonators. Science 362, 918–922.

Draper, B.E., Anthony, S.N., Jarrold, M.F., 2018. The FUNPET-a new hybrid ion funnel-ion carpet atmospheric pressure interface for the simultaneous transmission of a broad mass range. J. Am. Soc. Mass Spectrom. 29, 2160–2172.

Engen, J.R., 2009. Analysis of protein conformation and dynamics by hydrogen/deuterium exchange MS. Anal. Chem. 81, 7870–7875.

Engen, J.R., Wales, T.E., 2015. Analytical aspects of hydrogen exchange mass spectrometry. Annu Rev Anal Chem (Palo Alto, Calif) 8, 127–148.

Englander, S.W., 2006. Hydrogen exchange and mass spectrometry: a historical perspective. J. Am. Soc. Mass Spectrom. 17, 1481–1489.

Eschweiler, J.D., Frank, A.T., Ruotolo, B.T., 2017. Coming to grips with ambiguity: ion mobility-mass spectrometry for protein quaternary structure assignment. J. Am. Soc. Mass Spectrom. 28 (10), 1991–2000.

Espino, J.A., Mali, V.S., Jones, L.M., 2015. In cell footprinting coupled with mass spectrometry for the structural analysis of proteins in live cells. Anal. Chem. 87, 7971–7978.

Espino, J.A., Jones, L.M., 2019. Illuminating biological interactions with in vivo protein footprinting. Anal. Chem. 91, 6577–6584.

Faini, M., Stengel, F., Aebersold, R., 2016. The evolving contribution of mass spectrometry to integrative structural biology. J. Am. Soc. Mass Spectrom. 27, 966–974.

Fenn, J.B., Mann, M., Meng, C.K., Wong, S.F., Whitehouse, C.M., 1989. Electrospray ionization for mass-spectrometry of large biomolecules. Science 246, 64–71.

Fibriansah, G., Ng, T.S., Kostyuchenko, V.A., Lee, J., Lee, S., Wang, J.Q., Lok, S.M., 2013. Structural changes in dengue virus when exposed to a temperature of 37 degrees C. J. Virol. 87, 7585–7592.

Fort, K.L., Van De Waterbeemd, M., Boll, D., Reinhardt-Szyba, M., Belov, M.E., Sasaki, E., Zschoche, R., Hilvert, D., Makarov, A.A., Heck, A.J.R., 2017. Expanding the structural analysis capabilities on an Orbitrap-based mass spectrometer for large macromolecular complexes. Analyst 143, 100–105.

Fowler, M.L., McPhail, J.A., Jenkins, M.L., Masson, G.R., Rutaganira, F.U., Shokat, K.M., Williams, R.L., Burke, J.E., 2016. Using hydrogen deuterium exchange mass spectrometry to engineer optimized constructs for crystallization of protein complexes: case study of PI4KIIIbeta with Rab11. Protein Sci. 25, 826–839.

Fuerstenau, S.D., Benner, W.H., 1995. Molecular weight determination of megadalton DNA electrospray ions using charge detection time-of-flight mass spectrometry. Rapid Commun. Mass Spectrom. 9, 1528–1538.

Fuerstenau, S.D., Benner, W.H., Thomas, J.J., Brugidou, C., Bothner, B., Siuzdak, G., 2001. Mass spectrometry of an intact virus. Angew. Chem. Int. Ed. Eng. 40, 541–544.

Gan, J., Ben-Nissan, G., Arkind, G., Tarnavsky, M., Trudeau, D., Noda Garcia, L., Tawfik, D.S., Sharon, M., 2017. Native mass spectrometry of recombinant proteins from crude cell lysates. Anal. Chem. 89, 4398–4404.

Garcia-Moreno, M., Noerenberg, M., Ni, S., Jarvelin, A.I., Gonzalez-Almela, E., Lenz, C.E., Bach-Pages, M., Cox, V., Avolio, R., Davis, T., Hester, S., Sohier, T.J.M., Li, B., Heikel, G., Michlewski, G., Sanz, M.A., Carrasco, L., Ricci, E.P., Pelechano, V., Davis, I., Fischer, B., Mohammed, S., Castello, A., 2019. System-wide profiling of RNA-binding proteins uncovers key regulators of virus infection. Mol. Cell 74 (1), 196–211.e11.

Garcia, N.K., Guttman, M., Ebner, J.L., Lee, K.K., 2015. Dynamic changes during acid-induced activation of influenza hemagglutinin. Structure 23, 665–676.
Garcia, N.K., Lee, K.K., 2016. Dynamic viral glycoprotein machines: approaches for probing transient states that Drive membrane fusion. Viruses 8 (1), 15.
Gasilova, N., Nazabal, A., 2012. Monitoring ligand modulation of protein-protein interactions by chemical cross-linking and high-mass MALDI mass spectrometry. Methods Mol. Biol. 803, 219–229.
Gil-Santos, E., Ramos, D., Martinez, J., Fernandez-Regulez, M., Garcia, R., San Paulo, A., Calleja, M., Tamayo, J., 2010. Nanomechanical mass sensing and stiffness spectrometry based on two-dimensional vibrations of resonant nanowires. Nat. Nanotechnol. 5, 641–645.
Gingras, A.C., Gstaiger, M., Raught, B., Aebersold, R., 2007. Analysis of protein complexes using mass spectrometry. Nat. Rev. Mol. Cell Biol. 8, 645–654.
Goth, M., Pagel, K., 2017. Ion mobility-mass spectrometry as a tool to investigate protein-ligand interactions. Anal. Bioanal. Chem. 409, 4305–4310.
Guttman, M., Lee, K.K., 2016. Isotope labeling of biomolecules: structural analysis of viruses by HDX-MS. Methods Enzymol. 566, 405–426.
Guttman, M., Scian, M., Lee, K.K., 2011. Tracking hydrogen/deuterium exchange at glycan sites in glycoproteins by mass spectrometry. Anal. Chem. 83, 7492–7499.
Haberkant, P., Raijmakers, R., Wildwater, M., Sachsenheimer, T., Brügger, B., Maeda, K., Houweling, M., Gavin, A.C., Schultz, C., van Meer, G., Heck, A.J., Holthuis, J.C., 2013. In vivo profiling and visualization of cellular protein–lipid interactions using bifunctional fatty acids. Angew. Chem. Int. Ed. Engl. 52 (14), 4033–4038. https://doi.org/10.1002/anie.201210178.
Hambly, D.M., Gross, M.L., 2005. Laser flash photolysis of hydrogen peroxide to oxidize protein solvent-accessible residues on the microsecond timescale. J. Am. Soc. Mass Spectrom. 16, 2057–2063.
Harris, M.J., Raghavan, D., Borysik, A.J., 2019. Quantitative evaluation of native protein folds and assemblies by hydrogen deuterium exchange mass spectrometry (HDX-MS). J. Am. Soc. Mass Spectrom. 30, 58–66.
Haupl, B., Ihling, C.H., Sinz, A., 2017. Combining affinity enrichment, cross-linking with photo-amino acids, and mass spectrometry for probing protein kinase D2 interactions. Proteomics 17 (10), e1600459.
Hoffman, E.A., Frey, B.L., Smith, L.M., Auble, D.T., 2015. Formaldehyde crosslinking: a tool for the study of chromatin complexes. J. Biol. Chem. 290, 26404–26411.
Hogan Jr., C.J., Kettleson, E.M., Ramaswami, B., Chen, D.R., Biswas, P., 2006. Charge reduced electrospray size spectrometry of mega- and gigadalton complexes: whole viruses and virus fragments. Anal. Chem. 78, 844–852.
Holding, A.N., 2015. XL-MS: protein cross-linking coupled with mass spectrometry. Methods 89, 54–63.
Holm, T., Kopicki, J.D., Busch, C., Olschewski, S., Rosenthal, M., Uetrecht, C., Gunther, S., Reindl, S., 2018. Biochemical and structural studies reveal differences and commonalities among cap-snatching endonucleases from segmented negative-strand RNA viruses. J. Biol. Chem.
Holmes, K., Shepherd, D.A., Ashcroft, A.E., Whelan, M., Rowlands, D.J., Stonehouse, N.J., 2015. Assembly pathway of hepatitis B Core virus-like particles from genetically fused dimers. J. Biol. Chem. 290, 16238–16245.
Houde, D., Arndt, J., Domeier, W., Berkowitz, S., Engen, J.R., 2009. Characterization of IgG1 conformation and conformational dynamics by hydrogen/deuterium exchange mass spectrometry. Anal. Chem. 81, 2644–2651.
Hung, C.J., Hu, C.C., Lin, N.S., Lee, Y.C., Meng, M.S., Tsai, C.H., Hsu, Y.H., 2014. Two key arginine residues in the coat protein of Bamboo mosaic virus differentially affect the accumulation of viral genomic and subgenomic RNAs. Mol. Plant Pathol. 15, 196–210.

Jensen, P.F., Comamala, G., Trelle, M.B., Madsen, J.B., Jorgensen, T.J., Rand, K.D., 2016. Removal of N-linked glycosylations at acidic pH by PNGase A facilitates hydrogen/deuterium exchange mass spectrometry analysis of N-linked glycoproteins. Anal. Chem. 88, 12479–12488.
Kaddis, C.S., Lomeli, S.H., Yin, S., Berhane, B., Apostol, M.I., Kickhoefer, V.A., Rome, L.H., Loo, J.A., 2007. Sizing large proteins and protein complexes by electrospray ionization mass spectrometry and ion mobility. J. Am. Soc. Mass Spectrom. 18, 1206–1216.
Kaddis, C.S., Loo, J.A., 2007. Native protein MS and ion mobility large flying proteins with ESI. Anal. Chem. 79, 1778–1784.
Kadek, A., Mrazek, H., Halada, P., Rey, M., Schriemer, D.C., Man, P., 2014. Aspartic protease nepenthesin-1 as a tool for digestion in hydrogen/deuterium exchange mass spectrometry. Anal. Chem. 86, 4287–4294.
Kang, S., Hawkridge, A.M., Johnson, K.L., Muddiman, D.C., Prevelige Jr., P.E., 2006. Identification of subunit-subunit interactions in bacteriophage P22 procapsids by chemical cross-linking and mass spectrometry. J. Proteome Res. 5, 370–377.
Karaca, E., Melquiond, A.S., De Vries, S.J., Kastritis, P.L., Bonvin, A.M., 2010. Building macromolecular assemblies by information-driven docking: introducing the HADDOCK multibody docking server. Mol. Cell. Proteomics 9, 1784–1794.
Karaca, E., Rodrigues, J., Graziadei, A., Bonvin, A., Carlomagno, T., 2017. M3: an integrative framework for structure determination of molecular machines. Nat. Methods 14, 897–902.
Karas, M., Bachmann, D., Bahr, U., Hillenkamp, F., 1987. Matrix-assisted ultraviolet laser desorption of non-volatile compounds. Int. J. Mass Spectrom. Ion Process. 78, 53–68.
Katta, V., Chait, B.T., 1991. Conformational-changes in proteins probed by hydrogen-exchange electrospray-ionization mass-spectrometry. Rapid Commun. Mass Spectrom. 5, 214–217.
Kaur, U., Johnson, D.T., Chea, E.E., Deredge, D., Espino, J.A., Jones, L.M., 2019. Evolution of structural biology through the lens of mass spectrometry. Anal. Chem. 91 (1), 142–155.
Keifer, D.Z., Motwani, T., Teschke, C.M., Jarrold, M.F., 2016a. Acquiring structural information on virus particles with charge detection mass spectrometry. J. Am. Soc. Mass Spectrom. 27, 1028–1036.
Keifer, D.Z., Motwani, T., Teschke, C.M., Jarrold, M.F., 2016b. Measurement of the accurate mass of a 50 MDa infectious virus. Rapid Commun. Mass Spectrom. 30, 1957–1962.
Keifer, D.Z., Pierson, E.E., Jarrold, M.F., 2017. Charge detection mass spectrometry: weighing heavier things. Analyst 142, 1654–1671.
Klykov, O., Steigenberger, B., Pektas, S., Fasci, D., Heck, A.J.R., Scheltema, R.A., 2018. Efficient and robust proteome-wide approaches for cross-linking mass spectrometry. Nat. Protoc. 13, 2964–2990.
Knapman, T.W., Morton, V.L., Stonehouse, N.J., Stockley, P.G., Ashcroft, A.E., 2010. Determining the topology of virus assembly intermediates using ion mobility spectrometry-mass spectrometry. Rapid Commun. Mass Spectrom. 24, 3033–3042.
Kondylis, P., Schlicksup, C.J., Zlotnick, A., Jacobson, S.C., 2019. Analytical techniques to characterize the structure, properties, and assembly of virus capsids. Anal. Chem. 91, 622–636.
Konermann, L., 2017. Addressing a common misconception: ammonium acetate as neutral pH "buffer" for native electrospray mass spectrometry. J. Am. Soc. Mass Spectrom. 28, 1827–1835.
Konermann, L., Ahadi, E., Rodriguez, A.D., Vahidi, S., 2013. Unraveling the mechanism of electrospray ionization. Anal. Chem. 85, 2–9.
Konermann, L., Pan, J., Liu, Y.H., 2011. Hydrogen exchange mass spectrometry for studying protein structure and dynamics. Chem. Soc. Rev. 40, 1224–1234.

Konermann, L., Rodriguez, A.D., Sowole, M.A., 2014a. Type 1 and Type 2 scenarios in hydrogen exchange mass spectrometry studies on protein-ligand complexes. Analyst 139, 6078–6087.

Konermann, L., Vahidi, S., Sowole, M.A., 2014b. Mass spectrometry methods for studying structure and dynamics of biological macromolecules. Anal. Chem. 86, 213–232.

Konijnenberg, A., Bannwarth, L., Yilmaz, D., Kocer, A., Venien-Bryan, C., Sobott, F., 2015. Top-down mass spectrometry of intact membrane protein complexes reveals oligomeric state and sequence information in a single experiment. Protein Sci. 24, 1292–1300.

Kool, J., Jonker, N., Irth, H., Niessen, W.M., 2011. Studying protein-protein affinity and immobilized ligand-protein affinity interactions using MS-based methods. Anal. Bioanal. Chem. 401, 1109–1125.

Kurta, R.P., Donatelli, J.J., Yoon, C.H., Berntsen, P., Bielecki, J., Daurer, B.J., Demirci, H., Fromme, P., Hantke, M.F., Maia, F., Munke, A., Nettelblad, C., Pande, K., Reddy, H.K.N., Sellberg, J.A., Sierra, R.G., Svenda, M., Van Der Schot, G., Vartanyants, I.A., Williams, G.J., Xavier, P.L., Aquila, A., Zwart, P.H., Mancuso, A.P., 2017. Correlations in scattered X-ray laser pulses reveal nanoscale structural features of viruses. Phys. Rev. Lett. 119, 158102.

Kwong, P.D., Doyle, M.L., Casper, D.J., Cicala, C., Leavitt, S.A., Majeed, S., Steenbeke, T.D., Venturi, M., Chaiken, I., Fung, M., Katinger, H., Parren, P.W.L.H., Robinson, J., Van Ryk, D., Wang, L.P., Burton, D.R., Freire, E., Wyatt, R., Sodroski, J., Hendrickson, W.A., Arthos, J., 2002. HIV-1 evades antibody-mediated neutralization through conformational masking of receptor-binding sites. Nature 420, 678–682.

Leitner, A., Joachimiak, L.A., Unverdorben, P., Walzthoeni, T., Frydman, J., Forster, F., Aebersold, R., 2014. Chemical cross-linking/mass spectrometry targeting acidic residues in proteins and protein complexes. Proc. Natl. Acad. Sci. U. S. A. 111, 9455–9460.

Lenarcic, E.M., Landry, D.M., Greco, T.M., Cristea, I.M., Thompson, S.R., 2013. Thiouracil cross-linking mass spectrometry: a cell-based method to identify host factors involved in viral amplification. J. Virol. 87, 8697–8712.

Leney, A.C., Heck, A.J.R., 2017. Native mass spectrometry: what is in the name? J. Am. Soc. Mass Spectrom. 28, 5–13.

Lenz, C., Kuhn-Holsken, E., Urlaub, H., 2007. Detection of protein-RNA crosslinks by NanoLC-ESI-MS/MS using precursor ion scanning and multiple reaction monitoring (MRM) experiments. J. Am. Soc. Mass Spectrom. 18, 869–881.

Lermyte, F., Valkenborg, D., Loo, J.A., Sobott, F., 2018. Radical solutions: principles and application of electron-based dissociation in mass spectrometry-based analysis of protein structure. Mass Spectrom. Rev. 37, 750–771.

Leurs, U., Mistarz, U.H., Rand, K.D., 2015. Getting to the core of protein pharmaceuticals–Comprehensive structure analysis by mass spectrometry. Eur. J. Pharm. Biopharm. 93, 95–109.

Liang, S.Y., Patil, A.A., Han, C.H., Chou, S.W., Chang, W., Soo, P.C., Chang, H.C., Peng, W.P., 2018. Ionization of submicrometer-sized particles by laser-induced radiofrequency plasma for mass spectrometric analysis. Anal. Chem. 90, 13236–13242.

Liang, Y., Guttman, M., Davenport, T.M., Hu, S.L., Lee, K.K., 2016. Probing the impact of local structural dynamics of conformational epitopes on antibody recognition. Biochemistry 55, 2197–2213.

Liko, I., Allison, T.M., Hopper, J.T., Robinson, C.V., 2016. Mass spectrometry guided structural biology. Curr. Opin. Struct. Biol. 40, 136–144.

Lim, X.X., Chandramohan, A., Lim, X.Y., Bag, N., Sharma, K.K., Wirawan, M., Wohland, T., Lok, S.M., Anand, G.S., 2017. Conformational changes in intact dengue virus reveal serotype-specific expansion. Nat. Commun. 8, 14339.

Liu, F., Lossl, P., Scheltema, R., Viner, R., Heck, A.J.R., 2017. Optimized fragmentation schemes and data analysis strategies for proteome-wide cross-link identification. Nat. Commun. 8, 15473.
Longchamp, J.N., Rauschenbach, S., Abb, S., Escher, C., Latychevskaia, T., Kern, K., Fink, H.W., 2017. Imaging proteins at the single-molecule level. Proc. Natl. Acad. Sci. U. S. A. 114, 1474–1479.
Loo, J.A., 1995. Observation of large subunit protein complexes by electrospray-ionization mass-spectrometry. J. Mass Spectrom. 30, 180–183.
Lossl, P., Van De Waterbeemd, M., Heck, A.J.R., 2016. The diverse and expanding role of mass spectrometry in structural and molecular biology. EMBO J. 35, 2634–2657.
Lum, K.K., Cristea, I.M., 2016. Proteomic approaches to uncovering virus-host protein interactions during the progression of viral infection. Expert Rev. Proteomics 13, 325–340.
Luo, Y., Muesing, M.A., 2014. Mass spectrometry-based proteomic approaches for discovery of HIV-host interactions. Futur. Virol. 9, 979–992.
Lutomski, C.A., Lyktey, N.A., Pierson, E.E., Zhao, Z., Zlotnick, A., Jarrold, M.F., 2018. Multiple pathways in capsid assembly. J. Am. Chem. Soc. 140, 5784–5790.
Lutomski, C.A., Lyktey, N.A., Zhao, Z., Pierson, E.E., Zlotnick, A., Jarrold, M.F., 2017. Hepatitis B virus capsid completion occurs through error correction. J. Am. Chem. Soc. 139, 16932–16938.
Maleknia, S.D., Ralston, C.Y., Brenowitz, M.D., Downard, K.M., Chance, M.R., 2001. Determination of macromolecular folding and structure by synchrotron X-ray radiolysis techniques. Anal. Biochem. 289, 103–115.
Mallagaray, A., Creutznacher, R., Dülfer, J., Mayer, P.H.O., Grimm, L.L., Orduna, J.M., Trabjerg, E., Stehle, T., Rand, K.D., Blaum, B.S., Uetrecht, C., Peters, T., 2019. A post-translational modification of human Norovirus capsid protein attenuates glycan binding. Nat. Commun. 10, 1320.
Malvar, O., Ruz, J.J., Kosaka, P.M., Dominguez, C.M., Gil-Santos, E., Calleja, M., Tamayo, J., 2016. Mass and stiffness spectrometry of nanoparticles and whole intact bacteria by multimode nanomechanical resonators. Nat. Commun. 7, 13452.
Marcsisin, S.R., Engen, J.R., 2010. Hydrogen exchange mass spectrometry: what is it and what can it tell us? Anal. Bioanal. Chem. 397, 967–972.
Marklund, E.G., Benesch, J.L., 2019. Weighing-up protein dynamics: the combination of native mass spectrometry and molecular dynamics simulations. Curr. Opin. Struct. Biol. 54, 50–58.
Marklund, E.G., Degiacomi, M.T., Robinson, C.V., Baldwin, A.J., Benesch, J.L., 2015. Collision cross sections for structural proteomics. Structure 23, 791–799.
Masson, G.R., Jenkins, M.L., Burke, J.E., 2017. An overview of hydrogen deuterium exchange mass spectrometry (HDX-MS) in drug discovery. Expert Opin. Drug Discovery 12, 981–994.
McAllister, R.G., Metwally, H., Sun, Y., Konermann, L., 2015. Release of native-like gaseous proteins from electrospray droplets via the charged residue mechanism: insights from molecular dynamics simulations. J. Am. Chem. Soc. 137, 12667–12676.
Merk, A., Bartesaghi, A., Banerjee, S., Falconieri, V., Rao, P., Davis, M.I., Pragani, R., Boxer, M.B., Earl, L.A., Milne, J.L.S., Subramaniam, S., 2016. Breaking cryo-EM resolution barriers to facilitate drug discovery. Cell 165, 1698–1707.
Mikhailov, V.A., Mize, T.H., Benesch, J.L.P., Robinson, C.V., 2014. Mass-selective soft-landing of protein assemblies with controlled landing energies. Anal. Chem. 86, 8321–8328.
Mistarz, U.H., Bellina, B., Jensen, P.F., Brown, J.M., Barran, P.E., Rand, K.D., 2018. UV photodissociation mass spectrometry accurately localize sites of backbone deuteration in peptides. Anal. Chem. 90, 1077–1080.

Morgan, C.R., Engen, J.R., 2009. Investigating solution-phase protein structure and dynamics by hydrogen exchange mass spectrometry. Curr. Protoc. Protein Sci. 58 (1), 17.6.1–17.6.17.

Morton, V.L., Burkitt, W., O'Connor, G., Stonehouse, N.J., Stockley, P.G., Ashcroft, A.E., 2010. RNA-induced conformational changes in a viral coat protein studied by hydrogen/deuterium exchange mass spectrometry. Phys. Chem. Chem. Phys. 12, 13468–13475.

Mukhopadhyay, S., Kuhn, R.J., Rossmann, M.G., 2005. A structural perspective of the flavivirus lifecycle. Nat. Rev. Microbiol. 3, 13–22.

Neutze, R., Wouts, R., Van Der Spoel, D., Weckert, E., Hajdu, J., 2000. Potential for biomolecular imaging with femtosecond X-ray pulses. Nature 406, 752–757.

Nguyen, T.T., Sabat, G., Sussman, M.R., 2018. In vivo cross-linking supports a head-to-tail mechanism for regulation of the plant plasma membrane P-type H(+)-ATPase. J. Biol. Chem. 293, 17095–17106.

Novak, P., Kruppa, G.H., 2008. Intra-molecular cross-linking of acidic residues for protein structure studies. Eur. J. Mass Spectrom. (Chichester, Eng.) 14, 355–365.

O'Reilly, F.J., Rappsilber, J., 2018. Cross-linking mass spectrometry: methods and applications in structural, molecular and systems biology. Nat. Struct. Mol. Biol. 25, 1000–1008.

Pagel, K., Hyung, S.J., Ruotolo, B.T., Robinson, C.V., 2010. Alternate dissociation pathways identified in charge-reduced protein complex ions. Anal. Chem. 82, 5363–5372.

Parker, C.E., Mocanu, V., Mocanu, M., Dicheva, N., Warren, M.R., 2010. Mass spectrometry for post-translational modifications. In: Alzate, O. (Ed.), Neuroproteomics. Boca Raton (FL).

Peetz, O., Hellwig, N., Henrich, E., Mezhyrova, J., Dötsch, V., Bernhard, F., Morgner, N., 2019. LILBID and nESI: different native mass spectrometry techniques as tools in structural biology. J. Am. Soc. Mass Spectrom. 30, 181–191.

Peng, W.P., Chou, S.W., Patil, A.A., 2014. Measuring masses of large biomolecules and bioparticles using mass spectrometric techniques. Analyst 139, 3507–3523.

Peng, W.P., Yang, Y.C., Kang, M.W., Tzeng, Y.K., Nie, Z., Chang, H.C., Chang, W., Chen, C.H., 2006. Laser-induced acoustic desorption mass spectrometry of single bioparticles. Angew. Chem. Int. Ed. Eng. 45, 1423–1426.

Perez-Vargas, J., Vaughan, R.C., Houser, C., Hastie, K.M., Kao, C.C., Nemerow, G.R., 2014. Isolation and characterization of the DNA and protein binding activities of adenovirus core protein V. J. Virol. 88, 9287–9296.

Phillips, S.L., Soderblom, E.J., Bradrick, S.S., Garcia-Blanco, M.A., 2016. Identification of proteins bound to dengue viral RNA in vivo reveals new host proteins important for virus replication. MBio 7, e01865-15.

Pierson, E.E., Keifer, D.Z., Selzer, L., Lee, L.S., Contino, N.C., Wang, J.C., Zlotnick, A., Jarrold, M.F., 2014. Detection of late intermediates in virus capsid assembly by charge detection mass spectrometry. J. Am. Chem. Soc. 136, 3536–3541.

Piotrowski, C., Sinz, A., 2018. Structural investigation of proteins and protein complexes by chemical cross-linking/mass spectrometry. Adv. Exp. Med. Biol. 1105, 101–121.

Pogan, R., Dulfer, J., Uetrecht, C., 2018a. Norovirus assembly and stability. Curr Opin Virol 31, 59–65.

Pogan, R., Schneider, C., Reimer, R., Hansman, G., Uetrecht, C., 2018b. Norovirus-like VP1 particles exhibit isolate dependent stability profiles. J. Phys. Condens. Matter 30, 064006.

Politis, A., Borysik, A.J., 2015. Assembling the pieces of macromolecular complexes: hybrid structural biology approaches. Proteomics 15, 2792–2803.

Politis, A., Schmidt, C., 2018. Structural characterisation of medically relevant protein assemblies by integrating mass spectrometry with computational modelling. J. Proteome 175, 34–41.

Politis, A., Stengel, F., Hall, Z., Hernandez, H., Leitner, A., Walzthoeni, T., Robinson, C.V., Aebersold, R., 2014. A mass spectrometry-based hybrid method for structural modeling of protein complexes. Nat. Methods 11, 403–406.

Poor, T.A., Jones, L.M., Sood, A., Leser, G.P., Plasencia, M.D., Rempel, D.L., Jardetzky, T.S., Woods, R.J., Gross, M.L., Lamb, R.A., 2014. Probing the paramyxovirus fusion (F) protein-refolding event from pre- to postfusion by oxidative footprinting. Proc. Natl. Acad. Sci. U. S. A. 111, E2596–E2605.

Rand, K.D., 2013. Pinpointing changes in higher-order protein structure by hydrogen/deuterium exchange coupled to electron transfer dissociation mass spectrometry. Int. J. Mass Spectrom. 338, 2–10.

Rey, M., Man, P., Brandolin, G., Forest, E., Pelosi, L., 2009. Recombinant immobilized rhizopuspepsin as a new tool for protein digestion in hydrogen/deuterium exchange mass spectrometry. Rapid Commun. Mass Spectrom. 23, 3431–3438.

Root, K., Wittwer, Y., Barylyuk, K., Anders, U., Zenobi, R., 2017. Insight into signal response of protein ions in native ESI-MS from the analysis of model mixtures of covalently linked protein oligomers. J. Am. Soc. Mass Spectrom. 28 (9), 1863–1875.

Rose, M., Bobkov, S., Ayyer, K., Kurta, R.P., Dzhigaev, D., Kim, Y.Y., Morgan, A.J., Yoon, C.H., Westphal, D., Bielecki, J., Sellberg, J.A., Williams, G., Maia, F.R.N.C., Yefanov, O.M., Ilyin, V., Mancuso, A.P., Chapman, H.N., Hogue, B.G., Aquila, A., Barty, A., Vartanyants, I.A., 2018. Single-particle imaging without symmetry constraints at an X-ray free-electron laser. IUCrJ 5, 727–736.

Rose, R.J., Labrijn, A.F., Van Den Bremer, E.T., Loverix, S., Lasters, I., Van Berkel, P.H., Van De Winkel, J.G., Schuurman, J., Parren, P.W., Heck, A.J., 2011. Quantitative analysis of the interaction strength and dynamics of human IgG4 half molecules by native mass spectrometry. Structure 19, 1274–1282.

Rossi, V., Gaboriaud, C., Lacroix, M., Ulrich, J., Fontecillacamps, J.C., Gagnon, J., Arlaud, G.J., 1995. Structure of the catalytic region of human-complement protease C(1)-S—study by chemical cross-linking and 3-dimensional homology modeling. Biochemistry 34, 7311–7321.

Rostom, A.A., Robinson, C.V., 1999. Detection of the intact GroEL chaperonin assembly by mass spectrometry. J. Am. Chem. Soc. 121, 4718–4719.

Rozbesky, D., Rosulek, M., Kukacka, Z., Chmelik, J., Man, P., Novak, P., 2018. Impact of chemical cross-linking on protein structure and function. Anal. Chem. 90, 1104–1113.

Ruotolo, B.T., Giles, K., Campuzano, I., Sandercock, A.M., Bateman, R.H., Robinson, C.V., 2005. Evidence for macromolecular protein rings in the absence of bulk water. Science 310, 1658–1661.

Russel, D., Lasker, K., Webb, B., Velazquez-Muriel, J., Tjioe, E., Schneidman-Duhovny, D., Peterson, B., Sali, A., 2012. Putting the pieces together: integrative modeling platform software for structure determination of macromolecular assemblies. PLoS Biol. 10, e1001244.

Ruz, J.J., Tamayo, J., Pini, V., Kosaka, P.M., Calleja, M., 2014. Physics of nanomechanical spectrometry of viruses. Sci. Rep. 4, 6051.

Scalabrin, M., Dixit, S.M., Makshood, M.M., Krzemien, C.E., Fabris, D., 2018. Bifunctional cross-linking approaches for mass spectrometry-based investigation of nucleic acids and protein-nucleic acid assemblies. Methods 144, 64–78.

Schmidt, C., Urlaub, H., 2017. Combining cryo-electron microscopy (cryo-EM) and cross-linking mass spectrometry (CX-MS) for structural elucidation of large protein assemblies. Curr. Opin. Struct. Biol. 46, 157–168.

Schneidman-Duhovny, D., Pellarin, R., Sali, A., 2014. Uncertainty in integrative structural modeling. Curr. Opin. Struct. Biol. 28, 96–104.

Schulz, J., Bari, S., Buck, J., Uetrecht, C., 2013. Sample refreshment schemes for high repetition rate FEL experiments. In: Proceedings of SPIE, Advances in X-Ray Free-Electron Lasers Ii: instrumentation. The International Society for Optical Engineering, p. 8778.

Schwarz, R., Tanzler, D., Ihling, C.H., Sinz, A., 2016. Monitoring solution structures of peroxisome proliferator-activated receptor beta/delta upon ligand binding. PLoS One 11, e0151412.

Seo, J., Hoffmann, W., Warnke, S., Bowers, M.T., Pagel, K., Von Helden, G., 2016. Retention of native protein structures in the absence of solvent: a coupled ion mobility and spectroscopic study. Angew. Chem. Int. Ed. Eng. 55, 14173–14176.

Shakeel, S., Dykeman, E.C., White, S.J., Ora, A., Cockburn, J.J.B., Butcher, S.J., Stockley, P.G., Twarock, R., 2017. Genomic RNA folding mediates assembly of human parechovirus. Nat. Commun. 8, 5.

Shakeel, S., Evans, J.D., Hazelbaker, M., Kao, C.C., Vaughan, R.C., Butcher, S.J., 2018. Intrinsically-disordered N-termini in human parechovirus 1 capsid proteins bind encapsidated RNA. Sci. Rep. 8.

Sharon, M., Sinz, A., 2015. Studying protein–protein interactions by combining native mass spectrometry and chemical cross-linking. In: Kool, J., Niessen, W.M. (Eds.), Analyzing Biomolecular Interactions by Mass Spectrometry. Wiley-VCH, Weinheim.

Shoemaker, G.K., Van Duijn, E., Crawford, S.E., Uetrecht, C., Baclayon, M., Roos, W.H., Wuite, G.J., Estes, M.K., Prasad, B.V., Heck, A.J., 2010. Norwalk virus assembly and stability monitored by mass spectrometry. Mol. Cell. Proteomics 9, 1742–1751.

Shukla, A.K., Westfield, G.H., Xiao, K., Reis, R.I., Huang, L.Y., Tripathi-Shukla, P., Qian, J., Li, S., Blanc, A., Oleskie, A.N., Dosey, A.M., Su, M., Liang, C.R., Gu, L.L., Shan, J.M., Chen, X., Hanna, R., Choi, M., Yao, X.J., Klink, B.U., Kahsai, A.W., Sidhu, S.S., Koide, S., Penczek, P.A., Kossiakoff, A.A., Woods Jr., V.L., Kobilka, B.K., Skiniotis, G., Lefkowitz, R.J., 2014. Visualization of arrestin recruitment by a G-protein-coupled receptor. Nature 512, 218–222.

Singh, P., Nakatani, E., Goodlett, D.R., Catalano, C.E., 2013. A pseudo-atomic model for the capsid shell of bacteriophage lambda using chemical cross-linking/mass spectrometry and molecular modeling. J. Mol. Biol. 425, 3378–3388.

Sinz, A., Arlt, C., Chorev, D., Sharon, M., 2015. Chemical cross-linking and native mass spectrometry: a fruitful combination for structural biology. Protein Sci. 24, 1193–1209.

Siuzdak, G., Bothner, B., Yeager, M., Brugidou, C., Fauquet, C.M., Hoey, K., Chang, C.M., 1996. Mass spectrometry and viral analysis. Chem. Biol. 3, 45–48.

Snijder, J., Benevento, M., Moyer, C.L., Reddy, V., Nemerow, G.R., Heck, A.J., 2014a. The cleaved N-terminus of pVI binds peripentonal hexons in mature adenovirus. J. Mol. Biol. 426, 1971–1979.

Snijder, J., Burnley, R.J., Wiegard, A., Melquiond, A.S., Bonvin, A.M., Axmann, I.M., Heck, A.J., 2014b. Insight into cyanobacterial circadian timing from structural details of the KaiB-KaiC interaction. Proc. Natl. Acad. Sci. U. S. A. 111, 1379–1384.

Snijder, J., Heck, A.J., 2014. Analytical approaches for size and mass analysis of large protein assemblies. Annu Rev Anal Chem (Palo Alto, Calif) 7, 43–64.

Snijder, J., Kononova, O., Barbu, I.M., Uetrecht, C., Rurup, W.F., Burnley, R.J., Koay, M.S., Cornelissen, J.J., Roos, W.H., Barsegov, V., Wuite, G.J., Heck, A.J., 2016. Assembly and mechanical properties of the cargo-free and cargo-loaded bacterial nanocompartment encapsulin. Biomacromolecules 17, 2522–2529.

Snijder, J., Schuller, J.M., Wiegard, A., Lossl, P., Schmelling, N., Axmann, I.M., Plitzko, J.M., Forster, F., Heck, A.J., 2017. Structures of the cyanobacterial circadian oscillator frozen in a fully assembled state. Science 355, 1181–1184.

Snijder, J., Uetrecht, C., Rose, R., Sanchez-Eugenia, R., Marti, G.A., Agirre, J., Guérin, D., Wuite, G., Heck, A., Roos, W., 2013. Probing the biophysical interplay between a viral genome and its capsid. Nat. Chem. 5, 502.

Snijder, J., Van De Waterbeemd, M., Damoc, E., Denisov, E., Grinfeld, D., Bennett, A., Agbandje-Mckenna, M., Makarov, A., Heck, A.J., 2014c. Defining the stoichiometry and cargo load of viral and bacterial nanoparticles by Orbitrap mass spectrometry. J. Am. Chem. Soc. 136, 7295–7299.
Stiving, A.Q., Vanaernum, Z.L., Busch, F., Harvey, S.R., Sarni, S.H., Wysocki, V.H., 2019. Surface-induced dissociation: an effective method for characterization of protein quaternary structure. Anal. Chem. 91, 190–209.
Stuchfield, D., Barran, P., 2018. Unique insights to intrinsically disordered proteins provided by ion mobility mass spectrometry. Curr. Opin. Chem. Biol. 42, 177–185.
Susa, A.C., Xia, Z., Williams, E.R., 2017. Small emitter tips for native mass spectrometry of proteins and protein complexes from nonvolatile buffers that mimic the intracellular environment. Anal. Chem. 89, 3116–3122.
Tabb, D.L., 2012. Evaluating protein interactions through cross-linking mass spectrometry. Nat. Methods 9, 879–881.
Tanaka, K., Waki, H., Ido, Y., Akita, S., Yoshida, Y., Yoshida, T., Matsuo, T., 1988. Protein and polymer analyses up to m/z 100 000 by laser ionization time-of-flight mass spectrometry. Rapid Commun. Mass Spectrom. 2, 151–153.
Tang, X., Bruce, J.E., 2010. A new cross-linking strategy: protein interaction reporter (PIR) technology for protein-protein interaction studies. Mol. BioSyst. 6, 939–947.
Taube, S., Mallagaray, A., Peters, T., 2018. Norovirus, glycans and attachment. Curr. Opin. Virol. 31, 33–42.
Tinnefeld, V., Venne, A.S., Sickmann, A., Zahedi, R.P., 2017. Enrichment of cross-linked peptides using charge-based fractional diagonal chromatography (ChaFRADIC). J. Proteome Res. 16, 459–469.
Trabjerg, E., Nazari, Z.E., Rand, K.D., 2018. Conformational analysis of complex protein states by hydrogen/deuterium exchange mass spectrometry (HDX-MS): challenges and emerging solutions. TrAC Trends Anal. Chem. 106, 125–138.
Uetrecht, C., Barbu, I.M., Shoemaker, G.K., Van Duijn, E., Heck, A.J., 2011. Interrogating viral capsid assembly with ion mobility-mass spectrometry. Nat. Chem. 3, 126–132.
Uetrecht, C., Heck, A.J., 2011. Modern biomolecular mass spectrometry and its role in studying virus structure, dynamics, and assembly. Angew. Chem. Int. Ed. Eng. 50, 8248–8262.
Uetrecht, C., Lorenzen, K., Kitel, M., Heidemann, J., Spencer, J.H.R., Schlüter, H., Schulz, J., 2019. Native mass spectrometry provides sufficient ion flux for XFEL single particle imaging. J. Synchrotron Radiat. 26, 653–659.
Uetrecht, C., Rose, R.J., Van Duijn, E., Lorenzen, K., Heck, A.J., 2010. Ion mobility mass spectrometry of proteins and protein assemblies. Chem. Soc. Rev. 39, 1633–1655.
Uetrecht, C., Versluis, C., Watts, N.R., Roos, W.H., Wuite, G.J., Wingfield, P.T., Steven, A.C., Heck, A.J., 2008a. High-resolution mass spectrometry of viral assemblies: molecular composition and stability of dimorphic hepatitis B virus capsids. Proc. Natl. Acad. Sci. 105, 9216–9220.
Uetrecht, C., Versluis, C., Watts, N.R., Wingfield, P.T., Steven, A.C., Heck, A.J., 2008b. Stability and shape of hepatitis B virus capsids in vacuo. Angew. Chem. Int. Ed. Eng. 47, 6247–6251.
Vahidi, S., Konermann, L., 2016. Probing the time scale of FPOP (Fast photochemical oxidation of proteins): radical reactions extend over tens of milliseconds. J. Am. Soc. Mass Spectrom. 27, 1156–1164.
Van Berkel, W.J.H., Van Den Heuvel, R.H.H., Versluis, C., Heck, A.J.R., 2000. Detection of intact megaDalton protein assemblies of vanillyl-alcohol oxidase by mass spectrometry. Protein Sci. 9, 435–439.

Van De Waterbeemd, M., Llauro, A., Snijder, J., Valbuena, A., Rodriguez-Huete, A., Fuertes, M.A., De Pablo, P.J., Mateu, M.G., Heck, A.J., 2017. Structural analysis of a temperature-induced transition in a viral capsid probed by HDX-MS. Biophys. J. 112, 1157–1165.

Van De Waterbeemd, M., Snijder, J., Tsvetkova, I.B., Dragnea, B.G., Cornelissen, J.J., Heck, A.J., 2016. Examining the heterogeneous genome content of multipartite viruses BMV and CCMV by native mass spectrometry. J. Am. Soc. Mass Spectrom. 27, 1000–1009.

Van Schooten, J., Van Gils, M.J., 2018. HIV-1 immunogens and strategies to drive antibody responses towards neutralization breadth. Retrovirology 15, 74.

Wang, L., Pan, H., Smith, D.L., 2002. Hydrogen exchange-mass spectrometry. Mol. Cell. Proteomics 1, 132–138.

Ward, A.B., Sali, A., Wilson, I.A., 2013. Biochemistry. Integrative structural biology. Science 339, 913–915.

Webb, B., Viswanath, S., Bonomi, M., Pellarin, R., Greenberg, C.H., Saltzberg, D., Sali, A., 2018. Integrative structure modeling with the integrative modeling platform. Protein Sci. 27, 245–258.

Weis, D.D., Wales, T.E., Engen, J.R., Hotchko, M., Ten Eyck, L.F., 2006. Identification and characterization of EX1 kinetics in H/D exchange mass spectrometry by peak width analysis. J. Am. Soc. Mass Spectrom. 17, 1498–1509.

Weiss, V.U., Bereszcazk, J.Z., Havlik, M., Kallinger, P., Gösler, I., Kumar, M., Blaas, D., Marchetti-Deschmann, M., Heck, A.J., Szymanski, W.W., 2015. Analysis of a common cold virus and its subviral particles by gas-phase electrophoretic mobility molecular analysis and native mass spectrometry. Anal. Chem. 87, 8709–8717.

Wijesinghe, K.J., Urata, S., Bhattarai, N., Kooijman, E.E., Gerstman, B.S., Chapagain, P.P., Li, S., Stahelin, R.V., 2017. Detection of lipid-induced structural changes of the Marburg virus matrix protein VP40 using hydrogen/deuterium exchange-mass spectrometry. J. Biol. Chem. 292, 6108–6122.

Yamashita, M., Fenn, J.B., 1984. Electrospray ion-source—another variation on the free-jet theme. J. Phys. Chem. 88, 4451–4459.

Yang, M.L., Hoeppner, M., Rey, M., Kadek, A., Man, P., Schriemer, D.C., 2015. Recombinant nepenthesin II for hydrogen/deuterium exchange mass spectrometry. Anal. Chem. 87, 6681–6687.

Yates, J.R., Ruse, C.I., Nakorchevsky, A., 2009. Proteomics by mass spectrometry: approaches, advances, and applications. Annu. Rev. Biomed. Eng. 11, 49–79.

Yoon, C.H., Yurkov, M.V., Schneidmiller, E.A., Samoylova, L., Buzmakov, A., Jurek, Z., Ziaja, B., Santra, R., Loh, N.D., Tschentscher, T., Mancuso, A.P., 2016. A comprehensive simulation framework for imaging single particles and biomolecules at the European X-ray free-electron laser. Sci. Rep. 6, 24791.

Zanotto, L., Heerdt, G., Souza, P.C.T., Araujo, G., Skaf, M.S., 2018. High performance collision cross section calculation-HPCCS. J. Comput. Chem. 39, 1675–1681.

Zelter, A., Hoopmann, M.R., Vernon, R., Baker, D., MacCoss, M.J., Davis, T.N., 2010. Isotope signatures allow identification of chemically cross-linked peptides by mass spectrometry: a novel method to determine interresidue distances in protein structures through cross-linking. J. Proteome Res. 9, 3583–3589.

Zhang, B.J., Cheng, M., Rempel, D., Gross, M.L., 2018. Implementing fast photochemical oxidation of proteins (FPOP) as a footprinting approach to solve diverse problems in structural biology. Methods 144, 94–103.

CHAPTER EIGHT

Illuminating the virus life cycle with single-molecule FRET imaging

Maolin Lu*, Xiaochu Ma, Walther Mothes*
Department of Microbial Pathogenesis, Yale University School of Medicine, New Haven, CT, United States
*Corresponding authors: e-mail address: maolin.lu@yale.edu; walther.mothes@yale.edu

Contents

Abstract

Single-molecule Förster resonance energy transfer (smFRET) imaging has emerged as a powerful tool to probe conformational dynamics of viral proteins, identify novel structural intermediates that are hiding in averaging population-based measurements, permit access to the energetics of transitions and as such to the precise molecular mechanisms of viral replication. One strength of smFRET is the capability of characterizing biological molecules in their fully hydrated/native state, which are not necessarily available to other structural methods. Elegant experimental design for physiologically

Advances in Virus Research, Volume 105
ISSN 0065-3527
https://doi.org/10.1016/bs.aivir.2019.07.004

relevant conditions, such as intact virions, has permitted the detection of previously unknown conformational states of viral glycoproteins, revealed asymmetric intermediates, and allowed access to the real-time imaging of conformational changes during viral fusion. As more laboratories are applying smFRET, our understanding of the molecular mechanisms and the dynamic nature of viral proteins throughout the virus life cycle are predicted to improve and assist the development of novel antiviral therapies and vaccine design.

Abbreviations

5′UTR	5′ untranslated region
ADCC	antibody-dependent cellular cytotoxicity
AIDS	acquired immunodeficiency syndrome
APOBEC3	apolipoprotein B editing complex
bNAbs	broadly neutralizing antibodies
CA	capsid
CD4mc	CD4 mimetics
cryoET	cryo-electron tomography
CT	cytoplasmic tail
DIS	dimerization initiation site
Env	envelope glycoprotein
FCS	fluorescence correlation spectroscopy
gp41	glycoprotein 41, transmembrane domain of HIV-1 Env
gp120	glycoprotein 120, receptor/coreceptor binding domain of HIV-1 Env
gp160	glycoprotein 160, the precursor of HIV-1 Env
HA	influenza A hemagglutinin
HA_0	the precursor of HA
HA_1	the receptor sialic acid binding domain of HA
HA_2	the transmembrane domain of HA
HIV-1	human immunodeficiency virus 1
HR1, HR2	heptad repeat 1, heptad repeat 2
MA	matrix
MPER	membrane proximal extracellular region
NC	nucleocapsid
NMR	nuclear magnetic resonance
PIP_2	phosphatidylinositol 4,5-bisphosphate
Pol	polymerase
Psi	packaging signal
RNase H	ribonuclease H
RT	reverse transcriptase
smFRET	single-molecule Förster resonance energy transfer
SOSIP.664	SOS for a disulfide bridge, IP for an I559P mutation, 664 for the truncation side at residue 664
TIRFM	total internal reflection fluorescence microscope
TM	transmembrane region
tRNA	transfer RNA
VLPs	viral-like particles
vRNA	viral RNA

1. Introduction

Decades of research have led to a detailed understanding of the viral life cycle of most clinically relevant viruses. Molecular virologists have identified the function of individual viral proteins and structural biologists have provided structural snapshots of most viral enzymes, capsids and viral glycoproteins (Das and Arnold, 2013; Harrison, 2008; Mukhopadhyay et al., 2005; Vaney and Rey, 2011). These studies have driven the development of antiviral compounds, many of which have reached the clinic and become available treatments (Cohen et al., 2011; Das and Arnold, 2013). Cell biologists have extensively studied virus entry and egress (Marsh and Helenius, 2006; Sundquist and Krausslich, 2012). For almost all viruses, the cellular receptors have been determined, and genome-wide screens have identified various host dependence factors (Bushman et al., 2009; Carette et al., 2011; Hunter, 1997). Single-particle imaging studies have been powerful in elucidating virus entry, uncoating, nuclear entry, virus assembly and egress, as well as cell-to-cell transmission (Brandenburg and Zhuang, 2007; Campbell and Hope, 2008; Jouvenet et al., 2008; Melikyan, 2014; Sattentau, 2008). Correspondingly, we have a good understanding of the recognition of viral capsids and/or viral genomes by innate sensors and restriction factors, and of the adaptive immune responses, including humoral and cell-mediated immunity (Blanco-Melo et al., 2012; McFadden et al., 2017). While there are a few gaps in the understanding of individual steps of the life cycle in most viruses, two particular areas remain poorly understood. Macroscopically, we know very little about virus dissemination and pathogenesis at the organismal level (Uchil et al., 2019). And microscopically, we know little about the dynamics involved when viral proteins undergo conformational changes to accomplish a specific function throughout the viral life cycle.

Single-molecule Förster resonance energy transfer (smFRET) imaging is uniquely positioned to provide access to the microscopic dimension (Roy et al., 2008), specifically to monitoring the conformational dynamics and conformational changes of viral proteins as they carry out their functions. smFRET detects large conformational changes and domain movements that, along the time scale of biomolecular motions, occur in the milliseconds-to-seconds range (Fig. 1). Theoretically, smFRET measures the distance-dependent energy transfer from an excited donor to a nearby acceptor fluorophore in real time. The energy transfer efficiency (E_{FRET}) depends on the inter-dye distance between donor and acceptor (r) and is

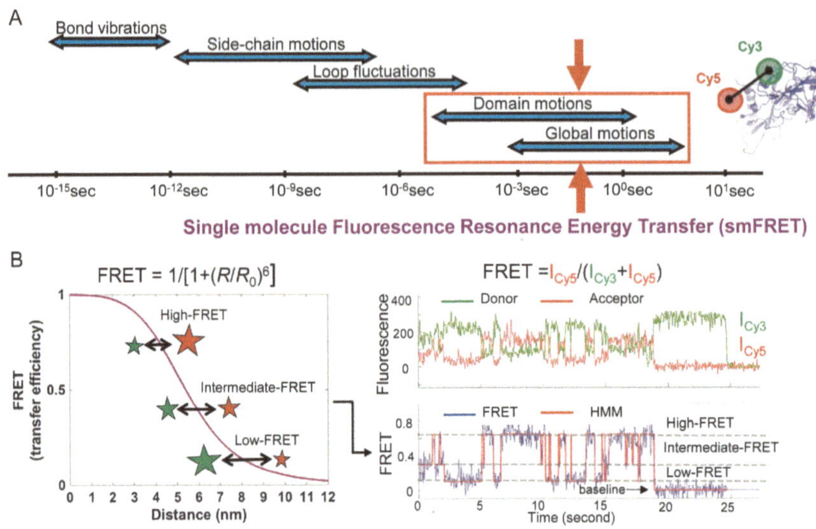

Fig. 1 smFRET imaging allows for real-time observation of conformational dynamics of biomolecules. (A) Along the timescale of biomolecular dynamics, large domain motions and global motions of proteins lie within the range of millisecond to a few seconds (Henzler-Wildman and Kern, 2007). (B) Diagrams depicting the distance dependence of the energy transfer from donor to acceptor fluorophore (left), and an example smFRET analysis of conformational dynamics for a labeled HIV-1 Env molecule (right). (Left) Energy transfer efficiency (FRET) between donor and acceptor depends on the distance. When they are close to each other, energy transfer efficiency is high, or vice versa. By recording donor (such as Cy3) and acceptor (such as Cy5) fluorescence, corresponding energy transfer efficiency is derived to monitor the relative distance changes between two labeled sites on a biomolecule of interest. At single-molecule level, distinct conformational states and the transitions between distinct states are directly visible based on FRET value and FRET trace in real time, respectively. (Right) For instance, single HIV-1 gp120 molecules within native trimers are dynamic in time observed from a FRET trace, and it constantly samples three conformational states characterized as low-FRET, intermediate-FRET, and high-FRET (Lu et al., 2019).

described as $E_{FRET} = (1 + (r/R_0)^6)^{-1}$, where R_0 is the Förster radius when $E_{FRET} = 50\%$. E_{FRET} can be determined experimentally as the ratio of acceptor intensity to total fluorescence intensity, according to $FRET = I_A/(I_A + I_D)$, where I_A is the intensity of acceptor fluorescence, and I_D is that of the donor. The cyanine dyes, Cy3, Cy5, and Cy7, are commercially available for smFRET owing to their brightness, photostability, and water solubility. The recent development of self-healing dyes has dramatically improved their lifespan and photon output while maintaining water solubility (Zheng et al., 2014). For smFRET, donor and acceptor fluorophores have to be introduced into the molecules of interest. This has historically been easier for RNA and DNA, which have dominated early studies (Abbondanzieri et al., 2008; Coey et al., 2016; Liu et al., 2008b, 2010). Recent advances in utilizing site specific labeling technologies as well as unnatural amino acids technologies have revolutionized our ability to introduce fluorophores at various positions into proteins of interest (Das et al., 2018; Lu et al., 2019; Plass et al., 2011; Sakin et al., 2017). smFRET imaging is preferentially performed on total internal reflection fluorescence microscopes (TIRFM) for their superior single-molecule sensitivity and high signal to noise ratio. smFRET has proven to be an insightful method for revealing the intrinsically dynamic and heterogeneous nature of biological systems, by identifying previously unknown conformational states and intermediates, as well as delineating the sequence of conformational transitions (Abbondanzieri et al., 2008; Akyuz et al., 2015; Dunkle et al., 2011; Gregorio et al., 2017; Lerner et al., 2018; Lu et al., 2019; Munro et al., 2010, 2014a; Roy et al., 2008; Zhao et al., 2010).

Adhering to the order of the virus life cycle, we will review recent insights gained from applying smFRET to the entry of the human immunodeficiency virus 1 (HIV-1) and influenza A into cells, reverse transcription, genome packaging, and assembly of HIV-1.

2. Viral glycoproteins as entry and immune evasion machines

Enveloped viruses, such as HIV-1 and influenza A, enter host cells by fusion of viral and cellular membranes (Harrison, 2008; Skehel and Wiley, 1998). While membrane fusion of small enveloped viruses with large cells is energetically favored, the electrostatic repulsion between membranes and the initial merging represents an energy barrier (Harrison, 2008). This barrier is overcome by viral fusion machines that bring membranes together for fusion. The lipid merger reaction proceeds through a hemi-fusion intermediate, and then a widening fusion pore that permits content mixing and the delivery of viral capsids into the cytoplasm of the host (Chernomordik and Kozlov, 2008). Viral fusion proteins are often also the only viral protein on the surface of viruses, and as such the target for neutralizing antibodies, which they evolve to evade. This dualism of functioning as viral fusion machines as well as immune evasion machines makes them interesting objects to study, and understanding these machines is vital for the rational design of small molecule inhibitors, immunogens for vaccines, and the development of strategies for a cure.

Enveloped viruses have evolved at least three distinct classes of viral fusion machines that can overcome this barrier and induce fusion (Harrison, 2015). The glycoproteins of HIV-1 and influenza A virus, the envelope glycoprotein (Env) and influenza A hemagglutinin (HA) belong to class 1 fusion machines. They are metastable trimeric proteins that respond to cellular cues, e.g. to receptor CD4 and coreceptor for HIV-1 Env, and to low pH within an endosome for influenza A HA, to initiate a series of conformational changes of loop-to-helix, as well as helix-to-loop transitions that lead to the generation of a highly stable hairpin conformation with a central alpha-helical coil-coiled in the post-fusion conformation (Bullough et al., 1994; Carr and Kim, 1993; Chan et al., 1997; Julien et al., 2013; Lyumkis et al., 2013; Mao et al., 2012; Pancera et al., 2014; Wilson et al., 1981). The energy released during this refolding process is believed to drive the fusion reaction. These metastable fusion proteins are generated from precursor proteins, HIV-1 gp160 trimers, and influenza A HA_0 trimers, and proteolytically primed into the metastable trigger-prone mature fusion proteins, trimers of HIV-1 gp120 and gp41 heterodimers, as well as trimers of influenza A HA_1 and HA_2 heterodimers, respectively. The actual fusion machines reside within the transmembrane subunits HIV-1 gp41 and influenza A HA_2. While structural snapshots of the precursor, mature and post-fusion conformations have been generated at low and high resolutions (Bullough et al., 1994; Carr and Kim, 1993; Julien et al., 2013; Kwong et al., 1998; Liu et al., 2008a; Lyumkis et al., 2013; Mao et al., 2012; Ozorowski et al., 2017; Wang et al., 2016; Wilson et al., 1981), smFRET has just recently provided real-time measurements of the conformational changes during fusion for influenza A HA_2 (Das et al., 2018).

In contrast, much of the immune evasion function is mediated by HIV-1 gp120 and influenza HA_1 that initially cover and protect the underlying fusion machines. Both are heavily glycosylated and are undergoing sequence variation. In the case of HIV-1, gp120 subunits are clearly dynamic, concealing the immunogenic functional receptor and coreceptor binding sites from antibody recognition (Barouch, 2008; Crispin et al., 2018; Julien et al., 2012; Kwong et al., 2002; Wilson et al., 1981; Wyatt et al., 1998). These immune evasion strategies are best understood for HIV-1 Env (Kwong et al., 2002, 2013; Wei et al., 2003). HIV-1 Env has a nearly perfect glycan shield as ~half of its molecular weight is contributed by sugars (Gristick et al., 2016; Stewart-Jones et al., 2016). Vaccine studies with soluble trimers containing systematically placed glycan holes have illustrated how much better the HIV-1 Env trimer is recognized by antibodies the moment holes are introduced into the glycan shield (Crooks et al., 2015; Ringe et al., 2019; Wagh et al., 2018). Due to the high mutation rate of reverse transcriptase (RT) and likely additional APOBEC3 mediated hyper-mutation, HIV-1 Env also undergoes rapid sequence variation, permitting epitope escape from neutralizing antibodies (Mulder et al., 2008). Last, Env is both structurally flexible and conformationally dynamic, and those features have also been elucidated by smFRET. This dynamic behavior enables concealing of vulnerable functional centers of Env by restricting antibody access to receptor and coreceptor binding sites. Critically for the understanding of HIV-1/AIDS is that the antibodies generated against the more open and highly immunogenic conformational states are ultimately non-neutralizing against primary HIV-1 isolates. In contrast, the broadly neutralizing antibodies (bNAbs) generated by few patients can recognize the glycan-shielded closed conformation. This model is further supported by the Tier classification of HIV-1 isolates (Montefiori et al., 2018). Tier 1 isolates are lab-adapted strains that have not been exposed to innate and adaptive immunity for a long time, and carry more open Env trimers that can be neutralized by "non-neutralizing" antibodies. In contrast, Env trimers from the primary Tier 2 and 3 clinical isolates are more closed and more resistant to most antibodies, but can be recognized by bNAbs (Montefiori et al., 2018). Clearly, structure-based HIV vaccine design that intends to elicit bNAbs focuses on developing immunogens that mimic the structure of glycan-shielded closed conformation of Env. The application of smFRET has begun to elucidate both aspects of these immune evasion and fusion machines (Das et al., 2018; Lu et al., 2019; Munro et al., 2014a).

3. smFRET imaging of viral glycoproteins in the context of intact virions

The establishment of smFRET methods for native virus particles has provided critical new insights into the conformational dynamics of HIV-1 Env and influenza A HA (Munro and Mothes, 2015; Munro et al., 2014a). In the case of Influenza HA, HA was imaged on the surface of HIV-1 particles (Das et al., 2018). The implementation of smFRET for these virus systems is based upon the utilization of site specific labeling as well as unnatural amino acids technologies (Das et al., 2018; Lu et al., 2019; Ma et al., 2018; Munro et al., 2014a). In the case of HIV-1 Env, positions that tolerate the incorporation of 6, 8 or 12 amino acid tags within the variable loops of gp120 were identified. These short peptide tags are site-specifically recognized by enzymes that catalyze the transfer of fluorophore-conjugated substrates onto single residues within these tags (Ma et al., 2018; Munro et al., 2014a). Tagged Env proteins must be validated to retain infectivity, Env synthesis, processing, virus incorporation and sensitivity to trimer specific neutralizing antibodies. For HIV-1 Env, tags were tolerated in V1, V4 and V5 (Munro et al., 2014a). Labeling of viruses must then be accomplished in a way that one virus particle, on average, only carries a single Cy3 and Cy5 pair. The other two gp120 protomers within the same trimer, and all other trimers on the surface of the same single virus, should remain unlabeled. This is accomplished by transfecting an excess of plasmid encoding the untagged wild-type Env over the tagged Env. It should also be noted that this method only works for viruses such as HIV-1, which carry a limited number of viral glycoproteins. For viruses with higher numbers of viral glycoproteins on their surface, the probability that two separate trimers carry dyes increases to a point where inter-molecular FRET signals can obscure the intra-molecular FRET signals of interest. Viruses purified from culture supernatants are then incubated with the labeling enzymes and their dye-conjugated substrates, purified to remove free dyes, and immobilized on quartz slides for TIRFM smFRET imaging (Fig. 2). If unnatural amino acids are used, clickable amino acids allow copper-free click chemistry to introduce the fluorophores (Das et al., 2018; Lu et al., 2019). The strengths of this approach are that the viral glycoproteins are studied within their natural virus environment using physiological conditions, avoiding the need for protein purification and/or use of detergents.

4. smFRET imaging reveals that the unliganded HIV-1 Env is dynamic, having access to three distinct prefusion conformational states

TIRFM imaging of HIV-1 Env carrying fluorophores in V1 and V4 of gp120 revealed that Env from three different isolates (laboratory-adapted NL4-3, clinical isolates JR-FL and BG505) (Ma et al., 2018; Munro et al., 2014a) shows anti-correlative behavior between donor and acceptor fluorescence intensities, indicative of energy transfer between both dyes, and suggesting that the unliganded Env is dynamic. Quantitative analysis showed that Env samples three distinct conformational states characterized by low-FRET, intermediate-FRET, and high-FRET values (Lu et al., 2019; Ma et al., 2018; Munro et al., 2014a) (Fig. 2). The low-FRET conformation

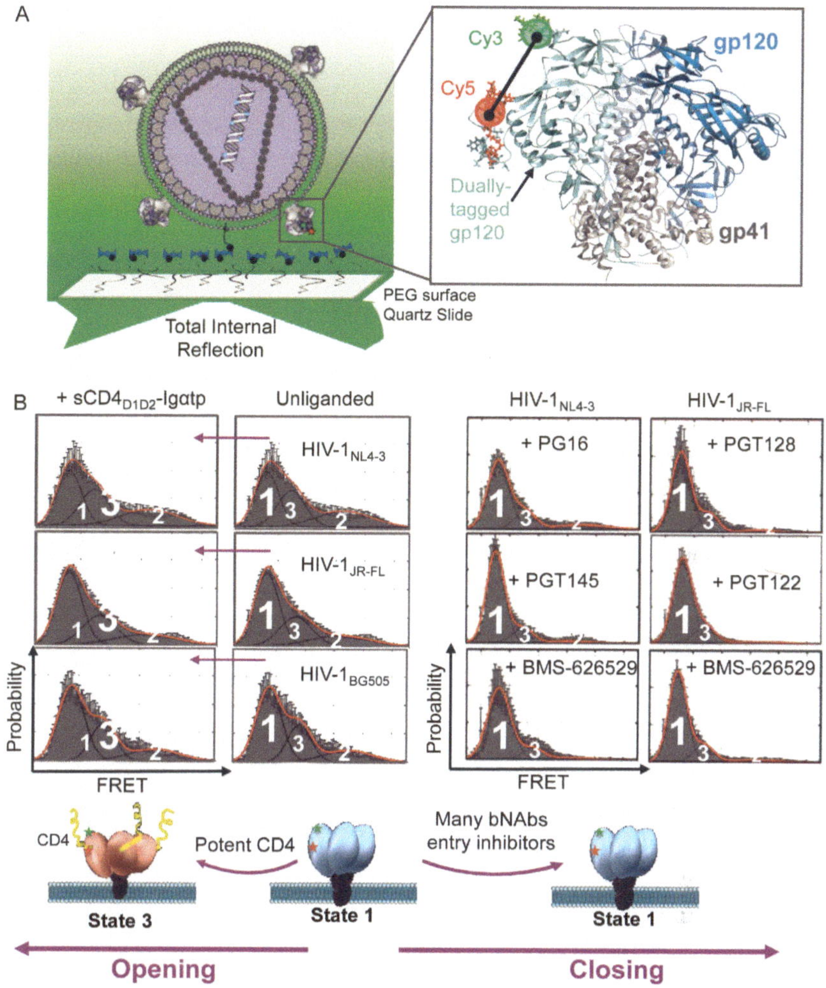

Fig. 2 smFRET imaging of native HIV-1 Env on the surface of intact viruses reveals three distinct conformational states. (A) Individual HIV-1 virus carrying single Cy3/Cy3 FRET dyes labeled gp120 within a trimer in the context of otherwise wild-type trimers was immobilized and imaged via total internal reflection microscopy. High-resolution structure of Env trimer shows the positions where dyes are labeled on gp120 (cyan) within other wild-type gp120/gp41 (blue/gray) trimer (structure was adapted from PDB: 4NCO). (B) Observation of the opening and further closing of HIV-1 Env trimer via FRET population histograms. Native functional HIV-1 Env (HIV-1 isolates NL4-3, JR-FL, and BG505) predominately resides in pre-triggered State 1, and has inherent access to State 3, through one necessary intermediate (State 2). HIV-1 Env opens into State 3 upon binding of the potent 12xCD4 (left panel). bNAbs PG16, PGT145 (NL4-3), PGT128, PGT122 (JR-FL), and entry inhibitor BMS reduce the occupancy of State 2 and State 3, further stabilizing State 1 (right panel). *Adapted from Ma, X.C., Lu, M.L., Gorman, J., Terry, D.S., Hong, X.Y., Zhou, Z., Zhao, H., Altman, R.B., Arthoe, J., Blanchard, S.C., Kwong, P.D., Munro, J.B., Mothes, W., 2018. HIV-1 Env trimer opens through an asymmetric intermediate in which individual protomers adopt distinct conformations. eLife 7, e34271; Munro, J.B., Gorman, J., Ma, X., Zhou, Z., Arthos, J., Burton, D. R., Koff, W.C., Courter, J.R., Smith 3rd, A.B., Kwong, P.D., Blanchard, S.C., Mothes, W., 2014. Conformational dynamics of single HIV-1 envelope trimers on the surface of native virions. Science 346, 759–763, with permission from AAAS.*

was the most populated state in all three cases, and represents the ground-state conformation of the prefusion Env, which is expected to correspond to the most immune evasive closed conformation. Consistent with this observation, the Tier 2 HIV-1 isolates JR-FL and BG505 exhibited greater occupancy of the low-FRET conformational state as compared to the lab-adapted NL4-3, which is known to be more open (Ma et al., 2018). Further statistical analyses of the transitions revealed that Env transitions from the low-FRET to the high-FRET and back, as well as between the high-FRET and the intermediate-FRET state. Transitions between low-FRET and intermediate-FRET states were rare, indicating that the high-FRET state is a necessary intermediate between low-FRET and intermediate-FRET. Based on this biophysical analysis, these three FRET states were designated as pre-triggered State 1 (low-FRET), intermediate State 2 (high-FRET), and State 3 (intermediate-FRET) (Herschhorn et al., 2016; Ma et al., 2018).

5. Ligands identify the nature of HIV-1 Env conformational states observed by smFRET

To identify the nature behind the three observed conformational states, we tested the effects of various ligands including soluble CD4 (sCD4) and numerous antibodies as well as Env mutants (Lu et al., 2019; Ma et al., 2018; Munro et al., 2014a). Interestingly, many bNAbs increased the occupancy of State 1 by reducing that of State 2 and State 3. For example, bNAbs directed against different epitopes of Env including the trimer apex containing the variable loops V1V2 (PG9, PG16 and PGT145), the V3 glycan patch (10-1074, PGT122, PGT128, 2G12), and the CD4 binding site (VRC01, 3BNC117), were either compatible with or stabilized the State 1 conformation of Env (Lu et al., 2019; Munro et al., 2014a). A small molecule entry inhibitor BMS-626529 (Temsavir®) exhibited a similar conformational signature of a State 1 stabilization (Lu et al., 2019; Munro et al., 2014a).

In contrast, soluble receptor CD4 (sCD4) containing ligands, small molecule CD4 mimetics, as well as coreceptor mimicking antibodies such as 17b, stabilized State 3. Specifically, the potent oligomeric sCD4 ligand 12xCD4 ($sCD4_{D1D2}$-IgαIg) (Arthos et al., 2002), or the simultaneous presence of the sCD4 (D1D2 domain) and additional presence of the coreceptor-surrogate antibody 17b, which binds to a CD4-induced epitope, all stabilized Env in State 3 conformations (Ma et al., 2018). These experiments allowed an assignment of the observed FRET states as the pre-triggered State 1 conformation, and the three-CD4 bound State 3 open conformation of Env (Ma et al., 2018). As such, the smFRET assay demonstrated that individual Env molecules on native virus particles are dynamic, spontaneously transitioning between three distinct well-populated functional conformational states: a pre-triggered Env (State 1), a default intermediate (State 2), and a three-CD4-bound conformation (State 3) along the Env activation pathway. Thermodynamically, the interaction with CD4 remodels the conformational landscape of Env by stabilizing the State 3 conformation over States 1 and 2, and by lowering the energy barriers for transitions into State 3 (Ma et al., 2018).

6. Stepwise opening of HIV-1 Env by cellular receptor CD4 and coreceptor mimics

With States 1 and 3 identified, the nature of State 2 had initially remained unclear. Differences in the response of HIV-1 isolates to sCD4 suggested that this conformational intermediate might be linked to incomplete CD4 occupancy on the trimer (Ma et al., 2018). To test this hypothesis, Ma et al. engineered trimers that can only bind 1 or 2 CD4 molecules per trimer, and then placed the conformational state monitoring two fluorophores either into the gp120 protomer that can bind CD4, or into the adjacent protomer that cannot bind CD4 (Fig. 3). This was accomplished using the D368R substitution in the CD4 binding pocket that impairs CD4 binding without causing indirect allosteric effects. Two mixed trimers that can only bind a single CD4 molecule were designed. In mixed trimer 1, the Cy3/Cy5 fluorophores were placed into a single protomer that can bind sCD4. The two other protomers within the trimer carried the D368R mutation to prevent CD4 binding. In mixed trimer 2, the Cy3/Cy5 dyes were placed on one of the two CD4-binding incompetent protomers next to the single CD4-binding competent protomer. smFRET imaging revealed that a trimer engaging only a single CD4 protomer is asymmetric. The CD4-bound protomer adopts State 3, while the protomers positioned left and right to the CD4-bound protomer adopt the State 2 conformation. Finally, a trimer with a single CD4 molecule bound was also able to completely open into State 3 if a coreceptor mimicking antibody 17b bound was also introduced (Ma et al., 2018) (Figs. 3 and 4). These data suggest a model for the sequential opening of the HIV-1 Env trimer through an asymmetric trimer intermediate in which only a single CD4 engages the trimer, followed by complete opening into State 3 upon further recruitment

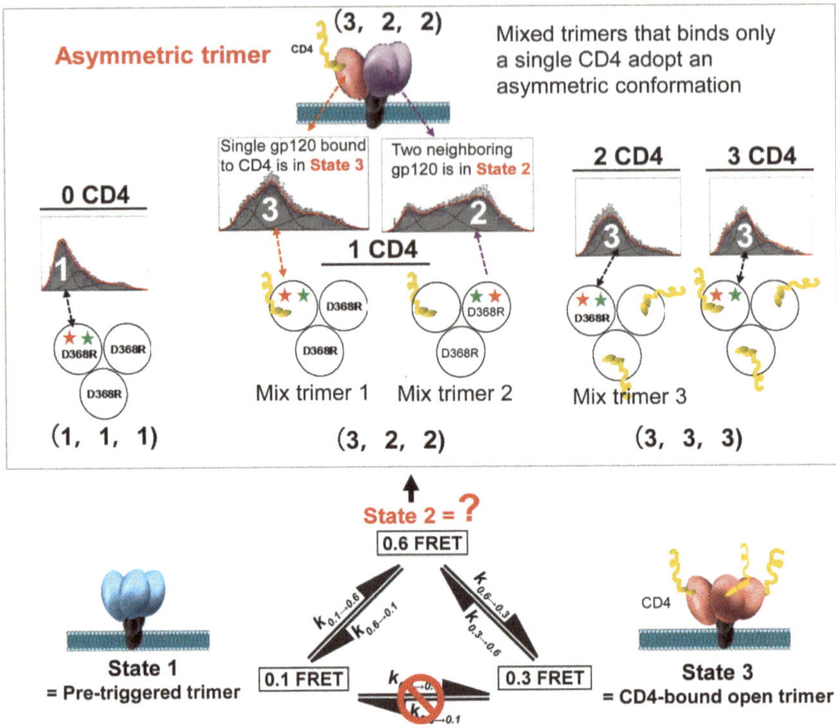

Fig. 3 Experimental scheme utilizing mixed trimers to test possible existence of asymmetric HIV-1 Env trimer intermediates. Env carrying the D368R substitution are incapable of binding CD4 (Ma et al., 2018). Taking advantage of D368R substitutions, trimers were engineered to engage none, a single CD4, two CD4, or three CD4 molecules, respectively. Within the unliganded trimer, all three gp120 protomers are in State 1 resulting in the trimer configuration (1, 1, 1). In contrast, the three CD4-bound trimer resides in a trimer configuration (3, 3, 3). In the case of single CD4-bound trimer, experimental data indicated that gp120 protomers binding the single CD4 adopted State 3. In contrast, protomers adjacent to the single bound CD4, adopted State 2. Thus, the Env trimer is in asymmetric configuration (3, 2, 2). Adding to this, when trimer binds a second CD4 molecule, all three gp120 subunits open into State 3, that is, the trimer is configured as (3, 3, 3). *Adapted from Ma, X.C., Lu, M.L., Gorman, J., Terry, D.S., Hong, X.Y., Zhou, Z., Zhao, H., Altman, R.B., Arthoe, J., Blanchard, S.C., Kwong, P.D., Munro, J.B., Mothes, W., 2018. HIV-1 Env trimer opens through an asymmetric intermediate in which individual protomers adopt distinct conformations. eLife 7, e34271.*

of either CD4 or coreceptor (Ma et al., 2018) (Fig. 4). State 3 within gp120 is likely the last conformational state within gp120 along the activation pathway that can be detected by smFRET. From there, a signal must be transmitted to the underlying gp41 fusion machine to drive membrane fusion (Fig. 4). As we will see below, smFRET analysis of the conformational

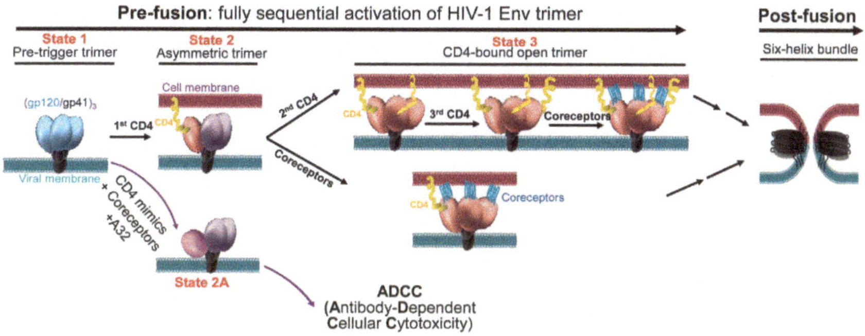

Fig. 4 A smFRET studies-derived model for the activation of HIV-1 Env by receptor and coreceptor. HIV-1 Env predominately resides in pre-triggered closed State 1 conformation, and opens through an asymmetric State 2 in which initially only a single CD4 molecule engages the trimer, to finally completely open into the three CD4-bound State 3 conformation in response to binding of two or three CD4 molecules or coreceptors. One off-pathway event is the formation of asymmetric State 2A that is vulnerable to ADCC. *Adapted from Ma, X.C., Lu, M.L., Gorman, J., Terry, D.S., Hong, X.Y., Zhou, Z., Zhao, H., Altman, R.B., Arthoe, J., Blanchard, S.C., Kwong, P.D., Munro, J.B., Mothes, W., 2018. HIV-1 Env trimer opens through an asymmetric intermediate in which individual protomers adopt distinct conformations. eLife 7, e34271.*

events within HA2, the influenza A equivalent of the HIV-1 gp41, has revealed conformational changes that correlate with membrane fusion (Das et al., 2018).

7. An asymmetric conformational state of HIV-1 Env highly vulnerable to ADCC

Clinical HIV-1 isolates evade the strong antibody response against the highly immunogenic open conformations by closing into the most immune evasive State 1 conformation. As a consequence, the success of bNAbs likely depends on learning to recognize State 1. However, the apparent need for Env to evade the antibody response means that the so-called "non-neutralizing" antibodies are actually potent in removing the vulnerable open conformations thereby forcing the selection of closed Env conformations. Recent work revealed that mechanistically the antibody-dependent cellular cytotoxicity (ADCC) is largely responsible for depleting Env variants that are more open and thus highly vulnerable (Alsahafi et al., 2019; Bournazos et al., 2014; Prevost et al., 2018; Richard et al., 2018; Veillette et al., 2014, 2015). In ADCC, antibodies bind Env expressed on the surface of infected cells, which can be recognized by Fc-receptor expressing cytotoxic CD8+

T cells or NK cells (Forthal and Finzi, 2018). Interestingly, the most ADCC vulnerable conformation of Env allows binding of anti-cluster A antibodies that recognize the inner domain of gp120 (Forthal and Finzi, 2018). This epitope is only accessible in the presence of sCD4 or a small molecule CD4 mimetic (CD4mc), coreceptor mimicking antibodies such as 17b, and anti-cluster A antibodies that recognize the inner domain (Alsahafi et al., 2019; Forthal and Finzi, 2018). Parallel smFRET analysis revealed that Env adopts a fourth conformational state that is connected to State 2, and hence designated State 2A (as shown in Fig. 4) (Alsahafi et al., 2019; Richard et al., 2016). Parallel cryo-electron tomography (cryoET) revealed an asymmetric trimer. Thus, these data suggest that State 2 and State 2A are asymmetric trimer configurations.

These experiments demonstrated the strengths of smFRET and cryo-ET to reveal asymmetric intermediates. Asymmetry is probably more common than previously thought. An asymmetric structural intermediate has also been observed by cryoET for the receptor-bound Env of the Murine Leukemia Virus (Riedel et al., 2017). During HIV-1 fusion, asymmetric activation was observed for the transmembrane gp41 subunit (Khasnis et al., 2016). Finally, molecular dynamic simulation also reveals that HIV-1 Env protomers within a trimer undergo scissoring movements which induce trimer asymmetry (Lemmin et al., 2017).

8. High-resolution structures of HIV-1 Env

Determination of mature cleaved HIV-1 Env trimeric structures have been extremely challenging, mainly due to the weak association between the gp120 and gp41 subunits, the heavy glycan shield, the variable loops, and conformational dynamics. An early breakthrough was the determination of the core gp120 subunit at atomic level in complex with soluble CD4 (sCD4, D1D2 domains) and coreceptor mimicking antibody 17b, followed by the determination of the post-fusion six-helix bundle of gp41 (Buzon et al., 2010; Chan et al., 1997; Kwong et al., 1998; Weissenhorn et al., 1997). The application of cryoET to complete virus particles provided a low resolution ~20 Å view of Env spikes and gradual opening in response to sCD4 and 17b (Liu et al., 2008a). A critical breakthrough in the characterization of the trimer was possible thanks to ~15 years of continued efforts by Rogier Sanders, James Binley and John Moore in collaboration with the groups of Ian Wilson and Andrew Ward that resulted in a stabilized soluble HIV-1 Env trimer (Binley et al., 2000; Ringe et al., 2013; Sanders et al., 2002, 2013). These trimers were designated SOSIPs, SOS for a disulfide bridge that creates a covalent link between the gp120 and the gp41, and IP for a I559P point mutation that destabilizes post-fusion conformations. These soluble trimers were also lacking the transmembrane region (TM), the membrane proximal extracellular region (MPER), were optimized for cleavage efficiency and were engineered using a naturally more stable HIV-1 Env isolate (Sanders et al., 2013). This design allowed the structural characterization of the HIV-1 Env trimer at atomic level by X-ray crystallography and by single-particle cryo-electron microscopy in its ligand-free, or antibody-bound, or small molecules-bound forms (Bartesaghi et al., 2013; Chuang et al., 2017; Julien et al., 2013; Kong et al., 2016; Kwon et al., 2015; Lee et al., 2016, 2017; Lyumkis et al., 2013; Pancera et al., 2014; Scharf et al., 2015; Scheid et al., 2016; Stewart-Jones et al., 2016). Advances in single-particle cryo-electron microscopy allowed the subsequent determination of the structure of the trans-membrane HIV-1_{JR-FL} Env containing the MPER and TM, only lacking the cytoplasmic tail, in complex with the bNAb PGT151 (Lee et al., 2016). Interestingly, all these Env structures determined by various groups and using different methodologies (X-ray crystallography or cryo-EM) revealed similar structural architectures, consolidating the current consensus model of the closed mature HIV-1 Env trimer (Fig. 5). In response to binding to sCD4, with or without the additional presence of 17b fab fragments, the sgp140 SOSIPs open into the CD4-bound conformation (Ozorowski et al., 2017; Wang et al., 2016, 2018). The V1V2 loops undergo a strong outward movement to align near sCD4 (Wang et al., 2016, 2018), and the fusion peptide folds into a new groove (Ozorowski et al., 2017; Wang et al., 2016).

smFRET analyses of the conformational effect of the antibody PGT122 that was used to stabilize the Env trimer by several groups revealed that it stabilized the Env trimer in State 1 (Julien et al., 2013; Pancera et al., 2014). By association, it was believed that these structures represent the pre-triggered closed State 1 conformation of HIV-1 Env seen by smFRET (Munro and Mothes, 2014, 2015; Munro et al., 2014a). On the other hand, several observations had raised questions about the similarity of sgp140 SOSIP.664 trimers to the functional Env on the surface of viruses and cells (Alsahafi et al., 2015, 2018; Kesavardhana and Varadarajan, 2014).

9. Associating the conformational states observed by smFRET with high-resolution structures of HIV-1 Env

To determine how the observed smFRET states relate to these high-resolution structures, donor and acceptor fluorophores were placed at exactly the same positions on both BG505 sgp140 SOSIP.664 and wild-type

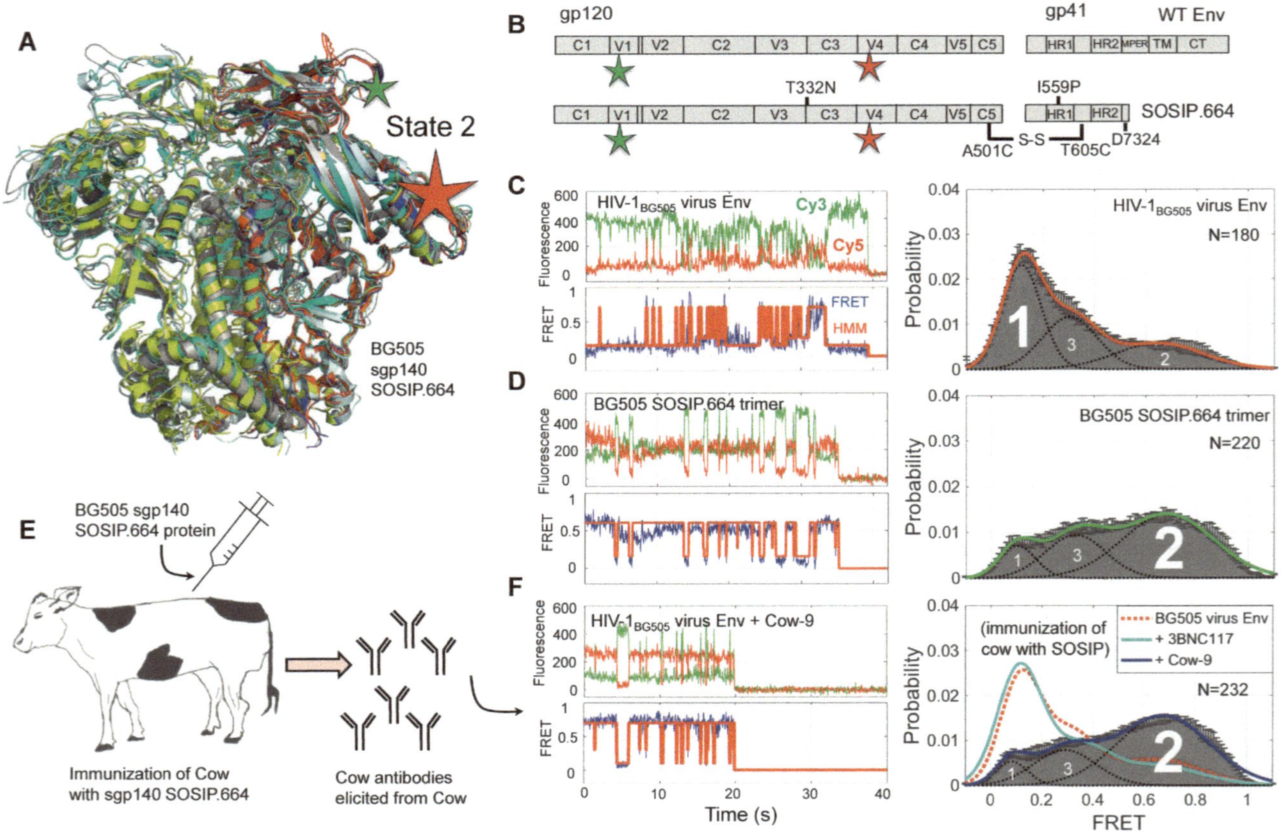

Fig. 5 See legend on opposite page.

BG505 Env on the surface of viral particles (Fig. 5) (Lu et al., 2019). Surprisingly, smFRET revealed that the conformational landscape of sgp140 SOSIP.664 differs from the State 1 Env conformation observed on the surface of virus particles. Instead, BG505 sgp140 SOSIP.664 closely resembles the downstream State 2 conformation (Fig. 5) (Lu et al., 2019). This large shift in the conformational landscape was confirmed for additional BG505 sgp140 SOSIP.664 proteins including further stabilized immunogens (Lu et al., 2019). Moreover, PGT151 used as a ligand in cryo-EM structures with the membrane-bound HIV-$1_{JR\text{-}FL}$ Env (Lee et al., 2016), also stabilized the State 2 on virion resident Env. These observations were verified for a second perspective where a dye was introduced into gp41 using amber suppression click labeling and the distance between gp41 and V4 in gp120 monitored by smFRET (Lu et al., 2019). Collectively, all data indicate that existing high-resolution structures are similar to the State 2 conformation of Env (Fig. 5). The high-resolution structure of State 1 remains unknown. These data are consistent with findings in other laboratories that have also observed differences between HIV-1 virus Env and sgp140 SOSIP.664. For instance, the processing of HIV-1 Env glycans differs (Cao et al., 2018; Go et al., 2017). And a recent report applying cross-linking followed by mass spectrometry also detected dramatic differences between both proteins, specifically within the trimer association domain, the C-terminus of gp120 subunit, and in the heptad repeat 1 (HR1) of gp41 (Castillo-Menendez et al., 2019).

Fig. 5 Existing high-resolution structures of HIV-1 Env trimer and derived immunogens resembles a State 2-like conformation. (A) High-resolution crystal and single-particle electron microscopy structures (PDB codes: 4ZMJ, 4NCO, 3J5M, 4TVP, 5I8H,5CJX, and 8V8L) were aligned and show similar structural architectures. (B) Schemes for comparing soluble gp140 SOSIP.664 with wild-type HIV-1 BG505 Env. Trimer stabilizing modifications SOS, IP, and gp41 truncations are illustrated in sgp140 SOSIP.664. (C and D) Representative FRET trajectories and FRET histograms of wild-type Env on virus surface and soluble gp140 SOSIP.664 indicating predominant State 1 or State 2 occupancies, respectively. (E and F) Immunizing cows with soluble gp140 SOSIP.664 as immunogens elicit cow antibodies that exhibit a preference for State 2. Note that the conformational landscape of HIV-1 Env on virus in the presence of antibody Cow-9 is indistinguishable to the State 2-dominated conformational landscape as that of immunogen itself SOSIP.664. In contrast, most bNAbs developed by patients against the native virus exhibit a preference for State 1. *Adapted from Lu, M., Ma, X., Castillo-Menendez, L.R., Gorman, J., Alsahafi, N., Ermel, U., Terry, D.S., Chambers, M., Peng, D., Zhang, B., Zhou, T., Reichard, N., Wang, K., Grover, J.R., Carmen, B.P., Gardner, M.R., Nikic-Spiegel, I., Sugawara, A., Arthos, J., Lemke, E.A., Smith, A.B., Farzan, M., Abrams, C., Munro, J.B., McDermott, A.B., Finzi, A., Kwong, P.D., Blanchard, S.C., Sodroski, J., Mothes, W., 2019. Associating HIV-1 envelope glycoprotein structures with states on virus observed by smFRET. Nature 568 (7752), 415–419, with permission from Nature.*

The interpretation that existing high-resolution structures correspond to the State 2 seen by smFRET is further supported by comparing the changes in distances between unliganded and CD4-bound sgp140 SOSIP with the corresponding FRET values (Stadtmueller et al., 2018). In the sgp140 SOSIP structures, the V1 and V4 positions where the fluorophores reside are very close. This is consistent with the observed high-FRET signal (Stadtmueller et al., 2018). Upon binding of sCD4, the V1V2 loops move out to align parallel to the bound sCD4 consistent with the lower FRET value (distance between fluorophores increasing, reducing energy transfer) (Ozorowski et al., 2017; Stadtmueller et al., 2018; Wang et al., 2016). Altogether, these observations are consistent with current sgp140 proteins being in State 2.

The smFRET approach revealed that soluble sgp140 SOSIP trimers including trimers further stabilized for immunogen design (Sanders and Moore, 2017), reside in a conformation that resembles State 2 (Lu et al., 2019). Importantly, many such stabilized sgp140 SOSIP proteins are now entering clinical vaccine trials for safety evaluation and to evaluate their potential to elicit bNAbs in patients. We thus tested the conformational preference of antibodies elicited using sgp140 SOSIP in earlier experimental vaccine studies in animals such as cows (Sok et al., 2017). Intriguingly, the sgp140 SOSIP elicited cow antibodies exhibited a preference for State 2 (Fig. 5). This preference is unlike many bNAbs such as 3BNC117 and 10-1074 (Caskey et al., 2017; Scheid et al., 2016) developed in some HIV-1-infected individuals fighting the native virus that exhibit a preference for the State 1 conformation (Lu et al., 2019; Munro et al., 2014a). These data suggest that State 2 specific immunogens elicit State 2-specific antibodies. Going forward, the characterization of a putative structure of State 1 becomes of great importance. A high-resolution structure would facilitate the design of State 1-specific immunogens that should be tested in vaccine settings.

10. smFRET analysis of conformational changes during membrane fusion for the influenza a hemagglutinin fusion glycoprotein (HA2)

smFRET application has elucidated the conformational dynamics of influenza A HA and provided dynamic insights into the fusion reaction of another class I fusion protein (Das et al., 2018). Similar to HIV-1 Env trimer, HA is synthesized as a precursor HA_0, then cleaved by cellular proteases into a metastable trimer of HA_1/HA_2 heterodimers: the receptor sialic acid binding domain HA_1 and transmembrane domain that harbors the fusion machine HA_2. In contrast to HIV-1 Env, influenza A HA_2 is activated by low pH within an endosome (Marsh and Helenius, 2006; White and Whittaker, 2016). Current models for the activation of HA_2 are based on structural characterizations of the uncleaved precursor, the mature protein and the low-pH activated form (Bullough et al., 1994; Carr and Kim, 1993; Stevens et al., 2006; Wilson et al., 1981). In this model, low pH induces a loop-to-helix conformational change within HA_2 that leads to the exposure and subsequent insertion of the fusion peptide into the target membrane, followed by a helix-to-loop transition that brings both membranes together for fusion (Skehel and Wiley, 2000). However, direct real-time measurements of single conformational changes during fusion have never been directly observed. Using unnatural amino acids (Sakin et al., 2017), Das et al. introduced donor and acceptor fluorophores before (near the fusion peptide) and toward the end of the switch regions within the influenza A fusion machine HA_2 (Das et al., 2018). At neutral pH, smFRET analysis revealed that the HA trimer is conformationally dynamic, sampling at least three conformational states: a most populated high-FRET (75% occupancy), an intermediate-FRET, and a low-FRET (Fig. 6). Lowering pH gradually shifts the conformational equilibrium of HA_2 from the prefusion high-FRET conformation to the low-FRET through a short-lived intermediate-FRET state (suggested to be a peptide-released intermediate in which the fusion peptide is positioned out of the hydrophobic pocket in HA_2). For a few minutes (up to 15), this conformational change is reversible and HA_2 can fully transition back to the high-FRET if the pH is neutralized. Interestingly, the presence of receptor sialic acid on target membranes was shown to increase the transitions into the low-FRET conformation even at neutral pH, and to accelerate the low pH induced conformational changes. The distinct features of the reversible low-FRET intermediate and the final irreversible low-FRET conformation suggest that two distinct conformational states are behind this FRET state, and that the pre-hairpin intermediate was not detectable due to the predicted long distance between the dyes (Fig. 6). Thus, smFRET indicates that HA_2 is activated through two structural intermediates and that these transitions are initially reversible. Reversibility would explain how HA is transported through the acidic Golgi lumen in infected cells without activation, and that only receptor and low pH on target cells most effectively activate HA_2 for fusion (Benhaim and Lee, 2018; Das et al., 2018). Going forward, monitoring additional perspectives to resolve these two conformational states, and

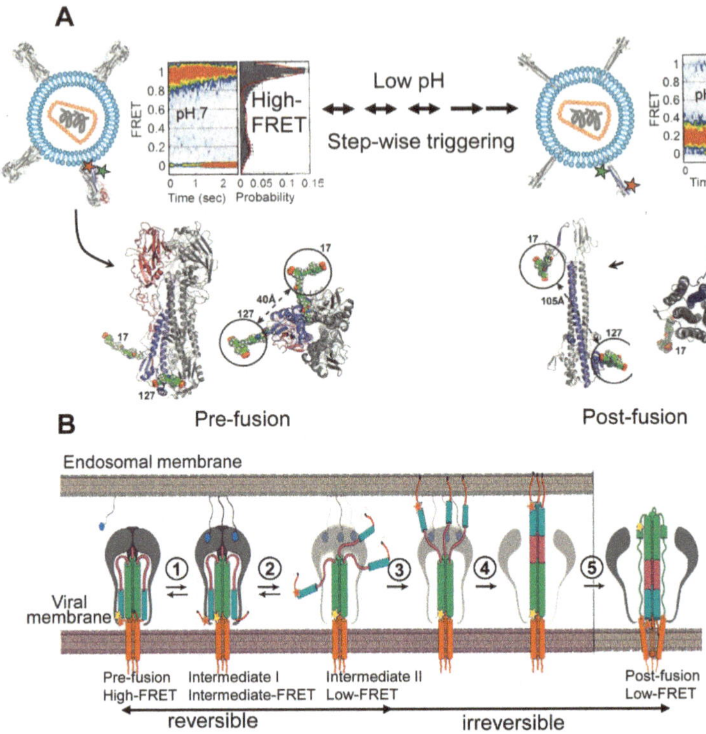

Fig. 6 smFRET imaging of influenza virus HA_2 reveals the step-wise conformational changes in HA_2 during receptor-binding, low pH triggering, and membrane fusion. (A) Low pH triggers conformational changes in HA_2. Conformational landscapes show large shifts from prefusion (high-FRET) to post-fusion (irreversible low-FRET) structures of HA_2 in response to lowering the pH. (B) Proposed model of HA-mediated influenza virus membrane fusion based on smFRET analysis. Facilitated by binding to sialic acid receptor, HA_2 initially reversibly transits from high-FRET pre-fusion state, through intermediate-FRET (peptide-released intermediate I), to reversible low-FRET intermediate II (steps 1 and 2). Accelerated by the combined presence of receptor and low pH, HA_2 transitions from the reversible low-FRET intermediate II to the irreversible low-FRET post-fusion state. *Adapted from Benhaim, M., Lee, K.K., 2018. Single-molecule analysis of a viral fusion protein illuminates a fusion-active intermediate state. Cell 174, 775–777; Das, D. K., Govindan, R., Nikic-Spiegel, I., Krammer, F., Lemke, E.A., Munro, J.B., 2018. Direct visualization of the conformational dynamics of single influenza hemagglutinin trimers. Cell 174, 926–937, with permission from Elsevier.*

correlating conformational changes within HA with the lipid mixing reaction that proceeds through hemi-fusion, flickering and widening fusion pores will result in additional crucial insights (Chernomordik and Kozlov, 2003, 2008; Melikyan, 2014; Melikyan et al., 2005).

11. Conformational dynamics during HIV-1 reverse transcription

Following virus entry, capsid uncoating and reverse transcription are the next steps in the viral life cycle of retroviruses. During reverse transcription, the reverse transcriptase (RT) reverse transcribes the viral RNA (vRNA) into double-stranded DNA. RT is a multifunctional enzyme that contains the polymerase (Pol) and ribonuclease H (RNase H) subunits. Structurally, RT is a heterodimer of p51 and p66 which share both a common amino terminus and four polymerase subdomains (fingers, palm, thumb, and connection) (Huang et al., 1998; Kohlstaedt et al., 1992; Sarafianos et al., 2009). RT begins with the formation of the HIV-1 reverse transcriptase initiation complex that includes the RT bound to primer (tRNA)-template vRNA complex (RT/tRNA/vRNA). The early phase of reverse transcription is a slow, non-processive process. It involves slow DNA polymerization, fast disassociation of RT and RNA complex, and frequent pauses of RT at different locations (Lanchy et al., 1996). While the first strand of DNA is synthesized, the RNA is cleaved from the RNA/DNA hybrid intermediate by the RNase H subunit. Leftover tighter bound RNA stretches then functions as primers for the synthesis of the second strand of DNA. It has been speculated that the initiation complex changes progressively in structural/conformational configurations along the RT catalytic pathway; however, it has been technically difficult to capture those dynamic features.

Several smFRET studies on RT have revealed surprising insights into the underlying mechanism of how RT (Pol and RNase H) interacts with the RNA and DNA templates using RNA and DNA primers to initiate Pol and RNase H activities, respectively. smFRET studies of labeling the duplex substrates with a dye in combination with another dye on RT p66 subunit at different locations (Fig. 7) enabled direct observations of enzyme-substrate associating/disassociating events of RT catalysis in real time (Abbondanzieri et al., 2008; Liu et al., 2008b, 2010). Using this elegant experimental design, orientational and translational dynamics of RT were further uncovered (Fig. 7) (Abbondanzieri et al., 2008; Liu et al., 2008b, 2010). RT was observed to adopt two horizontally-flipped binding orientations on nucleic acid duplexes substrates depending on whether a RNA or DNA primers were bound, directing its DNA synthesis or RNA hydrolysis activity, respectively (Fig. 7). RT switching between those two orientations was regulated by nucleotides and non-nucleoside reverse transcriptase inhibitors, a

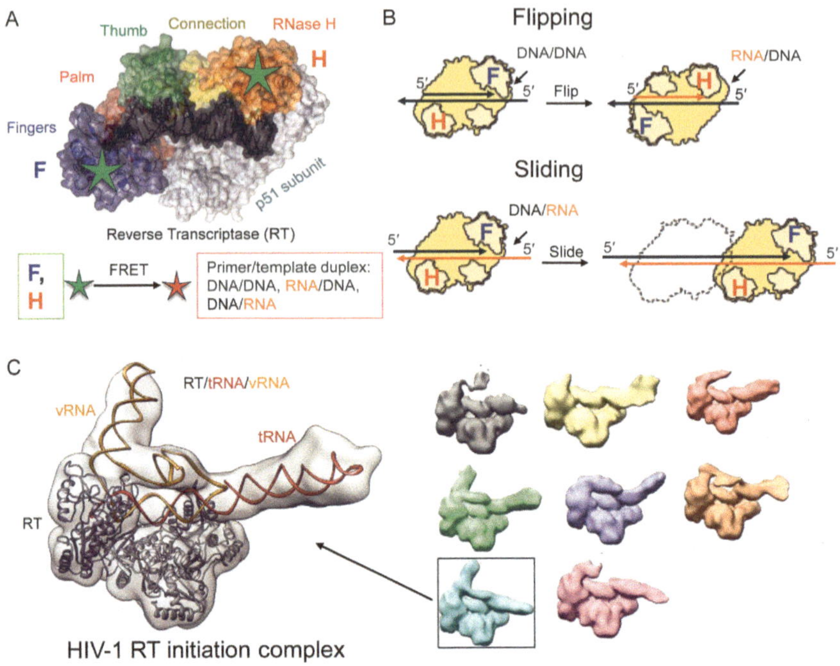

Fig. 7 Conformational variabilities in HIV-1 reverse transcriptase (RT) and RNA complex revealed by smFRET analysis. (A) Single-molecule FRET experimental design for monitoring the dynamics of RT. The overall structural density map of HIV-1 RT bound to a DNA/DNA duplex substrate. RT is labeled with Cy3, and primer/template duplex (DNA/DNA, RNA/DNA, DNA/RNA) are labeled with Cy5 dyes. Labeling sites for Cy3 on RT are located at either Fingers (F) or RNase H domains (H), highlighted by green stars. (B) RT exhibits orientational and sliding dynamics on primer/template nucleic acid substrates. smFRET reveals that RT is able to flip its orientation by engaging different substrates for adjusting its multifunctional catalysis, and RT can slide on nucleic acid substrates to facilitate polymerization site targeting during reverse transcription. (C) Single-particle cryo-EM high-resolution structure of HIV-1 RT initiation complex (RT/tRNA/vRNA). EM density map reveals that vRNA and tRNA are highly dynamic in conformations. *Adapted from Abbondanzieri, E.A., Bokinsky, G., Rausch, J.W., Zhang, J.X., Le Grice, S.F., Zhuang, X., 2008. Dynamic binding orientations direct activity of HIV reverse transcriptase. Nature 453, 184–189; Larsen, K.P., Mathiharan, Y.K., Kappel, K., Coey, A.T., Chen, D.H., Barrero, D., Madigan, L., Puglisi, J.D., Skiniotis, G., Puglisi, E.V., 2018. Architecture of an HIV-1 reverse transcriptase initiation complex. Nature 557, 118–122; Liu, S., Abbondanzieri, E.A., Rausch, J.W., Le Grice, S.F., Zhuang, X., 2008. Slide into action: dynamic shuttling of HIV reverse transcriptase on nucleic acid substrates. Science 322, 1092–1097, with permission from AAAS and Nature.*

major drug class of anti-retroviral therapy (Abbondanzieri et al., 2008). Extensive smFRET assay further revealed that those opposite orientations of RT are closely related to polymerization-competent and RT-paused polymerization-incompetent configurations of initiation complex in early phase of reverse transcription, respectively (Liu et al., 2010).

Besides this large-scale orientational motion (180° flipping), RT was also able to undergo long-range sliding on nucleic acid duplexes substrates, rapidly shuttling between two termini of substrates (Fig. 7) (Liu et al., 2008b). The sliding motion of RT play critical roles in multiple phase of reverse transcriptase: it shuttles RT to the tRNA terminus for DNA polymerization, translocates RT to multiple sites of RNA/DNA hybrid for vRNA degradation, and displaces non-template strand (Liu et al., 2008b). Flipping and sliding go hand-by-hand to effectively help RT carry out the catalysis of vRNA to double-stranded DNA.

RT is not the only component of RT/tRNA/vRNA complex involved in reverse transcription that is constantly dynamic. Indeed, tRNA/vRNA together or alone are also dynamic in real time. The combination of bulk NMR and smFRET allowed the observation of tRNA/vRNA hybrids adopting two or more conformations, suggesting conformational heterogeneity formed by conserved RNA sequences within HIV-1 reverse transcription initiation complex (Coey et al., 2016).

These smFRET applications highlighted the existence of multiple dynamically exchanging structures within the RT/tRNA/vRNA complex, and provide a framework of how various RT inhibitors shift the equilibrium between different states. And as in many areas, combining smFRET techniques with other structural methods such as single-particle cryo-electron microscopy is predicted to be particularly powerful. One such successful study describes the initiation complex determined by single-particle cryo-electron microscopy (Larsen et al., 2018). Strikingly, helices of highly dynamic primer/template (tRNA/vRNA) become visible at high-resolution. The overall structure (8 Å) shows RT being in an inactive polymerase conformation with open finger and thumb subdomain (4.5 Å), whereas the primer/template RNA complex moves away from the active sites (Fig. 7). Parallel smFRET studies helped to define the structures of highly dynamic and disordered parts of the complex. Altogether, smFRET applications in combination with additional structural techniques have allowed a detailed scrutiny into the slow initiation phase of RT, exposing its vulnerability to inhibition by RT inhibitors.

12. smFRET monitoring of conformational changes in HIV-1 genomes and capsid protein during HIV-1 Gag assembly

HIV-1 assembly is mediated by HIV-1 Gag polyprotein. The Gag polyprotein is synthesized as a Gag precursor that contains the matrix (MA), capsid (CA), nucleocapsid (NC) and p6 domain. The N-terminal

MA domain targets Gag to the plasma membrane through its specificity for *phosphatidylinositol 4,5*-bisphosphate (*PIP*$_2$) (Jouvenet et al., 2008; Mucksch et al., 2017; Ono et al., 2004). CA drives Gag multimerization into the immature viral core, NC recruits a dimeric RNA genome into an assembling particle through binding to the packaging signal (Psi), which is located in the 5′UTR of the RNA (Keane and Summers, 2016; Lu et al., 2011a) and the p6 domain recruits the endosomal sorting machinery to promote the release of the immature virus particle (Bell and Lever, 2013; Freed, 2015; Ganser-Pornillos et al., 2012; Mattei et al., 2016; Sundquist and Krausslich, 2012). Live cell imaging has revealed that assembly and genome packaging are coordinated whereby the genome stimulates the nucleation of capsid assembly (Jouvenet et al., 2009). Both Gag and RNA are conformationally dynamic and smFRET studies have provided insights into both processes (Beerens et al., 2013; Brigham et al., 2019; Coey et al., 2018; Jouvenet et al., 2009; Munro et al., 2014b).

Gag has been proposed to undergo major conformational changes, from a compact conformation in solution to the extended rod-like conformation observed within the final immature capsid (Datta et al., 2007; Ganser-Pornillos et al., 2012; Mattei et al., 2016; Sundquist and Krausslich, 2012). smFRET has allowed direct measurements of the conformational state of individual Gag molecules in solution as well within virus particles (Munro et al., 2014b). By site-specifically placing two dyes on different domains of the Gag polyprotein, smFRET analysis demonstrated that Gag monomers undergo transitions between a compact conformation and an extended linear conformation (Munro et al., 2014b). These two distinct Gag conformations form viral-like particles (VLPs) with different spherical morphology: in the compact conformation, Gag forms VLPs with an approximately ~30 nm in diameter in the presence of nucleic acid; with the presence of additional inositol hexaphosphate, Gag exhibits an extended rod-like conformation, and forms VLPs of ~100 nm in diameter (Munro et al., 2014b). Parallel fluorescence correlation spectroscopy (FCS) measurements combined with smFRET confirmed that the compact-to-extended conformational transition happens early when the Gag oligomers are still very small (Munro et al., 2014b). These experiments represent the first measurement of the conformational state of Gag within complete virus-like particles. Also, at that time, we speculated that inositol hexaphosphate may mimic the role of phosphoinositides in the plasma membrane to trigger this conformational change, but today we know that inositol hexaphosphate plays an integral role in stabilizing the HIV capsid (Dick et al., 2018; Mallery et al., 2018).

The packaging of a specifically dimeric RNA genome is incompletely understood. A dimerization initiation site (DIS) lies within the 5′UTR of the Psi packaging signal on the genomic RNA. The transition from RNA monomer to dimer is regulated by the U5 region. In the monomer, the U5 pairs with the DIS, thereby preventing dimer formation, while U5 pairing with a stem loop harboring the AUG Gag start codon, frees up the DIS for dimerization. The RNA is highly heterogeneous and competing models have been proposed for how monomers undergo conformational changes to form dimers that are specifically packaged into particles (Lu et al., 2011b). In the case of the murine leukemia virus, NC has a higher affinity for the dimerized genome (Miyazaki et al., 2010). In the case of HIV, smFRET and bulk in-gel FRET experiments have recently helped to elucidate how tRNA and NC promote transition to an extended dimer (Brigham et al., 2019). Intra-molecular conformational transitions within a single monomer were measured with fluorophores positioned at the termini of a 238-nucleotide RNA segment, which contains the Psi packaging signal as well as the primer binding site for the tRNA. Inter-molecular interactions between two RNA molecules were measured with fluorophores placed into separate RNA molecules. smFRET measurements indicated that the monomeric Psi can spontaneously sample the dimerization-competent and -incompetent forms. The annealing of tRNA to the monomer promoted the formation of the DIS exposed conformation, thereby shifting the equilibrium toward the kissing loop dimerization. NC binding then promoted the formation of the extended dimer, a conformation that facilitates packaging (Brigham et al., 2019). Thus, these experiments suggest that genome packaging proceeds through a series of events involving primer annealing, genome dimerization and packaging.

13. Concluding remarks

smFRET has previously provided critical insights into the dynamics of viral proteins throughout their life cycle (Abbondanzieri et al., 2008; Coey et al., 2016; Liu et al., 2008b, 2010). Here we focused on recent applications particularly in the Mothes as well as the Munro laboratories that have extended these applications to physiologically relevant conditions such as complete HIV virions (Brigham et al., 2019; Das et al., 2018; Lu et al., 2019; Ma et al., 2018; Munro et al., 2014a,b). smFRET identified a new conformational Env state on the surface of HIV-1 (Lu et al., 2019), detected and characterized novel structural intermediates in the activation of both HIV-1 and influenza A viral glycoproteins (Das et al., 2018; Ma et al., 2018), and identified a new conformational state of Env that is highly vulnerable to ADCC (Alsahafi et al., 2019). Going forward, monitoring additional perspectives to provide better structural overviews, and correlating conformational changes within envelope glycoproteins with the membrane fusion reaction that proceeds through hemi-fusion, flickering and widening fusion pores will result in additional crucial insights (Chernomordik and Kozlov, 2008; Melikyan, 2014). Since it is emerging that reverse transcription is also coupled to the uncoating of the viral capsid (Hulme et al., 2011), great insights can be expected from expanding RT studies to complete virions. Finally, most powerful approach would be a seamless integration of dynamic methods such as smFRET and NMR with single-particle and cryo-electron microscopy to provide a comprehensive characterization of the structure and dynamics of biological processes.

Of great help are recent advances in unnatural amino acid and click chemistry labeling technologies that are expanding our ability to place fluorophores into sterically restricted regions of proteins that don't tolerate enzymatic labeling tags or free cysteines. This technology has allowed access into conformational changes within tight viral fusion machines (Das et al., 2018; Lu et al., 2019). The generation of membrane permeable dyes will also make unnatural amino acid and click chemistry available to imaging in living cells, which at some point can be expected to permit smFRET studies in living cells. The continued application of smFRET methods to increasingly physiologically relevant experimental conditions will enrich our knowledge of the conformational events during the viral life cycle, and inform the design of antiviral inhibitors, as well as immunogens for vaccine design.

Acknowledgments

We thank Andrés Finzi, Jonathan R. Grover and James B. Munro for the critical reading of the manuscript, and James B. Munro for sharing unpublished results. Work in this area in the Mothes laboratory is supported by NIH grants RO1 GM116654, P01 AI150471, and P50 AI150464 to W.M., and by a Brown Coxe Fellowship to M.L.

Author contributions

M.L. and X.M. generated the figures, and M.L. and W.M. wrote the manuscript.

References

Abbondanzieri, E.A., Bokinsky, G., Rausch, J.W., Zhang, J.X., Le Grice, S.F., Zhuang, X., 2008. Dynamic binding orientations direct activity of HIV reverse transcriptase. Nature 453, 184–189.

Akyuz, N., Georgieva, E.R., Zhou, Z., Stolzenberg, S., Cuendet, M.A., Khelashvili, G., Altman, R.B., Terry, D.S., Freed, J.H., Weinstein, H., Boudker, O., Blanchard, S.C., 2015. Transport domain unlocking sets the uptake rate of an aspartate transporter. Nature 518, 68–73.

Alsahafi, N., Debbeche, O., Sodroski, J., Finzi, A., 2015. Effects of the I559P gp41 change on the conformation and function of the human immunodeficiency virus (HIV-1) membrane envelope glycoprotein trimer. PLoS One 10, e0129405.

Alsahafi, N., Anand, S.P., Castillo-Menendez, L., Verly, M.M., Medjahed, H., Prevost, J., Herschhorn, A., Richard, J., Schon, A., Melillo, B., Freire, E., Smith III, A.B., Sodroski, J., Finzi, A., 2018. SOSIP changes affect human immunodeficiency virus (HIV-1) envelope glycoprotein conformation and CD4 engagement. J. Virol. 92, e01080-18.

Alsahafi, N., Bakouche, N., Kazemi, M., Richard, J., Ding, S., Bhattacharyya, S., Das, D., Anand, S.P., Mothes, W., Lifson, J., Sodroski, J., Smith, A.B., Rouiller, I., Finzi, A., Munro, J.B., 2019. An asymmetric opening of HIV-1 Env is required for anti-cluster A antibody binding. Cell Host Microbe 25 (4), 578–587.

Arthos, J., Cicala, C., Steenbeke, T.D., Chun, T.W., Dela Cruz, C., Hanback, D.B., Khazanie, P., Nam, D., Schuck, P., Selig, S.M., Van Ryk, D., Chaikin, M.A., Fauci, A.S., 2002. Biochemical and biological characterization of a dodecameric CD4-Ig fusion protein: implications for therapeutic and vaccine strategies. J. Biol. Chem. 277, 11456–11464.

Barouch, D.H., 2008. Challenges in the development of an HIV-1 vaccine. Nature 455, 613–619.

Bartesaghi, A., Merk, A., Borgnia, M.J., Milne, J.L., Subramaniam, S., 2013. Prefusion structure of trimeric HIV-1 envelope glycoprotein determined by cryo-electron microscopy. Nat. Struct. Mol. Biol. 20, 1352–1357.

Beerens, N., Jepsen, M.D., Nechyporuk-Zloy, V., Kruger, A.C., Darlix, J.L., Kjems, J., Birkedal, V., 2013. Role of the primer activation signal in tRNA annealing onto the HIV-1 genome studied by single-molecule FRET microscopy. RNA 19, 517–526.

Bell, N.M., Lever, A.M., 2013. HIV Gag polyprotein: processing and early viral particle assembly. Trends Microbiol. 21, 136–144.

Benhaim, M., Lee, K.K., 2018. Single-molecule analysis of a viral fusion protein illuminates a fusion-active intermediate state. Cell 174, 775–777.

Binley, J.M., Sanders, R.W., Clas, B., Schuelke, N., Master, A., Guo, Y., Kajumo, F., Anselma, D.J., Maddon, P.J., Olson, W.C., Moore, J.P., 2000. A recombinant human immunodeficiency virus type 1 envelope glycoprotein complex stabilized by an intermolecular disulfide bond between the gp120 and gp41 subunits is an antigenic mimic of the trimeric virion-associated structure. J. Virol. 74, 627–643.

Blanco-Melo, D., Venkatesh, S., Bieniasz, P.D., 2012. Intrinsic cellular defenses against human immunodeficiency viruses. Immunity 37, 399–411.

Bournazos, S., Klein, F., Pietzsch, J., Seaman, M.S., Nussenzweig, M.C., Ravetch, J.V., 2014. Broadly neutralizing anti-HIV-1 antibodies require Fc effector functions for in vivo activity. Cell 158, 1243–1253.

Brandenburg, B., Zhuang, X., 2007. Virus trafficking—learning from single-virus tracking. Nat. Rev. Microbiol. 5, 197–208.

Brigham, B., Kitzrow, J., Reyes, J.P., Musier-Forsyth, K., Munro, J.B., 2019. Intrinsic conformational dynamics of the HIV-1 genomic RNA 5′ UTR. Proc. Natl. Acad. Sci. U. S. A. 116 (21), 10372–10381.

Bullough, P.A., Hughson, F.M., Skehel, J.J., Wiley, D.C., 1994. Structure of influenza haemagglutinin at the pH of membrane fusion. Nature 371, 37–43.

Bushman, F.D., Malani, N., Fernandes, J., D'Orso, I., Cagney, G., Diamond, T.L., Zhou, H., Hazuda, D.J., Espeseth, A.S., Konig, R., Bandyopadhyay, S., Ideker, T., Goff, S.P., Krogan, N.J., Frankel, A.D., Young, J.A., Chanda, S.K., 2009. Host cell factors in HIV replication: meta-analysis of genome-wide studies. PLoS Pathog. 5, e1000437.

Buzon, V., Natrajan, G., Schibli, D., Campelo, F., Kozlov, M.M., Weissenhorn, W., 2010. Crystal structure of HIV-1 gp41 including both fusion peptide and membrane proximal external regions. PLoS Pathog. 6, e1000880.

Campbell, E.M., Hope, T.J., 2008. Live cell imaging of the HIV-1 life cycle. Trends Microbiol. 16, 580–587.

Cao, L.W., Pauthner, M., Andrabi, R., Rantalainen, K., Berndsen, Z., Diedrich, J.K., Menis, S., Sok, D., Bastidas, R., Park, S.K.R., Delahunty, C.M., He, L., Guenaga, J., Wyatt, R.T., Schief, W.R., Ward, A.B., Yates, J.R., Burton, D.R., Paulson, J.C., 2018. Differential processing of HIV envelope glycans on the virus and soluble recombinant trimer. Nat. Commun. 9, 3693.

Carette, J.E., Raaben, M., Wong, A.C., Herbert, A.S., Obernosterer, G., Mulherkar, N., Kuehne, A.I., Kranzusch, P.J., Griffin, A.M., Ruthel, G., Dal Cin, P., Dye, J.M., Whelan, S.P., Chandran, K., Brummelkamp, T.R., 2011. Ebola virus entry requires the cholesterol transporter Niemann-Pick C1. Nature 477, 340–343.

Carr, C.M., Kim, P.S., 1993. A spring-loaded mechanism for the conformational change of influenza hemagglutinin. Cell 73, 823–832.

Caskey, M., Schoofs, T., Gruell, H., Settler, A., Karagounis, T., Kreider, E.F., Murrell, B., Pfeifer, N., Nogueira, L., Oliveira, T.Y., Learn, G.H., Cohen, Y.Z., Lehmann, C., Gillor, D., Shimeliovich, I., Unson-O'Brien, C., Weiland, D., Robles, A., Kummerle, T., Wyen, C., Levin, R., Witmer-Pack, M., Eren, K., Ignacio, C., Kiss, S., West Jr., A.P., Mouquet, H., Zingman, B.S., Gulick, R.M., Keler, T., Bjorkman, P.J., Seaman, M.S., Hahn, B.H., Fatkenheuer, G., Schlesinger, S.J., Nussenzweig, M.C., Klein, F., 2017. Antibody 10-1074 suppresses viremia in HIV-1-infected individuals. Nat. Med. 23, 185–191.

Castillo-Menendez, L.R., Nguyen, H.T., Sodroski, J., 2019. Conformational differences between functional human immunodeficiency virus envelope glycoprotein trimers and stabilized soluble trimers. J. Virol. 93, e01709–e01718.

Chan, D.C., Fass, D., Berger, J.M., Kim, P.S., 1997. Core structure of gp41 from the HIV envelope glycoprotein. Cell 89, 263–273.

Chernomordik, L.V., Kozlov, M.M., 2003. Protein-lipid interplay in fusion and fission of biological membranes. Annu. Rev. Biochem. 72, 175–207.

Chernomordik, L.V., Kozlov, M.M., 2008. Mechanics of membrane fusion. Nat. Struct. Mol. Biol. 15, 675–683.

Chuang, G.Y., Geng, H., Pancera, M., Xu, K., Cheng, C., Acharya, P., Chambers, M., Druz, A., Tsybovsky, Y., Wanninger, T.G., Yang, Y., Doria-Rose, N.A., Georgiev, I.S., Gorman, J., Joyce, M.G., O'Dell, S., Zhou, T., McDermott, A.B., Mascola, J.R., Kwong, P.D., 2017. Structure-based design of a soluble prefusion-closed HIV-1 Env trimer with reduced CD4 affinity and improved immunogenicity. J. Virol. 91, e02268-16.

Coey, A., Larsen, K., Puglisi, J.D., Viani Puglisi, E., 2016. Heterogeneous structures formed by conserved RNA sequences within the HIV reverse transcription initiation site. RNA 22, 1689–1698.

Coey, A.T., Larsen, K.P., Choi, J., Barrero, D.J., Puglisi, J.D., Puglisi, E.V., 2018. Dynamic interplay of RNA and protein in the human immunodeficiency virus-1 reverse transcription initiation complex. J. Mol. Biol. 430, 5137–5150.

Cohen, M.S., Chen, Y.Q., McCauley, M., Gamble, T., Hosseinipour, M.C., Kumarasamy, N., Hakim, J.G., Kumwenda, J., Grinsztejn, B., Pilotto, J.H., Godbole, S.V., Mehendale, S., Chariyalertsak, S., Santos, B.R., Mayer, K.H., Hoffman, I.F., Eshleman, S.H., Piwowar-Manning, E., Wang, L., Makhema, J., Mills, L.A., de Bruyn, G., Sanne, I., Eron, J., Gallant, J., Havlir, D., Swindells, S., Ribaudo, H., Elharrar, V., Burns, D., Taha, T.E., Nielsen-Saines, K., Celentano, D., Essex, M., Fleming, T.R., Team, H.S., 2011. Prevention of HIV-1 infection with early antiretroviral therapy. N. Engl. J. Med. 365, 493–505.

Crispin, M., Ward, A.B., Wilson, I.A., 2018. Structure and immune recognition of the HIV glycan shield. Annu. Rev. Biophys. 47, 499–523.

Crooks, E.T., Tong, T., Chakrabarti, B., Narayan, K., Georgiev, I.S., Menis, S., Huang, X., Kulp, D., Osawa, K., Muranaka, J., Stewart-Jones, G., Destefano, J., O'Dell, S., LaBranche, C., Robinson, J.E., Montefiori, D.C., McKee, K., Du, S.X., Doria-Rose, N., Kwong, P.D., Mascola, J.R., Zhu, P., Schief, W.R., Wyatt, R.T., Whalen, R.G., Binley, J.M., 2015. Vaccine-elicited tier 2 HIV-1 neutralizing antibodies bind to quaternary epitopes involving glycan-deficient patches proximal to the CD4 binding site. PLoS Pathog. 11, e1004932.

Das, K., Arnold, E., 2013. HIV-1 reverse transcriptase and antiviral drug resistance. Part 1. Curr. Opin. Virol. 3, 111–118.

Das, D.K., Govindan, R., Nikic-Spiegel, I., Krammer, F., Lemke, E.A., Munro, J.B., 2018. Direct visualization of the conformational dynamics of single influenza hemagglutinin trimers. Cell 174, 926–937.

Datta, S.A., Curtis, J.E., Ratcliff, W., Clark, P.K., Crist, R.M., Lebowitz, J., Krueger, S., Rein, A., 2007. Conformation of the HIV-1 Gag protein in solution. J. Mol. Biol. 365, 812–824.

Dick, R.A., Zadrozny, K.K., Xu, C., Schur, F.K.M., Lyddon, T.D., Ricana, C.L., Wagner, J.M., Perilla, J.R., Ganser-Pornillos, B.K., Johnson, M.C., Pornillos, O., Vogt, V.M., 2018. Inositol phosphates are assembly co-factors for HIV-1. Nature 560, 509–512.

Dunkle, J.A., Wang, L., Feldman, M.B., Pulk, A., Chen, V.B., Kapral, G.J., Noeske, J., Richardson, J.S., Blanchard, S.C., Cate, J.H., 2011. Structures of the bacterial ribosome in classical and hybrid states of tRNA binding. Science 332, 981–984.

Forthal, D.N., Finzi, A., 2018. Antibody-dependent cellular cytotoxicity in HIV infection. AIDS 32, 2439–2451.

Freed, E.O., 2015. HIV-1 assembly, release and maturation. Nat. Rev. Microbiol. 13, 484–496.

Ganser-Pornillos, B.K., Yeager, M., Pornillos, O., 2012. Assembly and architecture of HIV. Adv. Exp. Med. Biol. 726, 441–465.

Go, E.P., Ding, H., Zhang, S., Ringe, R.P., Nicely, N., Hua, D., Steinbock, R.T., Golabek, M., Alin, J., Alam, S.M., Cupo, A., Haynes, B.F., Kappes, J.C., Moore, J.P., Sodroski, J.G., Desaire, H., 2017. Glycosylation benchmark profile for HIV-1 envelope glycoprotein production based on eleven Env trimers. J. Virol. 91, e02428-16.

Gregorio, G.G., Masureel, M., Hilger, D., Terry, D.S., Juette, M., Zhao, H., Zhou, Z., Perez-Aguilar, J.M., Hauge, M., Mathiasen, S., Javitch, J.A., Weinstein, H., Kobilka, B.K., Blanchard, S.C., 2017. Single-molecule analysis of ligand efficacy in beta2AR-G-protein activation. Nature 547, 68–73.

Gristick, H.B., von Boehmer, L., West Jr., A.P., Schamber, M., Gazumyan, A., Golijanin, J., Seaman, M.S., Fatkenheuer, G., Klein, F., Nussenzweig, M.C., Bjorkman, P.J., 2016. Natively glycosylated HIV-1 Env structure reveals new mode for antibody recognition of the CD4-binding site. Nat. Struct. Mol. Biol. 23, 906–915.

Harrison, S.C., 2008. Viral membrane fusion. Nat. Struct. Mol. Biol. 15, 690–698.

Harrison, S.C., 2015. Viral membrane fusion. Virology 479-480, 498–507.

Henzler-Wildman, K., Kern, D., 2007. Dynamic personalities of proteins. Nature 450, 964–972.

Herschhorn, A., Ma, X., Gu, C., Ventura, J.D., Castillo-Menendez, L., Melillo, B., Terry, D.S., Smith 3rd, A.B., Blanchard, S.C., Munro, J.B., Mothes, W., Finzi, A., Sodroski, J., 2016. Release of gp120 restraints leads to an entry-competent intermediate state of the HIV-1 envelopeglycoproteins. mBio 7, e01598-16.

Huang, H., Chopra, R., Verdine, G.L., Harrison, S.C., 1998. Structure of a covalently trapped catalytic complex of HIV-1 reverse transcriptase: implications for drug resistance. Science 282, 1669–1675.

Hulme, A.E., Perez, O., Hope, T.J., 2011. Complementary assays reveal a relationship between HIV-1 uncoating and reverse transcription. Proc. Natl. Acad. Sci. U. S. A. 108, 9975–9980.

Hunter, E., 1997. Viral entry and receptors. In: Coffin, J.M., Hughes, S.H., Varmus, H.E. (Eds.), Retroviruses. Cold Spring Harbor, New York.

Jouvenet, N., Bieniasz, P.D., Simon, S.M., 2008. Imaging the biogenesis of individual HIV-1 virions in live cells. Nature 454, 236–240.

Jouvenet, N., Simon, S.M., Bieniasz, P.D., 2009. Imaging the interaction of HIV-1 genomes and Gag during assembly of individual viral particles. Proc. Natl. Acad. Sci. U. S. A. 106, 19114–19119.

Julien, J.P., Lee, P.S., Wilson, I.A., 2012. Structural insights into key sites of vulnerability on HIV-1 Env and influenza HA. Immunol. Rev. 250, 180–198.

Julien, J.P., Cupo, A., Sok, D., Stanfield, R.L., Lyumkis, D., Deller, M.C., Klasse, P.J., Burton, D.R., Sanders, R.W., Moore, J.P., Ward, A.B., Wilson, I.A., 2013. Crystal structure of a soluble cleaved HIV-1 envelope trimer. Science 342, 1477–1483.

Keane, S.C., Summers, M.F., 2016. NMR studies of the structure and function of the HIV-1 5'-leader. Viruses 8, 338.

Kesavardhana, S., Varadarajan, R., 2014. Stabilizing the native trimer of HIV-1 Env by destabilizing the heterodimeric interface of the gp41 postfusion six-helix bundle. J. Virol. 88, 9590–9604.

Khasnis, M.D., Halkidis, K., Bhardwaj, A., Root, M.J., 2016. Receptor activation of HIV-1 Env leads to asymmetric exposure of the gp41 trimer. PLoS Pathog. 12, e1006098.

Kohlstaedt, L., Wang, J., Friedman, J., Rice, P., Steitz, T., 1992. Crystal structure at 3.5 A resolution of HIV-1 reverse transcriptase complexed with an inhibitor. Science 256, 1783–1790.

Kong, R., Xu, K., Zhou, T.Q., Acharya, P., Lemmin, T., Liu, K., Ozorowski, G., Soto, C., Taft, J.D., Bailer, R.T., Cale, E.M., Chen, L., Choi, C.W., Chuang, G.Y., Doria-Rose, N.A., Druz, A., Georgiev, I.S., Gorman, J., Huang, J.H., Joyce, M.G., Louder, M.K., Ma, X.C., Mckee, K., O'Dell, S., Pancera, M., Yang, Y.P., Blanchard, S.C., Mothes, W., Burton, D.R., Koff, W.C., Connors, M., Ward, A.B., Kwong, P.D., Mascola, J.R., 2016. Fusion peptide of HIV-1 as a site of vulnerability to neutralizing antibody. Science 352, 828–833.

Kwon, Y.D., Pancera, M., Acharya, P., Georgiev, I.S., Crooks, E.T., Gorman, J., Joyce, M.G., Guttman, M., Ma, X., Narpala, S., Soto, C., Terry, D.S., Yang, Y., Zhou, T., Ahlsen, G., Bailer, R.T., Chambers, M., Chuang, G.Y., Doria-Rose, N.A., Druz, A., Hallen, M.A., Harned, A., Kirys, T., Louder, M.K., O'Dell, S., Ofek, G., Osawa, K., Prabhakaran, M., Sastry, M., Stewart-Jones, G.B., Stuckey, J., Thomas, P.V., Tittley, T., Williams, C., Zhang, B., Zhao, H., Zhou, Z., Donald, B.R., Lee, L.K., Zolla-Pazner, S., Baxa, U., Schon, A., Freire, E., Shapiro, L., Lee, K.K., Arthos, J., Munro, J.B., Blanchard, S.C., Mothes, W., Binley, J.M., McDermott, A.B., Mascola, J.R., Kwong, P.D., 2015. Crystal structure, conformational fixation and entry-related interactions of mature ligand-free HIV-1 Env. Nat. Struct. Mol. Biol. 22, 522–531.

Kwong, P.D., Wyatt, R., Robinson, J., Sweet, R.W., Sodroski, J., Hendrickson, W.A., 1998. Structure of an HIV gp120 envelope glycoprotein in complex with the CD4 receptor and a neutralizing human antibody. Nature 393, 648–659.

Kwong, P.D., Doyle, M.L., Casper, D.J., Cicala, C., Leavitt, S.A., Majeed, S., Steenbeke, T.D., Venturi, M., Chaiken, I., Fung, M., Katinger, H., Parren, P.W.L.H., Robinson, J., Van Ryk, D., Wang, L.P., Burton, D.R., Freire, E., Wyatt, R., Sodroski, J., Hendrickson, W.A., Arthos, J., 2002. HIV-1 evades antibody-mediated neutralization through conformational masking of receptor-binding sites. Nature 420, 678–682.

Kwong, P.D., Mascola, J.R., Nabel, G.J., 2013. Broadly neutralizing antibodies and the search for an HIV-1 vaccine: the end of the beginning. Nat. Rev. Immunol. 13, 693–701.

Lanchy, J.M., Ehresmann, C., Le Grice, S.F., Ehresmann, B., Marquet, R., 1996. Binding and kinetic properties of HIV-1 reverse transcriptase markedly differ during initiation and elongation of reverse transcription. EMBO J. 15, 7178–7187.

Larsen, K.P., Mathiharan, Y.K., Kappel, K., Coey, A.T., Chen, D.H., Barrero, D., Madigan, L., Puglisi, J.D., Skiniotis, G., Puglisi, E.V., 2018. Architecture of an HIV-1 reverse transcriptase initiation complex. Nature 557, 118–122.

Lee, J.H., Ozorowski, G., Ward, A.B., 2016. Cryo-EM structure of a native, fully glycosylated, cleaved HIV-1 envelope trimer. Science 351, 1043–1048.

Lee, J.H., Andrabi, R., Su, C.Y., Yasmeen, A., Julien, J.P., Kong, L., Wu, N.C., McBride, R., Sok, D., Pauthner, M., Cottrell, C.A., Nieusma, T., Blattner, C., Paulson, J.C., Klasse, P.J., Wilson, I.A., Burton, D.R., Ward, A.B., 2017. A broadly neutralizing antibody targets the dynamic HIV envelope trimer apex via a long, rigidified, and anionic beta-hairpin structure. Immunity 46, 690–702.

Lemmin, T., Soto, C., Stuckey, J., Kwong, P.D., 2017. Microsecond dynamics and network analysis of the HIV-1 SOSIP Env trimer reveal collective behavior and conserved microdomains of the glycan shield. Structure 25, 1631–1639.

Lerner, E., Cordes, T., Ingargiola, A., Alhadid, Y., Chung, S., Michalet, X., Weiss, S., 2018. Toward dynamic structural biology: two decades of single-molecule Forster resonance energy transfer. Science 359, eaan1133.

Liu, J., Bartesaghi, A., Borgnia, M.J., Sapiro, G., Subramaniam, S., 2008a. Molecular architecture of native HIV-1 gp120 trimers. Nature 455, 109–113.

Liu, S., Abbondanzieri, E.A., Rausch, J.W., Le Grice, S.F., Zhuang, X., 2008b. Slide into action: dynamic shuttling of HIV reverse transcriptase on nucleic acid substrates. Science 322, 1092–1097.

Liu, S., Harada, B.T., Miller, J.T., Le Grice, S.F., Zhuang, X., 2010. Initiation complex dynamics direct the transitions between distinct phases of early HIV reverse transcription. Nat. Struct. Mol. Biol. 17, 1453–1460.

Lu, K., Heng, X., Garyu, L., Monti, S., Garcia, E.L., Kharytonchyk, S., Dorjsuren, B., Kulandaivel, G., Jones, S., Hiremath, A., Divakaruni, S.S., LaCotti, C., Barton, S., Tummillo, D., Hosic, A., Edme, K., Albrecht, S., Telesnitsky, A., Summers, M.F., 2011a. NMR detection of structures in the HIV-1 5'-leader RNA that regulate genome packaging. Science 334, 242–245.

Lu, K., Heng, X., Summers, M.F., 2011b. Structural determinants and mechanism of HIV-1 genome packaging. J. Mol. Biol. 410, 609–633.

Lu, M., Ma, X., Castillo-Menendez, L.R., Gorman, J., Alsahafi, N., Ermel, U., Terry, D.S., Chambers, M., Peng, D., Zhang, B., Zhou, T., Reichard, N., Wang, K., Grover, J.R., Carmen, B.P., Gardner, M.R., Nikic-Spiegel, I., Sugawara, A., Arthos, J., Lemke, E.A., Smith, A.B., Farzan, M., Abrams, C., Munro, J.B., McDermott, A.B., Finzi, A., Kwong, P.D., Blanchard, S.C., Sodroski, J., Mothes, W., 2019. Associating HIV-1 envelope glycoprotein structures with states on virus observed by smFRET. Nature 568 (7752), 415–419.

Lyumkis, D., Julien, J.P., de Val, N., Cupo, A., Potter, C.S., Klasse, P.J., Burton, D.R., Sanders, R.W., Moore, J.P., Carragher, B., Wilson, I.A., Ward, A.B., 2013. Cryo-EM structure of a fully glycosylated soluble cleaved HIV-1 envelope trimer. Science 342, 1484–1490.

Ma, X.C., Lu, M.L., Gorman, J., Terry, D.S., Hong, X.Y., Zhou, Z., Zhao, H., Altman, R.B., Arthoe, J., Blanchard, S.C., Kwong, P.D., Munro, J.B., Mothes, W., 2018. HIV-1 Env trimer opens through an asymmetric intermediate in which individual protomers adopt distinct conformations. Elife 7, e34271.

Mallery, D.L., Marquez, C.L., McEwan, W.A., Dickson, C.F., Jacques, D.A., Anandapadamanaban, M., Bichel, K., Towers, G.J., Saiardi, A., Bocking, T., James, L.C., 2018. IP6 is an HIV pocket factor that prevents capsid collapse and promotes DNA synthesis. eLife 7, e35335.

Mao, Y., Wang, L., Gu, C., Herschhorn, A., Xiang, S.H., Haim, H., Yang, X., Sodroski, J., 2012. Subunit organization of the membrane-bound HIV-1 envelope glycoprotein trimer. Nat. Struct. Mol. Biol. 19, 893–899.

Marsh, M., Helenius, A., 2006. Virus entry: open sesame. Cell 124, 729–740.

Mattei, S., Schur, F.K., Briggs, J.A., 2016. Retrovirus maturation-an extraordinary structural transformation. Curr. Opin. Virol. 18, 27–35.

McFadden, M.J., Gokhale, N.S., Horner, S.M., 2017. Protect this house: cytosolic sensing of viruses. Curr. Opin. Virol. 22, 36–43.

Melikyan, G.B., 2014. HIV entry: a game of hide-and-fuse? Curr. Opin. Virol. 4, 1–7.

Melikyan, G.B., Barnard, R.J.O., Abrahamyan, L.G., Mothes, W., Young, J.A.T., 2005. Imaging individual retroviral fusion events: from hemifusion to pore formation and growth. Proc. Natl. Acad. Sci. U. S. A. 102, 8728–8733.

Miyazaki, Y., Garcia, E.L., King, S.R., Iyalla, K., Loeliger, K., Starck, P., Syed, S., Telesnitsky, A., Summers, M.F., 2010. An RNA structural switch regulates diploid genome packaging by Moloney murine leukemia virus. J. Mol. Biol. 396, 141–152.

Montefiori, D.C., Roederer, M., Morris, L., Seaman, M.S., 2018. Neutralization tiers of HIV-1. Curr. Opin. HIV AIDS 13, 128–136.

Mucksch, F., Laketa, V., Muller, B., Schultz, C., Krausslich, H.G., 2017. Synchronized HIV assembly by tunable PIP2 changes reveals PIP2 requirement for stable Gag anchoring. eLife 6, e25287.

Mukhopadhyay, S., Kuhn, R.J., Rossmann, M.G., 2005. A structural perspective of the flavivirus life cycle. Nat. Rev. Microbiol. 3, 13–22.

Mulder, L.C., Harari, A., Simon, V., 2008. Cytidine deamination induced HIV-1 drug resistance. Proc. Natl. Acad. Sci. U. S. A. 105, 5501–5506.

Munro, J.B., Mothes, W., 2014. The HIV-1 Env trimer in HD. Structure 22, 935–936.

Munro, J.B., Mothes, W., 2015. Structure and dynamics of the native HIV-1 Env trimer. J. Virol. 89, 5752–5755.

Munro, J.B., Altman, R.B., Tung, C.S., Cate, J.H.D., Sanbonmatsu, K.Y., Blanchard, S.C., 2010. Spontaneous formation of the unlocked state of the ribosome is a multistep process. Proc. Natl. Acad. Sci. U. S. A. 107, 709–714.

Munro, J.B., Gorman, J., Ma, X., Zhou, Z., Arthos, J., Burton, D.R., Koff, W.C., Courter, J.R., Smith 3rd, A.B., Kwong, P.D., Blanchard, S.C., Mothes, W., 2014a. Conformational dynamics of single HIV-1 envelope trimers on the surface of native virions. Science 346, 759–763.

Munro, J.B., Nath, A., Farber, M., Datta, S.A., Rein, A., Rhoades, E., Mothes, W., 2014b. A conformational transition observed in single HIV-1 Gag molecules during in vitro assembly of virus-like particles. J. Virol. 88, 3577–3585.

Ono, A., Ablan, S.D., Lockett, S.J., Nagashima, K., Freed, E.O., 2004. Phosphatidylinositol (4,5) bisphosphate regulates HIV-1 Gag targeting to the plasma membrane. Proc. Natl. Acad. Sci. U. S. A. 101, 14889–14894.

Ozorowski, G., Pallesen, J., de Val, N., Lyumkis, D., Cottrell, C.A., Torres, J.L., Copps, J., Stanfield, R.L., Cupo, A., Pugach, P., Moore, J.P., Wilson, I.A., Ward, A.B., 2017. Open and closed structures reveal allostery and pliability in the HIV-1 envelope spike. Nature 547, 360–363.

Pancera, M., Zhou, T., Druz, A., Georgiev, I.S., Soto, C., Gorman, J., Huang, J., Acharya, P., Chuang, G.Y., Ofek, G., Stewart-Jones, G.B., Stuckey, J., Bailer, R.T., Joyce, M.G., Louder, M.K., Tumba, N., Yang, Y., Zhang, B., Cohen, M.S., Haynes, B.F., Mascola, J.R., Morris, L., Munro, J.B., Blanchard, S.C., Mothes, W.,

Connors, M., Kwong, P.D., 2014. Structure and immune recognition of trimeric pre-fusion HIV-1 Env. Nature 514, 455–461.
Plass, T., Milles, S., Koehler, C., Schultz, C., Lemke, E.A., 2011. Genetically encoded copper-free click chemistry. Angew. Chem. Int. Ed. Engl. 50, 3878–3881.
Prevost, J., Richard, J., Ding, S., Pacheco, B., Charlebois, R., Hahn, B.H., Kaufmann, D.E., Finzi, A., 2018. Envelope glycoproteins sampling states 2/3 are susceptible to ADCC by sera from HIV-1-infected individuals. Virology 515, 38–45.
Richard, J., Pacheco, B., Gohain, N., Veillette, M., Ding, S.L., Alsahafi, N., Tolbert, W.D., Prevost, J., Chapleau, J.P., Coutu, M., Jia, M.X., Brassard, N., Park, J., Courter, J.R., Melillo, B., Martin, L., Tremblay, C., Hahn, B.H., Kaufmanna, D.E., Wu, X.L., Smith, A.B., Sodroski, J., Pazgier, M., Finzi, A., 2016. Co-receptor binding site antibodies enable CD4-mimetics to expose conserved anti-cluster A ADCC epitopes on HIV-1 envelope glycoproteins. EBioMedicine 12, 208–218.
Richard, J., Prevost, J., Alsahafi, N., Ding, S., Finzi, A., 2018. Impact of HIV-1 envelope conformation on ADCC responses. Trends Microbiol. 26, 253–265.
Riedel, C., Vasishtan, D., Siebert, C.A., Whittle, C., Lehmann, M.J., Mothes, W., Grunewald, K., 2017. Native structure of a retroviral envelope protein and its conformational change upon interaction with the target cell. J. Struct. Biol. 197, 172–180.
Ringe, R.P., Sanders, R.W., Yasmeen, A., Kim, H.J., Lee, J.H., Cupo, A., Korzun, J., Derking, R., van Montfort, T., Julien, J.P., Wilson, I.A., Klasse, P.J., Ward, A.B., Moore, J.P., 2013. Cleavage strongly influences whether soluble HIV-1 envelope glycoprotein trimers adopt a native-like conformation. Proc. Natl. Acad. Sci. U. S. A. 110, 18256–18261.
Ringe, R.P., Pugach, P., Cottrell, C.A., LaBranche, C.C., Seabright, G.E., Ketas, T.J., Ozorowski, G., Kumar, S., Schorcht, A., van Gils, M.J., Crispin, M., Montefiori, D.C., Wilson, I.A., Ward, A.B., Sanders, R.W., Klasse, P.J., Moore, J.P., 2019. Closing and opening holes in the glycan shield of HIV-1 envelope glycoprotein SOSIP trimers can redirect the neutralizing antibody response to the newly unmasked epitopes. J. Virol. 93, e01656-18.
Roy, R., Hohng, S., Ha, T., 2008. A practical guide to single-molecule FRET. Nat. Methods 5, 507–516.
Sakin, V., Hanne, J., Dunder, J., Anders-Osswein, M., Laketa, V., Nikic, I., Krausslich, H.G., Lemke, E.A., Muller, B., 2017. A versatile tool for live-cell imaging and super-resolution nanoscopy studies of HIV-1 Env distribution and mobility. Cell Chem. Biol. 24, 635–645.
Sanders, R.W., Moore, J.P., 2017. Native-like Env trimers as a platform for HIV-1 vaccine design. Immunol. Rev. 275, 161–182.
Sanders, R.W., Vesanen, M., Schuelke, N., Master, A., Schiffner, L., Kalyanaraman, R., Paluch, M., Berkhout, B., Maddon, P.J., Olson, W.C., Lu, M., Moore, J.P., 2002. Stabilization of the soluble, cleaved, trimeric form of the envelope glycoprotein complex of human immunodeficiency virus type 1. J. Virol. 76, 8875–8889.
Sanders, R.W., Derking, R., Cupo, A., Julien, J.P., Yasmeen, A., de Val, N., Kim, H.J., Blattner, C., de la Pena, A.T., Korzun, J., Golabek, M., de Los Reyes, K., Ketas, T.J., van Gils, M.J., King, C.R., Wilson, I.A., Ward, A.B., Klasse, P.J., Moore, J.P., 2013. A next-generation cleaved, soluble HIV-1 Env trimer, BG505 SOSIP.664 gp140, expresses multiple epitopes for broadly neutralizing but not non-neutralizing antibodies. PLoS Pathog. 9, e1003618.
Sarafianos, S.G., Marchand, B., Das, K., Himmel, D.M., Parniak, M.A., Hughes, S.H., Arnold, E., 2009. Structure and function of HIV-1 reverse transcriptase: molecular mechanisms of polymerization and inhibition. J. Mol. Biol. 385, 693–713.
Sattentau, Q., 2008. Avoiding the void: cell-to-cell spread of human viruses. Nat. Rev. Microbiol. 6, 815–826.

Scharf, L., Wang, H., Gao, H., Chen, S., McDowall, A.W., Bjorkman, P.J., 2015. Broadly neutralizing antibody 8ANC195 recognizes closed and open states of HIV-1 Env. Cell 162, 1379–1390.

Scheid, J.F., Horwitz, J.A., Bar-On, Y., Kreider, E.F., Lu, C.L., Lorenzi, J.C., Feldmann, A., Braunschweig, M., Nogueira, L., Oliveira, T., Shimeliovich, I., Patel, R., Burke, L., Cohen, Y.Z., Hadrigan, S., Settler, A., Witmer-Pack, M., West Jr., A.P., Juelg, B., Keler, T., Hawthorne, T., Zingman, B., Gulick, R.M., Pfeifer, N., Learn, G.H., Seaman, M.S., Bjorkman, P.J., Klein, F., Schlesinger, S.J., Walker, B.D., Hahn, B.H., Nussenzweig, M.C., Caskey, M., 2016. HIV-1 antibody 3BNC117 suppresses viral rebound in humans during treatment interruption. Nature 535, 556–560.

Skehel, J.J., Wiley, D.C., 1998. Coiled coils in both intracellular vesicle and viral membrane fusion. Cell 95, 871–874.

Skehel, J.J., Wiley, D.C., 2000. Receptor binding and membrane fusion in virus entry: the influenza hemagglutinin. Annu. Rev. Biochem. 69, 531–569.

Sok, D., Le, K.M., Vadnais, M., Saye-Francisco, K.L., Jardine, J.G., Torres, J.L., Berndsen, Z.T., Kong, L., Stanfield, R., Ruiz, J., Ramos, A., Liang, C.H., Chen, P.L., Criscitiello, M.F., Mwangi, W., Wilson, I.A., Ward, A.B., Smider, V.V., Burton, D.R., 2017. Rapid elicitation of broadly neutralizing antibodies to HIV by immunization in cows. Nature 548, 108–111.

Stadtmueller, B.M., Bridges, M.D., Dam, K.M., Lerch, M.T., Huey-Tubman, K.E., Hubbell, W.L., Bjorkman, P.J., 2018. DEER spectroscopy measurements reveal multiple conformations of HIV-1 SOSIP envelopes that show similarities with envelopes on native virions. Immunity 49, 235–246.

Stevens, J., Blixt, O., Tumpey, T.M., Taubenberger, J.K., Paulson, J.C., Wilson, I.A., 2006. Structure and receptor specificity of the hemagglutinin from an H5N1 influenza virus. Science 312, 404–410.

Stewart-Jones, G.B., Soto, C., Lemmin, T., Chuang, G.Y., Druz, A., Kong, R., Thomas, P.V., Wagh, K., Zhou, T., Behrens, A.J., Bylund, T., Choi, C.W., Davison, J.R., Georgiev, I.S., Joyce, M.G., Kwon, Y.D., Pancera, M., Taft, J., Yang, Y., Zhang, B., Shivatare, S.S., Shivatare, V.S., Lee, C.C., Wu, C.Y., Bewley, C.A., Burton, D.R., Koff, W.C., Connors, M., Crispin, M., Baxa, U., Korber, B.T., Wong, C.H., Mascola, J.R., Kwong, P.D., 2016. Trimeric HIV-1-Env structures define glycan shields from clades A, B, and G. Cell 165, 813–826.

Sundquist, W.I., Krausslich, H.G., 2012. HIV-1 assembly, budding, and maturation. Cold Spring Harb. Perspect. Med. 2, a006924.

Uchil, P.D., Haugh, K.D., Pi, R., Mothes, W., 2019. In vivo imaging driven approaches to study virus dissemination and pathogenesis. Annu. Rev. Virol. 6, 19.1–19.24.

Vaney, M.C., Rey, F.A., 2011. Class II enveloped viruses. Cell. Microbiol. 13, 1451–1459.

Veillette, M., Desormeaux, A., Medjahed, H., Gharsallah, N.E., Coutu, M., Baalwa, J., Guan, Y., Lewis, G., Ferrari, G., Hahn, B.H., Haynes, B.F., Robinson, J.E., Kaufmann, D.E., Bonsignori, M., Sodroski, J., Finzi, A., 2014. Interaction with cellular CD4 exposes HIV-1 envelope epitopes targeted by antibody-dependent cell-mediated cytotoxicity. J. Virol. 88, 2633–2644.

Veillette, M., Coutu, M., Richard, J., Batraville, L.A., Dagher, O., Bernard, N., Tremblay, C., Kaufmann, D.E., Roger, M., Finzi, A., 2015. The HIV-1 gp120 CD4-bound conformation is preferentially targeted by antibody-dependent cellular cytotoxicity-mediating antibodies in sera from HIV-1-infected individuals. J. Virol. 89, 545–551.

Wagh, K., Kreider, E.F., Li, Y., Barbian, H.J., Learn, G.H., Giorgi, E., Hraber, P.T., Decker, T.G., Smith, A.G., Gondim, M.V., Gillis, L., Wandzilak, J., Chuang, G.Y., Rawi, R., Cai, F., Pellegrino, P., Williams, I., Overbaugh, J., Gao, F., Kwong, P.D., Haynes, B.F., Shaw, G.M., Borrow, P., Seaman, M.S., Hahn, B.H., Korber, B.,

2018. Completeness of HIV-1 envelope glycan shield at transmission determines neutralization breadth. Cell Rep. 25, 893–908.

Wang, H., Cohen, A.A., Galimidi, R.P., Gristick, H.B., Jensen, G.J., Bjorkman, P.J., 2016. Cryo-EM structure of a CD4-bound open HIV-1 envelope trimer reveals structural rearrangements of the gp120 V1V2 loop. Proc. Natl. Acad. Sci. U. S. A. 113, E7151–E7158.

Wang, H., Barnes, C.O., Yang, Z., Nussenzweig, M.C., Bjorkman, P.J., 2018. Partially open HIV-1 envelope structures exhibit conformational changes relevant for coreceptor binding and fusion. Cell Host Microbe 24, 579–592.

Wei, X., Decker, J.M., Wang, S., Hui, H., Kappes, J.C., Wu, X., Salazar-Gonzalez, J.F., Salazar, M.G., Kilby, J.M., Saag, M.S., Komarova, N.L., Nowak, M.A., Hahn, B.H., Kwong, P.D., Shaw, G.M., 2003. Antibody neutralization and escape by HIV-1. Nature 422, 307–312.

Weissenhorn, W., Dessen, A., Harrison, S.C., Skehel, J.J., Wiley, D.C., 1997. Atomic structure of the ectodomain from HIV-1 gp41. Nature 387, 426–430.

White, J.M., Whittaker, G.R., 2016. Fusion of enveloped viruses in endosomes. Traffic 17, 593–614.

Wilson, I.A., Skehel, J.J., Wiley, D.C., 1981. Structure of the haemagglutinin membrane glycoprotein of influenza virus at 3 Å resolution. Nature 289, 366–373.

Wyatt, R., Kwong, P.D., Desjardins, E., Sweet, R.W., Robinson, J., Hendrickson, W.A., Sodroski, J.G., 1998. The antigenic structure of the HIV gp120 envelope glycoprotein. Nature 393, 705–711.

Zhao, Y.F., Terry, D., Shi, L., Weinstein, H., Blanchard, S.C., Javitch, J.A., 2010. Single-molecule dynamics of gating in a neurotransmitter transporter homologue. Nature 465, 188–U173.

Zheng, Q., Juette, M.F., Jockusch, S., Wasserman, M.R., Zhou, Z., Altman, R.B., Blanchard, S.C., 2014. Ultra-stable organic fluorophores for single-molecule research. Chem. Soc. Rev. 43, 1044–1056.

CHAPTER NINE

Polyhedra, spindles, phage nucleus and pyramids: Structural biology of viral superstructures

Fasséli Coulibaly*
Infection & Immunity Program, Department of Biochemistry and Molecular Biology, Biomedicine Discovery Institute, Monash University, Clayton, VIC, Australia
*Corresponding author: e-mail address: fasseli.coulibaly@monash.edu

Contents

Abstract

Viral infection causes comprehensive rearrangements of the cell that reflect as much host defense mechanisms as virus-induced structures assembled to facilitate infection. Regardless of their pro- or antiviral role, large intracellular structures are readily detectable by microscopy and often provide a signature characteristic of a specific viral

ISSN 0065-3527
https://doi.org/10.1016/bs.aivir.2019.08.001

infection. The structural features and localization of these assemblies have thus been commonly used for the diagnostic and classification of viruses since the early days of virology. More recently, characterization of viral superstructures using molecular and structural approaches have revealed very diverse organizations and roles, ranging from dynamic viral factories behaving like liquid organelles to ultra-stable crystals embedding and protecting virions.

This chapter reviews the structures, functions and biotechnological applications of virus-induced superstructures with a focus on assemblies that have a regular organization, for which detailed structural descriptions are available. Examples span viruses infecting all domains of life including the assembly of virions into crystalline arrays in eukaryotic and bacterial viruses, nucleus-like compartments involved in the replication of large bacteriophages, and pyramid-like structures mediating the egress of archaeal viruses. Among these superstructures, high-resolution structures are available for crystalline objects produced by insect viruses: viral polyhedra which function as the infectious form of occluded viruses, and spindles which are potent virulence factors of entomopoxviruses. In turn, some of these highly symmetrical objects have been used to develop and validate advanced structural approaches, pushing the boundary of structural biology.

Abbreviations

3D	tridimensional
AA10	auxiliary activity 10
bp	base pair
Bt	*Bacillus thuringiensis*
CBP21	chitin-binding protein 21
CSPAD	Cornell-SLAC pixel array detector
CT	C-terminal loop of fusolin
Cter	C-terminal loop of BV polyhedrin/granulin
C-terminus	carboxy terminus
DFFN	double flow focus nozzle
ds	double stranded
EM	electron microscopy
ER	endoplasmic reticulum
ES	embryonic stem cells
ESRF	European Synchrotron Radiation Facility
FGF-2	fibroblast growth factor-2
FIB	focused ion beam
Fn3	fibronectin type 3
GFP	green fluorescent protein
GP37	glycoprotein 37
ICTV	international committee on taxonomy of viruses
iPS	induced pluripotent stem cells
kDa	kilo Dalton
LCLS	Linac Coherent Light Source
LIF	leukemia inhibitory factor
LPMO	lytic polysaccharide monooxygenase

MCP	major capsid protein
MIR	multiple isomorphous replacement method
MTOC	microtubule-organization center
NIH/3T3	national institutes of health/"3-day transfer, inoculum 3×10^5 cells"
NMR	nuclear magnetic resonance
N-terminus	amino terminus
ODV	occlusion-derived virion
PDB	protein data bank
PhuZ	*Ph*age t*u*bulin/Fts*Z*
PISA	proteins, interfaces, structures and assemblies server
pK_a	logarithmic acid dissociation constant
PM	peritrophic membrane
PVAP	protein forming virus-associated pyramids
RMSD	root mean square deviation
SDS	sodium dodecyl sulfate
SEM	scanning electron microscopy
SFX	serial femtosecond crystallography
SLS	Swiss light source
SYN	synchrotron crystallography
T	triangulation
UV	ultraviolet
v/v	volume to volume
VAP	virus-associated pyramids
VP3	viral protein 3
XFEL	X-ray free electron laser

Virus names

201ϕ2-1	bacteriophage 201ϕ2-1
***Ac*MNPV**	*Autographa californica* multiple NPV
BmCPV	*Bombyx mori* CPV
BV	baculovirus
CIV	chilo iridovirus
***Cp*GV**	*Cydia pomonella* GV
CPV	cypovirus
EV	entomopoxvirus
GV	granulovirus
IBDV	infectious bursal disease virus
***Mm*EV**	*Melolontha melolontha* EV
NCLDV	nucleo-cytoplasmic large DNA viruses
NPV	nucleopolyhedrovirus
SIRV2	*Sulfolobus islandicus* rod-shaped virus 2
STIV1	*Sulfolobus* turreted icosahedral virus 1
TMV	tobacco mosaic virus
ϕKZ	bacteriophage ϕKZ
ϕPA3	bacteriophage ϕPA3

1. Introduction

Viruses have historically been defined by the small physical size of their infectious particles. While the discovery of giant viruses has blurred the boundary between viruses and cellular organisms (La Scola et al., 2003), the parasitic nature of viruses means that they can forgo many components that are essential to cellular life. In fact, most viruses heavily rely on genetic economy to maximize their replication efficiency and persist in a context of competition with the evolving host and other viral species. The broad use of self-assembly utilizing symmetrical arrangements of simple protein building blocks is perhaps one of the most striking ways of achieving such efficiency. This requirement was postulated by Watson and Crick from the very beginning of molecular biology to reconcile the limiting coding capacity of viral genomes with the comparatively large size of their capsids (Crick and Watson, 1956) and was supported by Caspar's analysis of X-ray diffraction data (Caspar, 1956). Indeed, exploration of the virosphere by pioneering structural biologists has never ceased to reveal elegant ways in which viruses use the icosahedral and helical symmetries to assemble virus particles of unique structures (Caspar and Klug, 1962; Rossmann, 2013).

In this pursuit of viral discovery and taxonomy, two concurrent trends balance each other. On one hand, modern large-scale metagenomics approaches have allowed an ever more comprehensive and unbiased description of the diversity of the virus world—moving away from an animal-centric view focused on pathogens directly affecting human activities and health (Paez-Espino et al., 2016; Shi et al., 2016, 2018; Simmonds et al., 2017). On the other hand, structural virology has established evolutionary connections between large groups of viruses by revealing long-distance evolutionary links where homology was undetectable at the primary DNA or protein sequence level (e.g., herpesviruses and tailed bacteriophages). These connections may be the result of common ancestry of the viruses as well as horizontal gene transfer. Together, these approaches have mapped a limited number of lineages that encompass a large part of the known virosphere (Abrescia et al., 2012; Krupovic and Bamford, 2010, 2011; Krupovic and Koonin, 2017).

Supported by constant innovation in structural biology, advances in the comprehension of viral architectures have been impressive. Excellent reviews summarize the current knowledge of virion structures (Prasad and Schmid, 2012; Rossmann, 2013). This chapter will focus on the elegant use of regular, and often symmetrical, superstructures that are not structural components of the virus particles. The review will not attempt to be exhaustive due to the remarkable diversity of strategies adopted by viruses across all domains of life, and the continuum between amorphous, ordered and highly symmetrical superstructures used by viruses. It will present select examples involved in key steps of the virus cycle from replication, assembly, egress and extracellular stability, to virulence enhancement.

2. Inclusion bodies, viroplasms and virion superassembly

With the exception of latent stages, most viruses cause vast changes in the biology of the cell. These changes commonly result in a cytopathic effect evidenced by a modification of the cell morphology including swelling and rounding. Another particularly striking symptom of viral infection is the formation of virus-induced intracellular bodies. These bodies are of very different nature depending on the specific virus or even different phases of the infectious cycle for a given virus. They can be separated into three broad classes: membrane-bound replication sites; virosomes or viral factories; and inclusion bodies. Membrane-bound replication sites are formed by remodeling of cellular compartments and commonly produced by positive-strand RNA viruses. Virosomes and viral factories refer to similar sites of replication concentrated in specific foci but that are not bound by an external membrane. This is well described for eukaryotic viruses as reviewed by Netherton and colleagues (Netherton and Wileman, 2011; Netherton et al., 2007) including large DNA viruses such as the Nucleo-Cytoplasmic Large DNA viruses (NCLDV) and herpesviruses; and small DNA viruses such as adenovirus and papillomavirus. More recently, membrane-less viral compartments have also been described as liquid organelles for negative-strand RNA viruses such as rabies (Nikolic et al., 2017) and borna disease viruses (Hirai et al., 2016). While such membrane-bound replication sites and viral factories have been comprehensively reviewed elsewhere (Jin et al., 2018; Moshe and Gorovits, 2012; Netherton and Wileman, 2011; Netherton et al., 2007), a bacterial example is discussed in Section 5.2. The term inclusion body is rather confusingly used interchangeably with viral factory and other virus-induced dense assemblies in many older publications. Here, we use the term inclusion bodies to refer specifically to membrane-less aggregates that are not associated with replication of the viral genome and typically form later in the infectious cycle. They may be formed by superstructures of the virions themselves or aggregates of viral proteins and/or nucleic acid. The functions of these inclusion bodies are diverse, ranging from dead-end products of antiviral cell responses to specific viral assembly sites essential to the infectious cycle.

While the structure and role of these assemblies are generally poorly understood, they have long been recognized as signatures of many pathogenic viruses. Their characterization has thus been particularly useful for the diagnostic and taxonomy of human (e.g., Negri bodies of rabies), insects (e.g., viral polyhedroses) and plant (e.g., geminivirus infections) diseases. Most of this section focuses on inclusion bodies that have a paracrystalline and crystalline structure.

2.1 Viral factories vs. inclusion bodies

Viral factories and inclusion bodies caused by viral infection of tobacco plants by tobacco mosaic virus (TMV) have been described by Iwanowski as early as 1903 (Iwanowski, 1903; Martelli and Russo, 1977). Similar objects were detected in the same year in cells infected by the rabies virus, starting a long list of virus-induced structures associated with animal diseases (Netherton et al., 2007; Netherton and Wileman, 2011). In animal viruses, most viral factories and inclusion bodies have initially been described as amorphous. They are typically associated with the microtubule-organization center (MTOC) through the action of dynein motors (Dohner et al., 2005; Wileman, 2006).

In the case of nuclear structures, detailed characterization revealed that some viral factories are liquid organelles segregating their content from the rest of the cell without an enclosing membrane. This is particularly seen in negative-strand viruses including rhabdoviruses (Heinrich et al., 2018; Nikolic et al., 2017) and influenza virus (Alenquer et al., 2019). The liquid behavior of these compartments is characterized by rapid responses to stimuli resulting in fusion/fission events, structural reorganization and material exchange. These compartments have the advantage of concentrating the factors required for viral replication, while shielding the RNA genome from recognition by the host antiviral systems (Heinrich et al., 2018; Nikolic et al., 2017). Similar liquid properties are found in protein- and nucleic acid-containing assemblies such as nuclear speckles, nucleolus, centrioles and stress granules.

By contrast, much of the virus-induced structures described in plant viruses have static properties characteristic of a solid assembly. For instance, infection by tobacco etch virus causes typical intranuclear inclusions and "cylindrical" inclusions that are characteristic of the virus (Edwardson, 1966). They can be detected by light microscopy on tissues dyed with orange-green combination or bromophenol blue during purification. Such inclusion bodies have largely been used for the classification and identification of plant viruses (Hiebert et al., 1984). These inclusion bodies are formed either by viral proteins or the virus particles, and may closely interact with cellular components.

2.2 Superassembly of virions

Many inclusion bodies are simply formed by the concentration of virions in a specific area of the cell. Because of their regular structures, some viruses may form highly ordered arrangements with a paracrystalline or crystalline structure. Because of their small size and labile nature, there are no high-resolution structures from *in cellula*-grown material. Structural descriptions of these superassemblies of virus particles result primarily from analysis by electron microscopy of thin cell sections.

2.2.1 Plants

In many plants the accumulation of virions results in the formation of inclusion bodies that are characteristic of the host-virus pair. These inclusions may not have a specific role and only result from the over-production of virions with high local concentrations. TMV and geminivirus are prime examples of this accumulation of virions. The case of TMV is probably the best characterized example of crystalline inclusion of virus particles. These intracellular crystals differ in their organization from Stanley's famous TMV preparations, which were produced *in vitro* and paracrystalline. TMV hexagonal crystals were isolated from infected cells and characterized by electron microscopy (EM) to demonstrate that they were composed of infectious particles (Rubio Huertos, 1954; Steere and Williams, 1953). TMV adopts several crystalline phases including liquid-crystalline phases.

2.2.2 Animals

Animal viruses also produce paracrystalline arrays and 3D crystals in infected cells (Fig. 1A–C). This has been well described for several adenovirus (Ad) types including Ad2, Ad4, Ad5, Ad6, Ad12 and Ad18 (Kjellen et al., 1955; Morgan et al., 1956; Weber and Liao, 1969). The virions of these viruses pack into cubic paracrystalline arrays in the nucleus of infected cells (Boulanger et al., 1974). Excess viral capsomers form a second type of

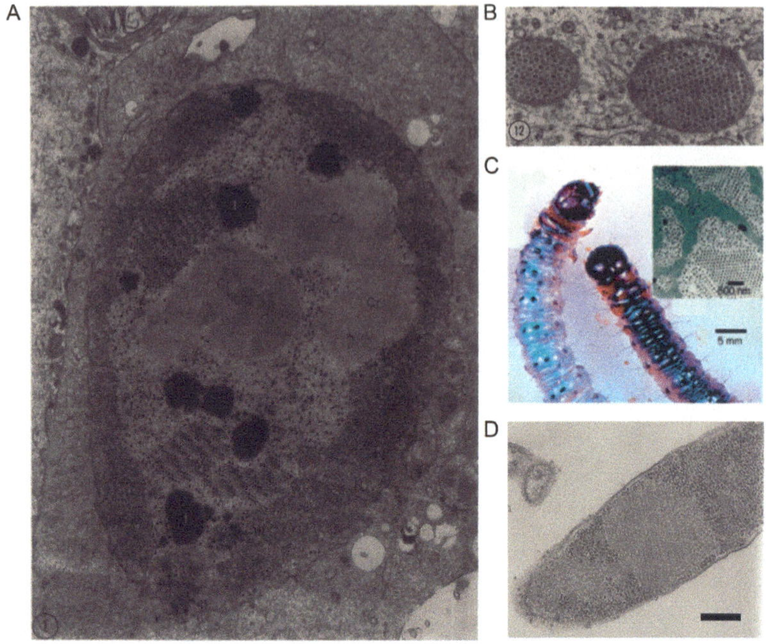

Fig. 1 Inclusions bodies and virion crystals. (A) Vero cells infected with adenovirus type 2. Intranuclear crystals (Cr), crystalline arrays of virions (Vp) and different amorphous inclusion bodies (I, II and III) are labeled within the nucleus (N). (B) Paracrystalline array of IBDV virions enclosed in a membrane envelope. (C) Larva infected with CIV (left) shows a global discoloration and a blue-red iridescence compared to a non-infected larva (right). The inset shows an EM image of a thinly-sliced cell with paracrystalline arrays of virions. Scale bar is 5 mm (800 nm in the inset). (D) *E. coli* C990 cell infected with the phage phiX174. Lysis was inhibited by mutation in the cellular slyD gene resulting in the formation of virion crystals. The scale bar is 200 nm. *Panel (A): Reproduced with permission from Henry C.J., Slifkin M., Merkow L.P., Pardo M., 1971. The ultrastructure and nature of adenovirus type 2-induced paracrystalline formations. Virology 44 (1), 215–218. Panel (B): Reproduced with permission from Lunger, P.D., Maddux, T.C., 1972. Fine-structure studies of the avian infectious bursal agent. I. In vivo viral morphogenesis. Avian Dis. 16 (4), 874–893. Panel (C): Reproduced with permission from Juhl, S.B., Chan, E.P., Ha, Y.-H., Maldovan, M., Brunton, J., Ward, V., Dokland, T., Kalmakoff, J., Farmer, B., Thomas, E.L., Vaia, R.A., 2006. Assembly of Wiseana iridovirus: viruses for colloidal photonic crystals. Adv. Funct. Mater. 16 (8), 1086–1094. Panel (D): From Duyvesteyn, H.M.E., Ginn, H.M., Pietilä M.K., Wagner, A., Hattne, J., Grimes, J.M., Hirvonen, E., Evans, G., Parsy, M.-L., Sauter, N.K., Brewster, A.S., Huiskonen, J.T., Stuart, D.I., Sutton, G., Bamford, D.H., 2018. Towards in cellulo virus crystallography. Sci. Rep. 1–7.*

intranuclear crystals of tubular structure (Fig. 1A). Early crystallographic analysis suggested a helical tubular arrangement (Lifchitz et al., 1975). However the 3D structure remains undetermined and the crystal composition has remained uncertain until a study combining mutant Ad5 viruses and immuno-labeling EM, which concluded that the crystals are made of penton capsomers in complex with the viral fiber (Franqueville et al., 2008). Intracellular crystals of virions have also been described in poliovirus (Dales et al., 1965) and the infectious bursal disease virus (IBDV, Fig. 1B) where it is thought to be a protection mechanism for virions at the center of the crystalline arrangement (Lunger and Maddux, 1972). The most striking example is provided by iridoviruses in the NCLDV group (Williams, 2010; Xeros, 1954). These large DNA viruses infect both vertebrates and invertebrates. They are the largest family of insect viruses after the occluded viruses, baculoviruses and cypoviruses described in Section 3. Infection of insects by iridoviruses results in a change of appearance, with a slight blue-green iridescence (Fig. 1C). This rainbow-like opalescence is caused by optical properties of the large virions arranged in paracrystalline arrays. In tissues of heavily infected larvae, such a regular spacing between virus particles causes optical interferences of visible light due to Bragg scattering. The iridovirus virions have large, icosahedral capsids. A lipid bilayer is present underneath the icosahedral shell of the major capsid protein (MCP). The capsids are decorated with long fibers of ~30 nm that prevent close packing of the virions and may modulate the nature and level of iridescence by affecting the inter-layer spacing. The function of these crystals in the virus cycle, if any, is unknown. The bright iridescence of the larvae may render the insect more easily detectable by predators and indirectly facilitate dissemination. Applications taking advantage of the optical properties of these crystals have been proposed to produce photonics crystals with tuneable properties (Juhl et al., 2006).

2.2.3 Bacteriophages

Most bacteriophages (or phages) do not form highly packed virion arrangements because they induce cell lysis in the late stages of infection. However, in cases where lysis is inhibited, virions artificially accumulate and 3D crystals have occasionally been observed. Infection of *E. coli* cells resistant to lysis by phage phiX174 provided an opportunity to analyze *in cellula* crystals of infectious virions (Fig. 1D). Here the cells were harvested and analyzed both by cryo-EM and X-ray free electron laser (XFEL) (Duyvesteyn et al., 2018). Nevertheless, this study is an example of how viral structures are used as model systems to develop structural biology approaches as described in more detail in Section 6. The poor diffracting power of the crystals has not yet permitted structure determination. The researchers could show that condensates are formed in 75% of the cells and some of these objects have a low resolution crystalline order (to ~50 Å).

3. Intracellular crystals of insect viruses

Polyhedroses are common diseases of invertebrates, collectively affecting over 600 species of insects and crustaceans. These diseases have been known for time immemorial as associated to the rearing of silk worm in China and more recently Europe (Vida, 1527), where they have first been described scientifically at the start of the 19th century (Cornalia, 1856; Nysten, 1808). The impact of the *Bombyx mori* polyhedrosis on the silk industry has driven initial research efforts on this large group of diseases. They are now best known for the beneficial uses of baculovirus for the biological control of insect pests and as an expression system of recombinant proteins in insect cells (Summers, 2006).

Polyhedroses are characterized by the production of intracellular crystals that adopt various polyhedral shapes depending on the virus/cell combination. These crystals have been the subject of much curiosity due to their striking crystalline structures observable by simple light microscopy and a remarkable stability. Because of their protein-rich composition, they have once been thought to be protein-based infectious agents akin to the more recently discovered prions. However, at the turn of the 1950s, the advent of electron microscopy established that these crystals contained virus particles, which were the actual vectors of the disease. The role of viral polyhedra was subsequently shown to be to encase virions in an intracellular protein matrix and protect their viability once released in the environment (Bergoin et al., 1971; Summers, 1971; Vago and Croissant, 1959). Thus, viral polyhedra are also called occlusion bodies by analogy with the occlusion of gas within metallic crystals. The role of these crystals in the viral cycle is summarized in Fig. 2. Infection by these viruses leads to overexpression of a major protein, the polyhedrin or granulin, at impressive levels (Rohrmann, 1986). The protein spontaneously crystallizes in the infected cell, a phenomenon that is often observed in proteins destined to be stored or transported in an inactive form such as insulin or the bacterial toxin of *Bacillus thuringiensis* (Bt) (Doye and Poon, 2006; Schönherr et al., 2018). Virions are embedded into the growing crystalline matrix through an active process that involves viral factors. Eventually, these crystals may represent a large proportion of the dry weight of the infected larvae, which excretes them during the rest of its life or releases them upon death. These crystals are remarkably stable biochemically, which results in a better preservation of the virus viability once released in the environment. *In vitro*, cypovirus polyhedra are resistant

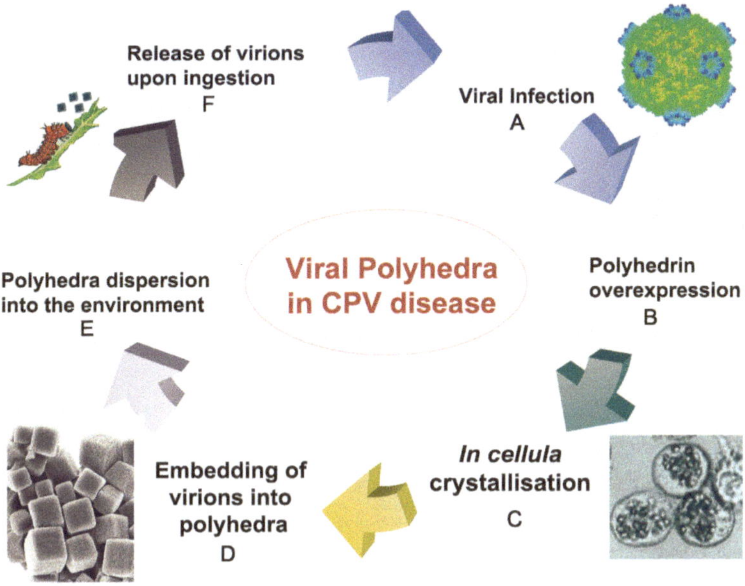

Fig. 2 Function of occlusion bodies in the infectious cycle. The role of occlusion bodies in the infectious cycle is illustrated with CPV polyhedra. (A and B) Viral infection of target cells leads to the overexpression of the CPV polyhedrin (baculovirus polyhedrin and granulin in the case of NPV and GV, respectively; spheroidin for EV). Virion image © Protein Data Bank Japan licensed under CC-BY-4.0 International; EMDB: 5256. (C) The viral protein assembles into intracellular crystals that are visible by light microscopy. (D) During the crystallization process, virus particles are targeted to the crystal and embedded into the dense protein matrix. (E) Upon release into the environment, the crystalline "armor" prolongs the persistence of the virus owing to the robustness of the occlusion bodies. (F) However, the crystals readily dissolve upon ingestion in the alkaline environment of the insect midguts, releasing the virus to initiate a new infectious cycle.

to extreme biochemical conditions including concentrated acids, detergents and solvents (Coulibaly et al., 2007). Like bacterial spores, these occlusion bodies allow the virus to persist in the absence of favorable growth conditions, a permissive host in this case. This trait represents a particularly important selective advantage for insect viruses, which hosts have seasonal or even multiannual cycles. Since most of these viruses preferentially infect early stages of their host development (Rohrmann, 2019), long-term viability in the environment is a bottleneck in the persistence of the viral population. By contrast with the robustness of polyhedra in a broad range of chemical environments, the crystals are very sensitive to alkaline pH even in otherwise mild conditions. Dissolution is observed within seconds in diluted solutions of sodium carbonate at pH above 10.5. *In vivo*, this striking example of metastability explains the rapid release of virions from viral polyhedra when they reach the alkaline midguts of larvae that have been feeding on contaminated material.

Two types of polyhedroses have been distinguished in early studies depending on the cytoplasmic or nuclear localization of the crystals. Insect poxviruses or entomopoxviruses (EV) also produce occlusion bodies but they do not have a polyhedral shape. These occlusion bodies are called spheroids (or spherules in earlier studies) reflecting their ovoid shape (Arif, 1995; Bergoin et al., 1971).

Nuclear crystals are characteristic of diseases caused by viruses of the *Baculoviridae* family (Miller, 2013; Rohrmann, 2019). These viruses have a double-stranded (ds) DNA genome replicated and packaged into the nucleocapsid in the nucleus of infected cells. Thus the viral progeny co-localizes with the site of crystal formation. The virion is rod-shaped presenting one or several helical nucleocapsids enveloped in a lipid membrane. Cytoplasmic crystals of polyhedral shape are typical of diseases caused by members of the cypovirus genus within the *Reoviridae* family. These viruses have a segmented dsRNA genome and a non-enveloped icosahedral capsid. Spheroids are also found in the cytoplasm where insect poxviruses replicate. The EVs have a large DNA genome encapsidated in a characteristic brick-shaped virion. EV particles are enveloped and present one lateral bodies of electron dense material.

More details will be provided in later sections but it is already evident that these viruses differ in most aspects of their infectious cycles, setting them apart in the virosphere (Fig. 3). The only features shared by these three classes of viruses are the use of intracellular crystals and their tropism for invertebrate hosts. Spheroids have been recognized as structurally and functionally different to polyhedra early on. However, the evolutionary relationship between nuclear and cytoplasmic polyhedra has been a conundrum for decades with shared functional and structural traits but the absence of significant sequence or biophysical similarities at the protein level (Anduleit et al., 2005).

Despite their ubiquitous distribution and widespread impact on human activities, the molecular organization of viral occlusion bodies has remained a puzzle for over 50 years due to the lack of structural information. Native, viral polyhedra isolated from infected insects posed technical challenges due to their small size. Advances in synchrotron capabilities and the development of new approaches in microcrystallography are described in Section 6. These have allowed structure determination of the two classes of polyhedra and significant advances in understanding their function in the viral cycle as described in the next sections.

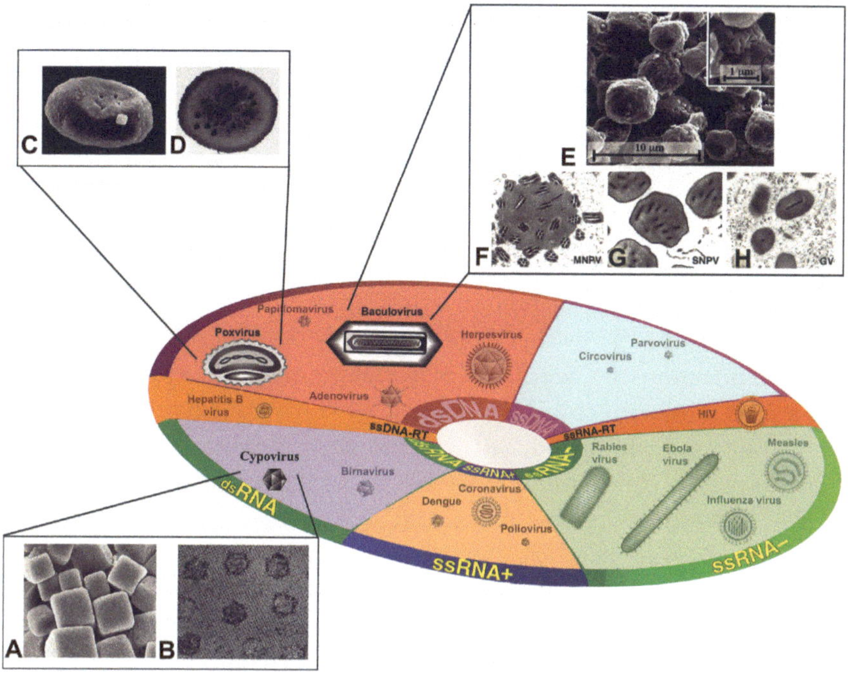

Fig. 3 Occlusion bodies produced by three unrelated viral families. Center: simplified representation of the virosphere adapted from the scheme of the International Committee for the Taxonomy of Viruses (Fauquet and Mayo, 2005). The three insets represent scanning EM (A: CPV, C: EV, E: NPV) and negative-strain EM of thin slices (B: CPV, D: EV, F: MNPV, G: SNPV, H: GV) of the main three classes of occlusion bodies. *Panel (B): Courtesy of J. Doye; panel (D): Courtesy of M. Bergoin; panel (E): Coulibaly, F., Chiu, E., Gutmann, S., Rajendran, C., Haebel, P.W., Ikeda, K., Mori, H., Ward, V.K., Schulze-Briese, C., Metcalf, P., 2009. The atomic structure of baculovirus polyhedra reveals the independent emergence of infectious crystals in DNA and RNA viruses. Proc. Natl. Acad. Sci. U. S. A. 106 (52), 22205–22210; panels (F)–(H): King et al. (2011) with permission.*

3.1 Occlusion bodies of RNA viruses: Cypovirus polyhedra

Cypoviruses (CPV) are insect viruses belonging to the *Reoviridae* family, which regroups segmented dsRNA viruses infecting hosts as diverse as fungi, plants, aquatic invertebrates, insects and vertebrates, including humans. Within this large family, cypovirus is the only genus producing polyhedra.

The genome of cypovirus is composed of 10 or 11 segments coding for 10–12 proteins. CPVs were isolated from over 250 species of insects (Mori and Metcalf, 2010) and have been classified in 16 CPV species recognized by the ICTV with 5 other species under consideration (Mertens et al., 2011).

CPVs have a non-enveloped icosahedral virion that represents the transcriptional unit of the virus. This replication strategy shields the genomic RNA from the potent antiviral cellular defenses based on recognition of dsRNA. The structure of the virion has been determined by cryo-electron microscopy to high resolution, revealing an atypical organization. Viral segments are enclosed in a single icosahedral shell of 60 dimers of the major capsid protein without the $T=13$ external shell characteristic of most other *Reoviridae*. Despite this seemingly simpler architecture, CPV virions are extremely stable owing to additional domains forming non-covalent cross-links of the capsid (Yu et al., 2011).

Upon infection, the virus induces overexpression of the polyhedrin protein, a 29 kDa protein coded by segment 10. This protein self-assembles in the cytoplasm and produces crystals that embed virions during crystallogenesis. The resulting matrix is extremely compact with a density of 1.28–1.30 g cm^{-3} (Lefkowitz et al., 2018), intermediate between the density of typical protein crystals of recombinant protein (1.1–1.2 g cm^{-3}) and that of pure protein, 1.35 g cm^{-3}. Due to their robustness and high density, these crystals are readily purified by mechanical disruption of the infected larvae, followed by differential centrifugation and a density gradient purification. Polyhedra-like crystals can also be produced in cell culture by overexpression of the CPV polyhedrin protein. Intracellular crystallization is achieved in a heterologous system using the baculovirus expression system (Mori et al., 1993). The initial structures of recombinant and infectious CPV polyhedra were determined using microcrystallography approaches (Coulibaly et al., 2007) at synchrotron X-ray beamlines specifically designed to address the challenges presented by crystals of such small size (typically 1–5 μm). In addition to their small size, the structure of native CPV polyhedra was one of the first structures determined from crystals produced *in vivo*, i.e., in infected insects. These approaches are described in more details in Section 6 and are now routinely available at most synchrotrons worldwide (Banerjee et al., 2018; Boudes et al., 2014; Smith et al., 2012).

The crystalline matrix of CPV polyhedra is composed of a compact arrangement of the polyhedrin protein following cubic symmetry (*I*23 space group, unit cell 101–106 Å, PDB ID: 2OH5). All molecules in the crystal have the same conformation—i.e., the asymmetric unit contain a single molecule—which can be described as a pointing left hand. A beta-sandwich forms the palm, while helices H1 and H4 represent the extended index finger and the thumb, respectively (Fig. 4). The C-terminus of the

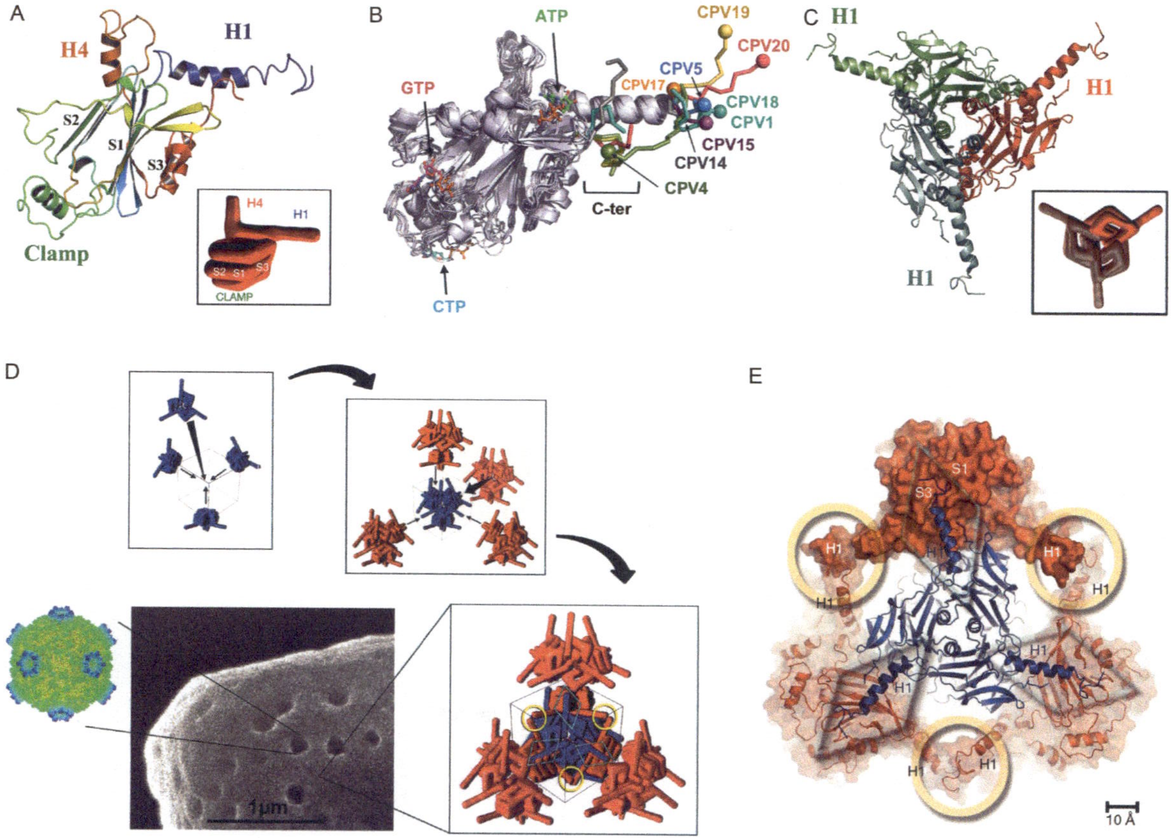

Fig. 4 See legend on next page.

polyhedrin protein forms a third layer to the beta-sandwich and comes into contact with the N-terminal helix. In the crystal, polyhedrin forms a trimer with the palm domain mediating most of the oligomeric interactions. The index and thumb structures project out of the trimer (Fig. 4C) and mediate inter-trimer interactions important for the stability of the crystalline lattice (Fig. 4D–E). The extensive buried surfaces and their hydrophobic nature suggest that the trimers are the building blocks of CPV polyhedra. These trimers form tetrahedral clusters at the center of the cubic unit cell and each lattice point (Fig. 4D). Each unit cell contains the equivalent of two clusters that are translationally related throughout the entire crystal. A 2 μm crystal is formed by ~200 million polyhedrin molecules and contains thousands of virus particles. The virions are distributed homogenously throughout the crystal but without any apparent symmetry. The resulting matrix is extremely dense with a solvent content of only 19% and no continuous channel allowing diffusion through the crystal.

In addition to the polyhedrin molecule and virions, the CPV1 polyhedra contain stoichiometric amounts of the nucleotides ATP, CTP and GTP. A CTP molecule is bound at the center of the tetrahedral cluster, and the ATP/GTP are located at the interface between clusters

Fig. 4 Structure-function of cypovirus polyhedra. (A) Ribbon representation of the CPV1 polyhedrin molecule with a blue-red gradient from N- to C-terminus (PDB ID: 2OH5). Inset: schematic representation of CPV polyhedrin as a left hand with H1 as the pointing finger; H4, the thumb; S1, S2 and S3, the fist. (B) Superposition of nine polyhedrin structures from CPV1 (2OH5), 2 (5A8T), 4 (5A8S), 5 (5A8U), 14, (5A96), 15 (5A98), 17 (4ZQX), 18 (5A9P), 19 (5A99), 20 (5A9A). PDB IDs are given in parentheses. The nucleotides ATP, GTP and CTP bound to CPV1 polyhedra are shown in a ball-and-stick representation. The N- and C-termini of the proteins are colored differently for each molecule. Labels are given next to the N-terminus represented by a colored sphere. (C) The polyhedrin trimer (inset: schematic as (A)). (D) Assembly of polyhedra using the schematic representation of (C). Trimers assemble in tetrahedral clusters (top), which themselves pack tightly in the crystals. The unit cell contains two tetrahedral clusters represented in blue and red (middle and bottom right). Circles and diamond indicate multiple interacting surfaces involving H1. The scale of a unit cell is shown on the central panel on a SEM image of a CPV1 polyhedron. Holes occupied by virion (© Protein Data Bank Japan licensed under CC-BY-4.0 International; EMDB: 5256) are evident on the crystal surface. (E) Similar representation as (D) with a central trimer shown as a ribbon in blue and three trimers colored in red shown as molecular surfaces. H1 interactions are highlighted as in (D). *Panels (A) and (C): From Coulibaly, F., 2012, Les polyèdres viraux: armures cristallines des virus d'insectes, Virologie 16 (3), 134–147. with permission; panel (E): From Coulibaly, F., Chiu, E., Ikeda, K., Gutmann, S., Haebel, P.W., Schulze-Briese, C., Mori, H., Metcalf, P., 2007. The molecular organization of cypovirus polyhedra. Nature 446 (7131), 97–101.*

(Coulibaly et al., 2007). The role of these nucleotides remains unclear but could participate to assembly and stability of the crystals. This site is occupied by purine nucleotides in six of the nine CPV types of known structure. However, only CPV1 and 18 have both ATP and GTP, while others have only one of the nucleotide and present very different binding modes. The CTP site seen in CPV1 is only occupied in one other type of polyhedra, CPV20 which binds an UTP molecule instead of CTP (Ji et al., 2015).

The CPV polyhedrin has evolved to self-assemble into a dense matrix that will protect the virions in the environment, and release them only once ingested by a new host. This overall role can be described as three distinct functions: the ability to crystallize intracellularly, to bind virions for their recruitment into the growing matrix, and to efficiently release virions in the midguts of larvae.

Self-assembly is driven by a very high self-complementarity between the trimeric building blocks of polyhedra. The surface complementarity factor is 0.70, which is comparable to antigen-antibody complexes (Wodak and Janin, 2002). Thus, these contacts resemble more closely specific protein-protein complexes than crystal contacts. This is also true in the proportion of buried surface. In CPV1 polyhedra, a total of ~75% of the total polyhedrin surface is buried in crystal contacts. A third factor contributing to the efficiency of assembly and stability is the extreme density of the crystal. It is held by such a dense network of interactions that the residual solvent content is only 19%, among the top 10 crystals by density in the Protein Data Bank. Helix H1—the extended index—was proposed to play an important role in the stability of the crystal by mediating multiple inter-trimer contacts (Fig. 4E) and perhaps helping incoming building blocks dock onto the existing crystal core (Coulibaly et al., 2007).

How polyhedrin binds to the virion is not well understood. The clamp domain—also called base domain—is one of the most variable in conformation and is dispensable for crystallization. In CPV1 polyhedrin, replacement of residues 71–103 by a flexible linker resulted in a cubic unit cell edge only ~0.9 Å longer, with an overall structure largely unaffected except for the missing clamp domain. It was proposed that this domain may recruit the virion to the growing crystal. Its location as the center of the tetrahedral cluster where two- and threefold symmetry axes intersect would mediate compatible interactions between the crystal and the icosahedral capsid, which also has 23 symmetry (Ji et al., 2015).

On the virion side, the VP3 protein, which forms a turret projecting out of the virion, has been first proposed as the polyhedrin-binding protein.

Fusion proteins with the green fluorescent protein (GFP) determined that residues 1–70 (Ikeda et al., 2006) or 42–93 (Mori et al., 2007) may represent the minimal domain for incorporation into polyhedrin crystals. However, structure determination of the virion by high-resolution cryo-EM revealed that this domain is buried and inaccessible for interaction with polyhedrin (Yu et al., 2011). Whether interaction with the polyhedrin is able to lock this domain in a different conformation or another component of the virion is responsible for its incorporation into polyhedra remains to be shown.

Tomography of polyhedra shows gaps within the matrix, which suggests that the A fiber that projects out from the viral turrets could drive the recruitment of virions into the crystal (Chen et al., 2011).

The third function of polyhedra is to efficiently deliver particles to the new host. For this, the crystals must rapidly release the virions upon ingestion. Release is achieved by crystal dissolution in response to the alkaline environment encountered in the midguts of most lepidopteran larvae. Serine proteases have also been proposed to be involved in this process but their identity is not known. Global destabilization of the polyhedrin trimer has been proposed to result from the deprotonation of tyrosine residues in conditions where the pH is higher than the pK_a of tyrosine (pK_a = 10.5 in solution). Deprotonated hydroxyl groups would represent clusters of negative charges repelling each other and contributing to the unfolding of the polyhedrin protein. In support to this hypothesis, polyhedrin is enriched in tyrosine residues that cluster in the palm and thumb of the left hand-shaped polyhedrin molecules, regions that are involved in a large number of crystal contacts (Coulibaly et al., 2007). While most tyrosine residues are variable between CPV strains, conservation of Tyr_{232} (CPV1 numbering) across all CPV polyhedrin structures—except CPV17—suggests that it may play a specific role by direct repulsion with glutamate residue E_{217}. Deprotonation of Tyr_{232} would affect the conformation of the C-terminal helix H5 and initiate disassembly (Ji et al., 2015).

Structure determination of polyhedrins from eight other CPV species (CPV2, 4, 5, 14, 15, 18, 19 and 20) revealed a strong conservation of the fold of the protein across the genus (Ji et al., 2015). Despite sequences exhibiting as little as 12% sequence identity, all secondary structure elements are conserved and the overall root mean square deviations are very low (0.2–1.5 Å over ~80% equivalent residues). Structural differences are primarily located in three regions: the N- and C-termini and the clamp/base domain. The N-terminal helix and C-terminal beta-sheet are functionally important and mediate multiple interactions in the crystal. Accordingly, their structures and positioning with regard to the main body of the molecule are conserved. The very N- and C-termini of polyhedrins of different CPV species have extensions of variable lengths and structures. These extensions appear to squeeze between adjacent molecules in the crystal without altering the overall cubic lattice organization and unit cell length. These additional crystal contacts may increase stability in several ways. First, they increase the long-distance crosslinking between subunits within polyhedra. This is particularly significant for CPV19 and CPV20 polyhedrins (Fig. 4B) which are two of the most robust crystals when exposed to an alkaline carbonate buffer. These proteins present extended structures at their N-termini just before H1, which contact one and two additional molecules, respectively. By comparison, the weakest polyhedrin crystals in these conditions are formed by CPV5 and CPV15, which lack these 4–5 residue extension. These N- and C-terminal extensions may also limit diffusion of destabilizing molecules by occlusion of voids within the crystals. There is, however, no direct evidence so far that this is an important factor since none of the polyhedra have solvent channels running through the entire crystal (Coulibaly et al., 2007). However, crystals modified to present continuous solvent channels by replacement of the clamp domain with a poly-(Gly-Ser) linker dissolve in conditions that do not affect their native counterparts (Ji et al., 2015). This suggests that the ability to limit diffusion of small molecules into the crystal is an important factor contributing to the robustness of CPV polyhedra.

3.2 Occlusion bodies of DNA viruses: Baculovirus polyhedra and granules

Baculoviruses form a large family of viruses historically divided into two distinct groups, the nucleopolyhedroviruses (NPV) and the granuloviruses (GV). NPV produce polyhedral micron-sized crystals that embed up to hundreds of virions. GV produce granules smaller than 1 μm in length that contain a single virus particle. Genomic analysis has revealed that baculoviruses form four distinct genera (Herniou and Jehle, 2007; Jehle et al., 2006). NPV and granuloviruses infecting Lepidoptera (e.g., butterflies and moths) form the alpha- and beta-baculovirus genera, respectively. The gamma- and delta-baculoviruses regroup viruses infecting Hymenoptera (e.g., sawflies but not bees, ants or wasps) and Diptera (e.g., flies), respectively. The genome of these viruses is composed of circular double-stranded DNA with a size range of 80–180 kbp (Rohrmann, 2019). The genome is packaged into a helical nucleocapsid, which is itself enveloped by a lipid membrane.

The structure of the virion has been revised in light of cryo-EM images that did not suffer from artifacts present in negative stained samples (Wang et al., 2016). In this model, the envelope is only loosely fitting around the inner nucleocapsid with contacts between the two layers only at the apical ends. The spikes formed by the attachment and fusion proteins are concentrated in these areas of contact around the virion tips. This study identified a tegument layer that occupies some of the space left between the nucleocapsid and the envelope. Some NPVs package multiple nucleocapsids—up to 17—into a single envelope. These viruses are called multicapsids such as the prototype of the family, the *Autographa californica* multiple NPV (*Ac*MNPV).

The infectious cycle of baculovirus involves a nuclear replication step ultimately leading to the production of two infectious forms, budded virions (BV) and occlusion-derived virions (ODV). ODV are released upon ingestion of the occlusion bodies and mediate infection of the midgut epithelium. BV are produced in a second phase and disseminate infection to other tissues within the host, which depending on the host/virus pair may result in chronic infection or acute, systemic infection and possibly death. Baculovirus occlusion bodies are called polyhedra for NPVs and granules for GVs. They consist of a crystalline matrix of a 29 kDa protein called polyhedrin or granulin (Rohrmann, 1986).

The gene that codes for polyhedrin/granulin is the most conserved across the genome, resulting in proteins with >90% sequence identity within alpha-baculoviruses, >60% within beta-baculoviruses and ~50% between alpha, beta and gamma baculoviruses (Herniou and Jehle, 2007). However, baculoviruses infecting mosquitoes (CuniNPV, delta-baculovirus) and prawns (unclassified NPV) produce occlusion bodies composed of proteins that appear to be completely unrelated in sequence, size and organization to polyhedrins/granulins. Structural analysis of these occlusion bodies may shed light on their origin and relationship with other occlusion body proteins.

While they share the same name as cypovirus polyhedra, baculovirus occlusion bodies differ fundamentally in several key aspects. First, baculovirus polyhedra form in the nucleus of infected cells, which corresponds to the site of replication and assembly of these DNA viruses. Second, baculovirus polyhedra are coated with an external layer called the calyx or polyhedra envelope. This envelope is composed of proteoglycans that form a smooth layer over the crystal. A structure of the N-terminal fragment of the envelope protein has been determined for CpGV (PDB ID: 4YE7, unpublished). This layer appears to have a protective role and polyhedra lacking the envelope have a pitted appearance resulting from the loss of virions at their surface (Summers and Arnott, 1969). Intracellularly, the envelope is associated with fibrils of the p10 protein. While these fibers are required for the formation of the polyhedra envelope, their exact role has not been identified yet. Third, baculovirus polyhedra are formed by *in cellula* crystallization of a protein called polyhedrin or granulin, depending on the viral genus, which bears no detectable sequence similarity with cypovirus polyhedrins.

These differences, and the evolutionary distance between the viral families, suggested that the proteins may have emerged independently at least on two occasions. However, the subsequent discovery of striking similarities at the level of the crystalline lattice—with the same space group and very similar unit cell parameters—were suggestive of a common origin for both types of viral polyhedra (Anduleit et al., 2005).

Structural analysis of baculovirus polyhedra has benefited from the approaches developed for cypovirus polyhedra but was complicated by a weaker diffracting power. Their structure determination required the combination of multiple datasets with the best 19 crystals combined in the final dataset of infectious NPV polyhedra (Coulibaly et al., 2009). This approach foreshadowed automated serial microcrystallography methodologies described in Section 6. The structure of NPV polyhedra was first determined using larger crystals produced by a *Ac*MNPV polyhedrin gene with the G25D mutation (Lin et al., 2000). These crystals are brick shaped, ~10–15 μm in maximum dimension and do not efficiently embed virions. The structure of native polyhedra containing infectious NPV was subsequently determined by molecular replacement (PDB ID: 3JW6 and 3JVB). Structures of virus-containing occlusion bodies are currently available for the alpha-baculoviruses *Wiseana porina* NPV (Coulibaly et al., 2009) and *Ac*MNPV (Coulibaly et al., 2009; Ji et al., 2010) and the beta-baculovirus of the codling moth (*Cydia pomonella* GV) (Gati et al., 2017).

As expected from their sequence similarity (69%), the structures of polyhedrin and granulin are very similar, with 0.634 Å RMSD over 173 equivalent Calpha atoms. They are described together here and differences are discussed in Section 6. These proteins adopt a jelly-roll fold decorated by extended N-terminal and C-terminal extensions (Fig. 5A). The beta-sandwich domain has the canonical jelly-roll topology found in many icosahedral capsid proteins and a similar wedge shape. Projecting away from the jelly-roll, the N-terminal extension forms an 8-turn α-helix (H1/2) capped by a beta-hairpin. The helix is bent in its middle, resulting in a ~45° kink.

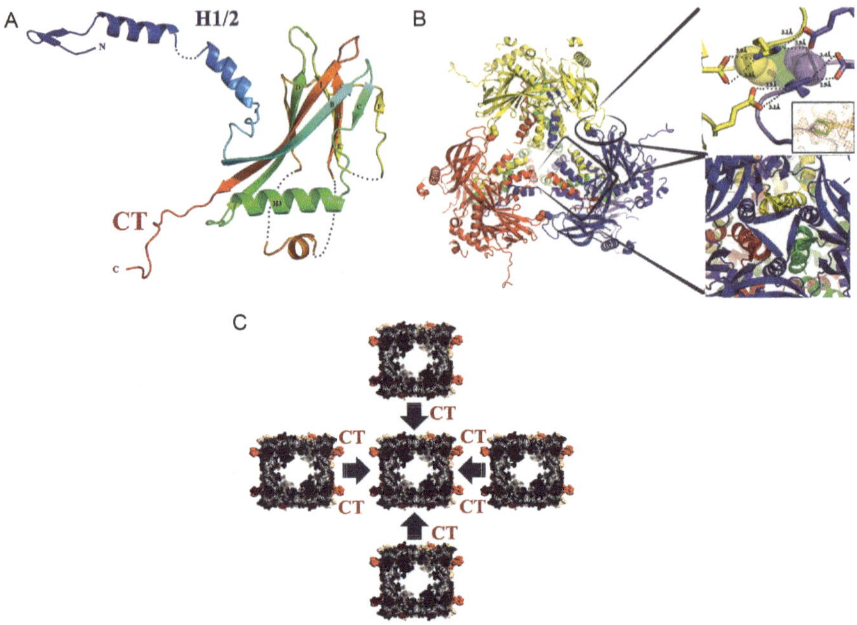

Fig. 5 Structure-function of baculovirus polyhedra and granules. (A) Ribbon representation of the WNPV polyhedrin molecule with a blue-red gradient from N- to C-terminus (PDB ID: 3JVB). Missing segments are indicated by dotted lines. (B) Dodecameric arrangement forming the putative building block of BV polyhedra. Trimeric assemblies are crosslinked in the building block by domain swapping mediated by the N-terminus H1/2 (bottom right), and intermolecular disulfide bonds (top right). (C) Two building blocks form a cubic unit cell shown as a brown/black molecular surface. Unit cells interlock like lego blocks through the C-terminal segment of polyhedrin, Cter (red projections). *Panels (A) and (B): From Coulibaly, F., Chiu, E., Gutmann, S., Rajendran, C., Haebel, P. W., Ikeda, K., Mori, H., Ward, V.K., Schulze-Briese, C., Metcalf, P., 2009. The atomic structure of baculovirus polyhedra reveals the independent emergence of infectious crystals in DNA and RNA viruses. Proc. Natl. Acad. Sci. U. S. A. 106 (52), 22205–22210.*

In BV polyhedra, the polyhedrin jelly-rolls assemble into hollow trimers involving interactions between the beta-sheets of two adjacent molecules. The gap between the jelly-rolls is filled by a 6-helix bundle of the N-terminal helix (Fig. 5B). Three of these helices come from three distal trimers forming a domain swap strengthened by the N-terminal beta-hairpin, which augments the largest jelly-roll beta-sheet (BIDG when beta-strands are named sequentially) in the primary sequence. This arrangement results in a domain-swapped dodecamer, which is likely to be the building block of the crystal. The four trimers in the dodecamer have a tetrahedral geometry in keeping with the cubic space group *I*23 (Fig. 5B).

Interestingly, trimers are covalently crosslinked within the dodecamer by an intermolecular disulfide bridge formed by a cysteine residue facing the twofold crystal symmetry axis (Fig. 5B). This cysteine—located at position 132 in the AcMPNV polyhedrin—is conserved in alpha- and beta-baculoviruses, suggesting that in all these viruses the dodecameric building block of the crystal is covalently locked into place for stability. Formation of the disulfide bond is assumed to occur in late phases of the viral cycle, when the normal reducing environment of the cytoplasm is disrupted.

The dodecamers pack tightly into the crystal resulting in a dense protein matrix with only 22% solvent. The C-terminal extension (Cter) forms a molecular arm projecting out of the dodecamer (Fig. 5). It does not contribute to the folding and stability of the trimers or to their interaction within the dodecamer. Instead, its role is to interlock neighboring dodecamers, clipping them together in all three dimensions (Fig. 5C).

Since the asymmetric unit of the crystal contains a single polyhedrin subunit, all molecules are in an identical environment within the crystalline part of polyhedra. Virions are homogenously distributed throughout the occlusion bodies. However, the virus particles do not follow the lattice symmetry and do not appear to have a regular organization within the crystal. Thus, if present, conformational differences of polyhedrin/granulin at the crystal-virion interface cannot be visualized by X-ray crystallography due to the departure from the crystal symmetry.

Helix H1/2 plays a special role in morphogenesis. By contrast with CPV polyhedrin, this N-terminal helix is buried inside the dodecameric building block and does not project outward. It contains sequences required for localization in the nucleus, where ODV assemble (Jarvis et al., 1991). Thus, it is possible that H1/2 interacts with importin(s) to prevent the formation of the polyhedrin dodecamers and their premature assembly within the cytoplasm.

Indeed, most mutations resulting in markedly altered crystal morphologies affect H1/2. For instance, mutations G25D (Lin et al., 2000), L43I (Cheng et al., 1998) and P58L (Carstens et al., 1986) lead to the formation of brick-shaped crystals of abnormally large size. These crystals also fail to embed virus particles. It is not known whether failure to incorporate virions results from a mislocalization in the cytoplasm due to disruption of the nuclear localization sequence present in H1/2 or because of direct disruption of the virion binding site.

Baculovirus polyhedra dissolve at alkaline pH. As for CPV polyhedrin, tyrosine residues have been proposed to play a role in the destabilization of the lattice in a chemical environment where the pH is higher than the theoretical pK_a of these residues. The high tyrosine content and their localization supports this hypothesis. Most tyrosine residues are present at the interface between trimers within the dodecameric building block, where the N-termini of the other three trimers insert in a central cavity to form a domain swap (Fig. 7B). Tyrosine residues are present on both partners of the domain swap. The Cter extension also present a terminal tyrosine as the last residue of the baculovirus polyhedrin in alpha- and beta-baculoviruses. In the case of disulfide-bonded polyhedra, reduction of these covalent bonds is anticipated to accompany dissolution due to a destabilization in alkaline conditions. Host proteases frequently co-purify with the occlusion bodies and accelerate dissolution, although the requirement for proteases *in vivo* has not been established (Rohrmann, 2019).

3.3 Crystals of insect poxviruses: Spheroids and spindles

3.3.1 Poxvirus spheroids

Poxvirus occlusion bodies are intracellular crystals of the spheroidin protein, a large protein of 942–1007 residues with no sequence homolog and no known functional or structural domains. The crystals are ovoid in shape and embed EV virions, often resulting in a pitted appearance in scanning EM (SEM) (Fig. 6A). The high proportion of cysteines (20–46) and the requirement for reducing agents to dissolve spheroids suggest the presence of cross-linking inter-subunit disulfide bonds. Spheroids contain an alkaline protease of unknown origin that contributes to virus release from the crystal in the midguts. Spheroids embed additional virulence factors such as the spindles, crystals of the fusolin protein, involved in disruption of the membrane lining the host midguts. Their structure presents a multi-domain protein unrelated to polyhedrin/granulin proteins (Coulibaly, unpublished).

3.3.2 Poxvirus spindles

Spindles are crystals of a single viral protein called fusolin. Contrary to occlusion bodies, they do not contain virus particles and are thus non-infectious. Instead, spindles are potent virulence factors that enhance the pathogenicity not only of their homologous virus but also of unrelated oral pathogens and toxins. Such an effect has been elegantly demonstrated in co-feeding experiments where larvae were fed with food containing both spindles and a viable baculovirus (Xu and Hukuhara, 1992) or spindles and the Bt toxin (Mitsuhashi et al., 2013).

In lepidopterans, oral administration of spindles has a drastic effect on the insect midgut. In the case of the silkworm *B. mori*, ingestion of spindles

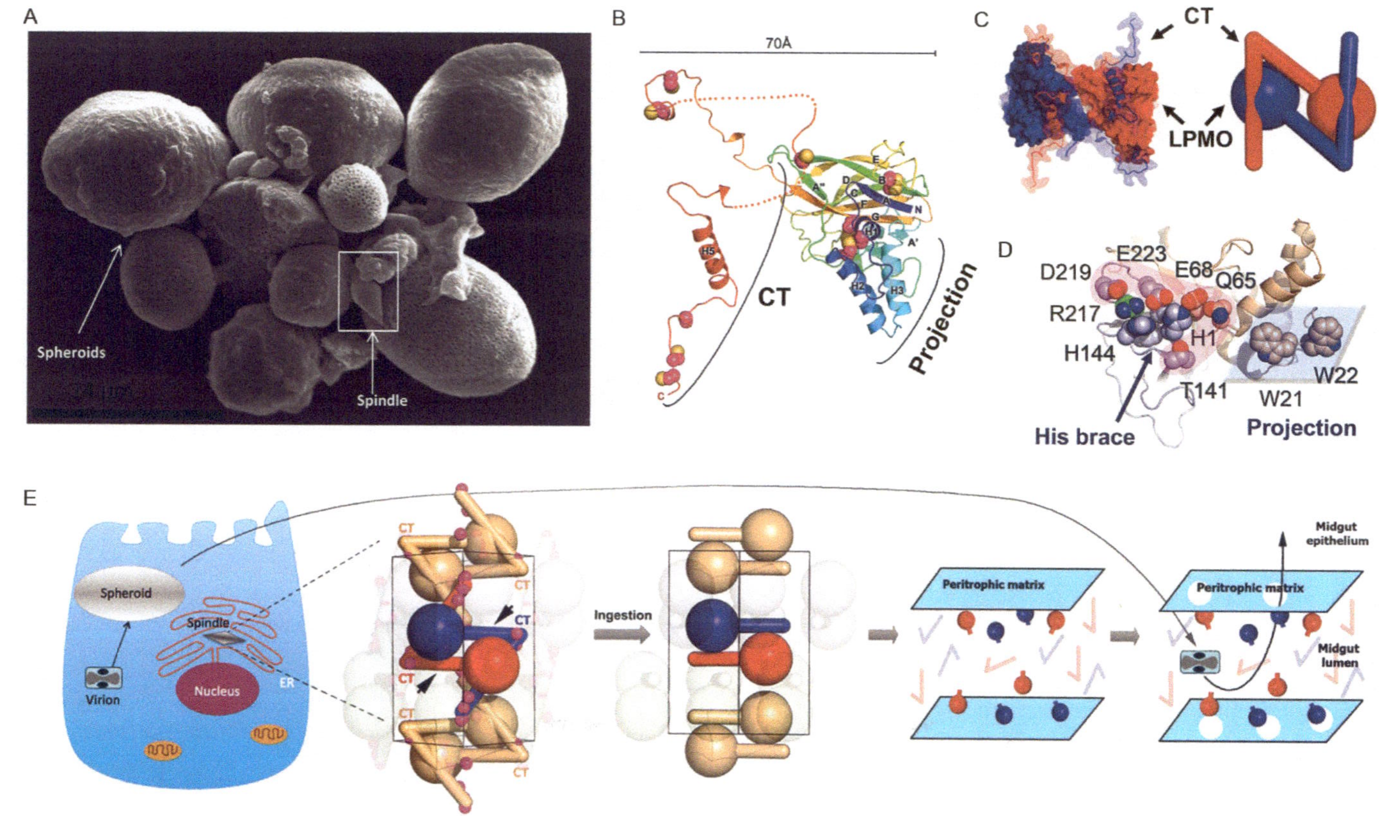

Fig. 6 See legend on opposite page.

produced by a coleopteran poxvirus resulted in the complete disappearance of the peritrophic membrane. This membrane is an acellular layer that forms a tube lining the midgut of the larvae (Mitsuhashi and Miyamoto, 2003). Without this protective layer, the larvae were more sensitive to the unrelated NPV and Bt toxin (Mitsuhashi et al., 2013; Xu and Hukuhara, 1992). In coleopterans, the phenotype is less pronounced but lesions of the peritrophic membrane were also demonstrated using SEM. The enhancement of virulence is specific of fusolin, which may be ingested as free spindles or spindles embedded into the larger spheroid crystals (Mitsuhashi et al., 2007). The crystalline structure does not seem to be required for activity since fusolin derived from spindles denatured in 5 M guanidium had comparable enhancing capacity to native spindles (Takemoto et al., 2008).

Spindles have a characteristic bipyramidal shape with maximum dimensions of ~1–3 μm (Fig. 6A). Contrary to spheroids, which often have a pitted appearance, the spindles surface is always smooth when imaged by SEM. In the cell, spindles are surrounded by an ER-derived lipid membrane (Gauthier et al., 1995). A study investigating resistance of spindles to various environmental factors such as UV, heat and chemicals also established the remarkable robustness of spindles (Mitsuhashi et al., 2008) that surpasses both classes of polyhedra.

Fig. 6 Structure-function of spheroids and spindles produced by insect poxviruses. (A) SEM image of several *Mm*EV spheroids and a spindle purified from infected cockchafers (Mm). Four spheroids have a smooth appearance while the others reveal holes occupied by EV infectious particles. (B) Ribbon representation of a fusolin molecule with a blue-red gradient from N- to C-terminus (PDB ID: 4OW5). The structure of fusolin was determined from intact spindles purified from infected larvae. The projection of the Fn3 domain and C-terminal domain (CT) are labeled. (C) Molecular and schematic representations of the fusolin dimers. (D) Representation of fusolin residues homologous to CBP21 residues involved in catalysis (light blue spheres) and substrate binding (light brown spheres). The metal binding site is formed by a histidine brace (His brace). Arginine 217 shown in green is the main difference with CBP21 in the active site. (E) Model of the role of spheroids and spindles in EV infection. From left to right: schematic representation of an infected cell, assembled fusolin dimers in spindles, fusolin arrangement following loss of CT by proteolytic digestion, dissolved spindles release the fusolin LPMO domain in the midguts of the insect (blue surfaces), EV virions are also released in the midguts and invade the host through holes in the peritrophic matrix that are generated by activated fusolin. *Panels (B)–(E): From Chiu, E., Hijnen, M., Bunker, R.D., Boudes, M., Rajendran, C., Aizel, K., Olieric, V., Schulze-Briese, C., Mitsuhashi, W., Young, V., Ward, V.K., Bergoin, M., Metcalf, P., Coulibaly, F., 2015. Structural basis for the enhancement of virulence by viral spindles and their in vivo crystallization. Proc. Natl. Acad. Sci. U. S. A. 112 (13), 3973–3978.*

Fusolin is a 350–380 residue glycoprotein. It is translocated into the endoplasmic reticulum through an N-terminal signal sequence. Most fusolin sequences have a single consensus site for N-glycosylation. This site has been shown to be used for the fusolin of *Anomala cuprea* EV and appears to be important for folding. The signal sequence is cleaved immediately before a strictly conserved His-Gly-Tyr motif. The mature protein is composed of two domains (Takemoto et al., 2008). The N-terminal domain represents ~2/3 of the protein and presents significant sequence and functional similarity with the GP37 protein produced by some baculoviruses (Arif, 1995; Li et al., 2003). The presence of homologous fusolin-like proteins in both EVs and baculoviruses probably results from horizontal gene transfer (Salvador et al., 2012). Both proteins are homologous to the bacterial protein CBP21 of *Serratia marcescens* and were shown to bind chitin *in vitro*.

Structure determination of spindles directly purified from infected insects used X-ray microcrystallography approaches similar to those described for polyhedra (Chiu et al., 2015). Despite the small size of the crystals, which were typically ~3 μm in length, spindles diffracted to high resolution (1.9 Å, PDB ID: 4OW5). Spindles are tetragonal crystals belonging to space group $P4_12_12$. The asymmetric unit of the crystal contains a single fusolin molecule but the building block is a dimer assembled around a crystallographic twofold axis.

3.3.2.1 Structure description

Fusolin has a two-domain organization with a compact N-terminal domain and an extended C-terminal domain (CT) composed of a long linker and an alpha-helix (Fig. 6B). The N-terminal domain has a modified fibronectin type 3 (Fn3) fold homologous to CBP21. A large projection sub-domain formed by helices H1–H3 gives an atypical wedge shape to the fusolin protein compared to CBP21. The CT extension projects ~30 Å away from the main body of the molecule. Its sequence greatly varies between EV species (76–134 residues) due to linkers differing in length and sequence. However, all CT extensions are predicted to have a central helix that mediates the formation of a domain-swapped dimer in spindles (Fig. 6C). In the dimer, this amphipathic helix fits into a hydrophobic groove of the neighboring Fn3 domain (Chiu et al., 2015).

The fusolin dimer is predicted to be stable by interface analysis, with a total buried surface of 3530 Å^2. A significant fraction of fusolin remains dimeric upon *in vitro* dissolution of spindles at pH 13, even in the presence of SDS. This stability is dependent on an intact CT arm, since its proteolytic removal by chymotrypsin abolishes the presence of dimers in those conditions. A fusolin molecule without CT is not predicted to form a stable dimer by interface analysis with the "Proteins, Interfaces, Structures and Assemblies" server (PISA) (Chiu et al., 2015).

3.3.2.2 The first viral representative of a lytic polysaccharide monooxygenase (LPMO)

The Fn3 domain is structurally homologous to bacterial proteins of the Auxiliary Activity 10 (AA10) family. These proteins are characterized by an ability to disrupt polysaccharides in their crystalline form. While they were initially only recognized as chitin-binding protein (hence their previous classification as Carbohydrate Binding Module 33), careful chemical analysis demonstrated their enzymatic activity (Vaaje-Kolstad et al., 2010). By contrast with classical glycosyl hydrolases, these enzymes function through an oxidative process. Like all AA10, fusolin has a highly conserved copper binding site (Courtade and Aachmann, 2019; Hemsworth et al., 2013). The nature of the metal natively bound to preparations of spindles used for structural analysis has not been determined. However, this site can be populated with various metallic cations, including Cu^{2+} (PDB ID: 4X27) (Chiu et al., 2015), which mediates catalysis in the case of CBP21. The metal binding site is located in a shallow groove on one side of the Fn3 beta-sandwich. It is formed by a so-called histidine brace composed of two histidines, one of which is the N-terminal residue (Fig. 6D). The metal is coordinated by the N-terminal amino group of His_1 (numbering of the mature protein after removal of the signal sequence) as well as nitrogen atoms from the imidazole side chain of His_1 and His_{144}. An alanine residue blocking access to the T-shape coordination sphere of the metal in AA10 LPMOs (Courtade and Aachmann, 2019; Forsberg et al., 2014; Frandsen et al., 2016; Vaaje-Kolstad et al., 2017) is also present in fusolin, supporting a functional role in catalysis.

Substrate binding involves a planar surface around the metal binding site extending onto the projection domain. This domain is larger in fusolin compared to other AA10 proteins. The actual substrate specificity of fusolin has not been determined. Based on the fact that chitin binding has been demonstrated for fusolin and GP37 proteins, it was proposed that fusolin acts on chitin, which appears to correlate with phylogenic sequence analysis (Chiu et al., 2015). The planar surface formed by the projection domain is well suited to bind and disrupt crystalline chitin. While most AA10 proteins lack aromatic residues at the substrate binding interface, fusolin presents prominent tryptophans that provide a favorable surface for polysaccharide binding.

Taken together, fusolin has all the hallmarks of an active LPMO. However, demonstrating activity of a protein naturally present in a crystalline form on a crystalline substrate presents unusual challenges. We have only recently obtained preliminary experimental evidence of the LPMO activity (Coulibaly and Vaaje-Kolstad, unpublished). An obstacle to activity measurement comes from the fact that fusolin molecules within the spindle crystals are inactive. This is due to a complete occlusion of the active site by the CT arm of a neighboring molecule in the crystal. In addition to covering the substrate binding surface, this CT arm inserts the side chain of a glutamate residue into the active site. The carboxylate moiety of this residue is precisely positioned where molecular oxygen normally binds Cu(II) to produce reactive oxygen species that will eventually cleave the glycosidic bond. Thus, fusolin is only functional at the surface of spindles or as a soluble factor once released from the crystal. In addition to the structural data, the strongest support for the importance of the LPMO activity on the fusolin enhancement of virulence comes from mutagenesis of recombinant fusolin, which showed an absolute requirement for the Ile-His-Glu motif, which includes His_{144} from the metal binding site (Takemoto et al., 2008). Similarly, fusion proteins where the N-terminal amino group of His_1 is blocked are nonfunctional.

3.3.2.3 The first crystalline LPMO

Fusolin and some of its GP37 homologs are the only LPMOs known to form functional, intracellular crystals. This feature seems to be linked to the presence of the CT extension, which is found in all fusolins and the four baculovirus GP37 that assemble into spindle-like crystals (Chiu et al., 2015). CT sequences differ in length and have no significant sequence similarity, yet they have the same structural role in all three spindle structures determined so far. In addition to its role in dimerization described previously, CT mediates most of the inter-dimer interactions that allow the formation of a stable tridimensional crystal. In the absence of CT, fusolin is not predicted to form stable crystals by interface analysis with the PISA server (Fig. 6E). Thus, CT appears to have evolved as an extension to the AA10 domain allowing it to crystallize intracellularly. This crystalline form of the LPMO is ultra-stable, resisting to exposure to temperatures up to 95 °C, UV-B irradiation and chemicals such as 3% (v/v) formaldehyde for at least 30 min (Mitsuhashi et al., 2008, 2013). Structural analysis revealed that the molecular basis of this remarkable stability lies in the ability of CT to covalently crosslink all fusolin dimers within the crystal. In *Melolontha melolontha* EV (*Mm*EV), 12 inter-dimer disulfide bonds crosslink any dimer in the crystalline matrix to six neighboring dimers. Thus, each spindle could be described as a single molecule of a fusolin polymer. The cross-linking cysteines are all present in the CT arm before and after the domain-swapped helix (magenta spheres in Figs. 6E and 7). This feature has only been visualized in the structure of *Mm*EV spindles but other fusolin proteins also have

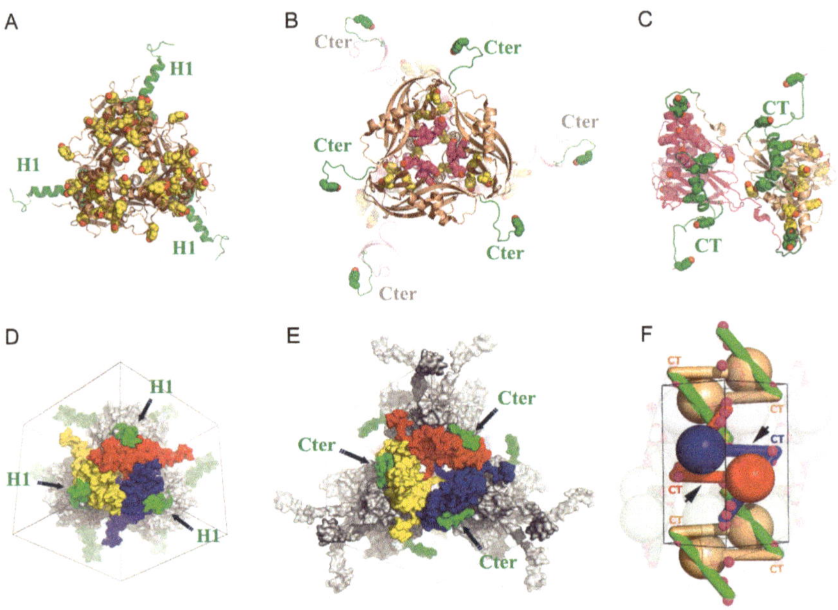

Fig. 7 Functional convergences in assembly and dissolution between 3 classes of viral intracellular crystals. (A–C) Ribbon representation of the putative building blocks of CPV polyhedra (PDB ID: 2OH5) (A), BV polyhedra/granules (PDB ID: 3JVB) (B) and EV spindles (PDB ID: 4OW5) (C). Only one trimer is clearly visible for BV polyhedra for clarity. Molecular arms important for assembly and stability of the crystals are colored in green. Sphere representation of tyrosine side chains with oxygen atoms colored in red and carbon atoms colored in yellow, magenta (domain-swapped molecules) or green (molecular arms). (D-F) Surface representation focused on one trimer (CPV polyhedrin, D and BV polyhedrin, E) or dimer (EV fusolin, F) colored in blue, red and yellow. Neighboring molecules are in gray or light brown except for the molecular arms highlighted in green as in (A)-(C). *Panels (D) and (E): From Chiu, E., Hijnen, M., Bunker, R.D., Boudes, M., Rajendran, C., Aizel, K., Olieric, V., Schulze-Briese, C., Mitsuhashi, W., Young, V., Ward, V.K., Bergoin, M., Metcalf, P., Coulibaly, F., 2015. Structural basis for the enhancement of virulence by viral spindles and their in vivo crystallization. Proc. Natl. Acad. Sci. U. S. A. 112 (13), 3973–3978.*

multiple cysteines in disordered and repetitive CT sequences expected to play a similar structural role. Poxviruses encode enzymes that manipulate the redox potential of the cytoplasm and catalyze the formation of disulfide bonds. Interestingly, a similar 3D network of disulfide bonds has been proposed to stabilize other *in vivo* crystals involved in insect-specific diseases, such as the parasporal Bt toxins and EV spheroids. In all cases, these crystals are ultra-stable in the environment but need to dissolve to release the active component of the crystal as described in the following paragraph.

3.3.2.4 Release of the LPMO domain from spindles

Studies in insect larvae have shown that the AA10 domain is sufficient for the activity of fusolin in virulence enhancement. When spindles are incubated in digestive juices, a host serine protease cleaves off a C-terminal sequence equivalent to the CT arm (Takemoto et al., 2008). Proteolytic release of the AA10 domain has been reproduced *in vitro* using chymotrypsin and results in a destabilization of the domain-swapped dimer. Accordingly, when CT is removed in silico, most inter-dimer contacts are lost and the oligomer is no longer predicted to be stable in solution (Chiu et al., 2015). Indeed, removal of the CT arm is predicted to strongly destabilize the crystalline lattice. Efficient dissolution is only observed at alkaline pH. As in both classes of polyhedra, destabilization of spindles could be linked to the deprotonation of multiple tyrosines. CT is particularly rich in tyrosine residues representing ~15% of the residues for *Mm*EV fusolin. Tyrosines residues are found on either side of the domain-swapped helix but also at the interface between this helix and the neighboring subunit (Fig. 7). These two tyrosines (Tyr_{345} and Tyr_{348}) are among the few conserved residues in the CT of the six published sequences of fusolin. Specifically, deprotonation of Tyr_{348} introduces a negative charge in a strongly electronegative environment, which is likely to destabilize the dimeric interface. Dissociation of coleopteran EV spindles is unexpectedly inefficient in the digestive juice of lepidopteran species, *Bombyx mori* and *Spodoptera litura*, and the native hosts *Anomala cuprea* larvae (Mitsuhashi et al., 2019). The ultra-stability of spindles also represents an obstacle for optimum use as a synergistic agent in bioinsecticides, which could be surmounted by structure-based protein engineering.

3.3.2.5 Mode of action of spindles

The mode of action of spindles is summarized in Fig. 6E. Production of spindles in infected cells result in its crystallization within the ER.

These ultra-stable crystals are released upon cell lysis and shed in the environment during the course of infection or following the death of the larva. These crystals are ingested by a new host together with spheroids that contains infectious particles. Spheroids release virions in the midguts of the new host by a poorly understood mechanism. Spindles also dissolve in the midguts due to the proteolytic cleavage of the CT arm in alkaline conditions. The released AA10 domain is then activated to cleave chitin, a major constituent of the PM tube that protects the insect midguts. Damage to the PM enhances the ability of spheroid-derived virions to invade the new host and cause disease, ultimately increasing virulence.

3.3.2.6 A unique strategy of *in vivo* crystallization

Spindles assemble in a completely different way to polyhedra. The stability of polyhedra is primarily achieved through solvent exclusion (<20%) and high surface complementarity between polyhedrin molecules. The highly compact packing means that ~75% of the polyhedrin surface is involved in crystal contacts. By contrast, spindles contain 33–40% of solvent, which is not unusual for protein crystals. Most of the crystal contacts involve the CT arm, suggesting that addition of this extension to a soluble precursor has permitted crystallization. Analysis of sequence conservation supports this scenario. The Fn3 domain is clearly homologous to baculovirus GP37 proteins and the bacterial AA10 proteins, but CT arms are highly variable and only present in AA10 proteins that form intracellular crystals (i.e. fusolin and some GP37).

3.4 Four evolutionary and structurally distinct classes of functional viral crystals

Three distinct families of viruses produce occlusion bodies as one of their infectious forms: the *Reoviridae* (genus: cypovirus), *Baculoviridae* and *Poxviridae* (sub-family: *Entomopoxvirinae*). Viruses from these viral families differ in most aspects of their biology (Fig. 3). The only features in common between these three viruses are their invertebrate hosts and the use of occlusion bodies raising the question of whether this infectious form was disseminated through horizontal transfer from a common ancestral source or emerged independently in these three viral families.

Early structural studies revealed remarkable similarities between the lattices of both classes of polyhedra (Anduleit et al., 2005). CPV and BV polyhedra shared a space group *I*23 with very similar unit cell parameters, built from a 29 kDa proteins. Among the >152,000 structures currently available in the Protein Data Bank, 43 have an *I*23 space group with unit cell parameter 101.5–104 Å. All but 3 (Anisep protease PDB ID: 3wy8; putative phage capsid, PDB ID: 2R9I; Dotd protein, PDB ID: 3ady) are baculovirus or cypovirus polyhedra. These remarkable similarities suggested the possibility that, although protein sequences have diverged beyond recognition, the lattice could have been conserved from an ancestral virus-containing crystal.

Unexpectedly, the structures of both types of polyhedra have not provided support to this hypothesis. The cypovirus and baculovirus polyhedrins differ in their folds and the way they assemble into virus-containing crystals (Figs. 4, 5 and 7). In the case of cypovirus polyhedrin, compact trimers form the likely building blocks of the crystal. Upright beta-sandwiches interacting through much of the length of the beta-strands bury large molecular surfaces. By contrast, the jelly-roll domains of baculovirus polyhedrin have a looser, propeller-like arrangement in the trimer, which primarily involve the loops decorating the beta-sandwich. In baculoviruses, the building block is most likely the dodecamer stabilized both by domain swapping and disulfide crosslinks (Chiu et al., 2012).

Interestingly, in the three classes of crystals, extended structures forming molecular arms are important for the stability and perhaps morphogenesis of the crystals (Fig. 7). In cypovirus, the H1 helix braces the central trimers from the periphery and extends to neighboring unit cells. The Cter plays a similar role in baculovirus polyhedra, projecting outward of the dodecamer to mediate long-distance cross-linking interactions between neighboring unit cells. Fusolin uses a different CT extension that stabilizes the dimeric building block of the crystal and mediates most inter-dimer contacts, including all intermolecular disulfide bonds. These molecular arms are not evolutionary related. They differ in sequence, structure and location in the crystal lattice. While baculovirus polyhedrin also has an N-terminal helix, it is buried in the center of the unit cell to stabilize the core of the dodecamer rather than projecting outward as H1 in CPV. The convergence to similar stabilization principles suggests that molecular arms are important features for the formation of stable crystals *in vivo*. This may be relevant to engineer the propensity for *in vivo* crystallization as a path toward structure determination of recombinant proteins (Koopmann et al., 2012).

However, if the two classes of polyhedra emerged independently, it remains a mystery why they converged to adopt very similar crystalline organizations. Perhaps the high symmetry of the *I*23 space group allows to achieve the unique metastability of polyhedra characterized by an extreme stability in most chemical and environmental conditions but a very rapid dissolution in an alkaline environment. Organization into a crystalline lattice is by nature repetitive due to internal symmetries. Such a repetitive arrangement is more conducive to the amplification of localized changes in response to pH changes than amorphous aggregates. This would result in propagation of the dissociation signal more specifically and rapidly than in non-crystalline occlusion bodies.

4. Use of viral crystals in biotechnology applications

4.1 Biological control of insects

Given their natural role in regulating insect populations, the initial biotechnological applications of polyhedra, spheroids and spindles have focused on their use for the control of insect pests. Much of the research and developments have centered on baculoviruses, which infect insects—particularly Lepidoptera—relevant to agriculture and forestry. They are considered as safe, relatively easy to produce in large quantities and highly pathogenic to their specific host. Their occluded form makes them convenient to formulate, store and apply to crop. Compared to non-occluded viruses, baculoviruses have an extended persistence in the field due to slower inactivation by UV and dehydration (Lacey et al., 2015). Despite their promises, baculovirus-based insecticides have not been widely adopted (Lacey et al., 2015; Summers, 2006). Several factors have contribute to these setbacks. First, baculoviruses have a relatively narrow host range, especially compared to chemical insecticides such as neonicotinoid or organophosphates. While the ability to target specific pests and avoid off-target damage is attractive from an environmental viewpoint, this characteristic severely complicates the practical implementation of baculovirus-based pest management. Second, the relatively high-cost of production and slow killing limit their cost-effectiveness in many agricultural settings, particularly for low value/high volume crops. However, use of baculovirus in integrated pest management strategies has met some success in cases where insects rapidly develop resistance to chemical insecticides, such as *Helicoverpa* spp. and *Spodoptera* spp. The current status of these programs has been reviewed in details by Lacey et al. (2015). One of the most widely commercialized virus is a granulovirus infecting the codling moth, *Cydia pomonella* (*Cp*GV), which is now used worldwide for the protection of apple trees. As for most other viruses, the restricted persistence of the virus in the environment is a major limitation in the field (Lacey et al., 2008). Overcoming this limitation for *Cp*GV and other baculoviruses is not trivial but will be facilitated by the structural and molecular advances in our understanding of the metastability of baculovirus occlusion bodies described in Section 3.

To overcome the lack of potency of baculovirus-based insecticides, genetic engineering has been most commonly required. These studies have focused on the insertion or inactivation of viral and toxin genes to accelerate and enhance the insecticidal activity of the modified viruses. However, the use of genetically modified viruses is problematic due to obstacles in regulatory approval and public acceptance (Summers, 2006). An alternative to modifying the viruses has been to identify virulence factors that can be expressed in plants or co-delivered with the viral agent. Spindles and GP37 have been investigated as a potential synergistic factor to enhance the virulence of baculoviruses, EV and the effect of Bt toxin (Mitsuhashi et al., 2013). While their ingestion severely disrupts the PM of lepidopterans, spindles are not toxic by themselves. However, they strongly enhance the virulence of other oral pathogenic agents in co-feeding experiments. This effect has been well studied for the autologous EV but also heterologous pathogens such as the unrelated baculoviruses (Summers, 2006). In this context, reduction of the lethal dose of up to six orders of magnitude has been achieved. For large viruses, this potentiating effect can be easily explained by the removal of the physical barrier of the PM. Pores in the chitin-matrix layers of the PM have diverse geometry and sizes but are typically <200 nm (Hegedus et al., 2009; Kawakita et al., 2016), smaller than EV and baculovirus virions. The mechanism of enhancement of the Bt toxin effect is less clear since the protein should diffuse through the PM pores once crystals dissolve. It has been proposed that spindles reduce adsorption of Bt toxin to the PM but the mechanisms remain to be elucidated (Mitsuhashi et al., 2013). In case of chemical disruption of the PM by Calcofluor, no enhancement of the Bt toxin effect was observed, suggesting that spindle disruption is qualitatively different (Rees et al., 2009). In a practical application of this approach for pest management, expression of the *Pseudaletia separata* EV fusolin in transgenic rice combined with the baculovirus *Pseudaletia unipuncta* MNPV has been effective at providing some levels of improved protection from armyworm larvae but has not been developed further at this stage (Hukuhara et al., 1999).

4.2 Polyhedra as microcontainers for protein cargo

Independently of their role in the viral cycle, the remarkable stability of polyhedra has prompted their use in diverse biotechnological applications harnessing their packaging and protective abilities.

Engineering of baculovirus polyhedra has been recently reviewed by Lopez et al. (2018). Beyond their role as bioinsecticides, a modified polyhedra have been used in applications ranging from efficient protein production and purification to the development of diagnostic tests and vaccines. Most of these developments were designed without a knowledge of the polyhedra structure and we can expect novel and improved approaches in the future. Some factors, however, complicate the use of baculovirus polyhedra. These include the presence of infectious virions in the crystalline matrix, which can induce off-target biological effects, reactogenicity and represent a biosafety risk. In addition, baculovirus polyhedra are sealed by the polyhedron envelope or calyx, which may prevent access to surface proteins and antigens, a desired feature in many biotechnological applications. Finally, the polyhedra assemble in the nucleus, which can be incompatible with some cytoplasmic or secreted targets.

Cypovirus polyhedra are simpler, which may facilitate their engineering. They lack an external envelope, assemble in the cytoplasm and can be produced recombinantly in insect cells in the absence of the native virus. These crystals have been engineered to embed proteins of interest during the *in cellula* crystallization process by various approaches involving co-expression of the native polyhedrin with a tagged protein of interest (Fig. 8A). Initial engineering of such nanocontainers used the N-terminus of the cypovirus protein VP3 (residues 1–70 or 42–93), which had been proposed as the viral polyhedra-binding domain of the virion (Ikeda et al., 2006; Yu et al., 2011). The polyhedrin-binding role of VP3 has since been questioned by the high-resolution structure determination of CPV1 particles and cryo-EM, as discussed in Section 3.1 (Chen et al., 2011; Yu et al., 2011).

Based on the structure of polyhedra (Coulibaly et al., 2007), an alternative targeting strategy was designed that exploited the central role of the H1 helix in the crystalline organization. When fused to the protein of interest, the 30 residues of H1 direct the incorporation of the chimeric protein into crystals formed by the native polyhedrin. Co-expression of the fusion protein with the native polyhedrin resulted in the production of mixed crystals, where the cargo is found both embedded within the crystalline matrix and at its surface (Ijiri et al., 2009). This approach was used to produce crystals containing several growth factors to induce cell proliferation and differentiation in cell culture and *in vivo*. Fibroblast growth factor-2 (FGF-2) incorporated into CPV polyhedra induced the proliferation of NIH/3T3 cells (Fig. 8B) and human epidermal keratinocytes. In the latter case, the crystals could be deposited and air dried with little loss of activity. Further studies have shown

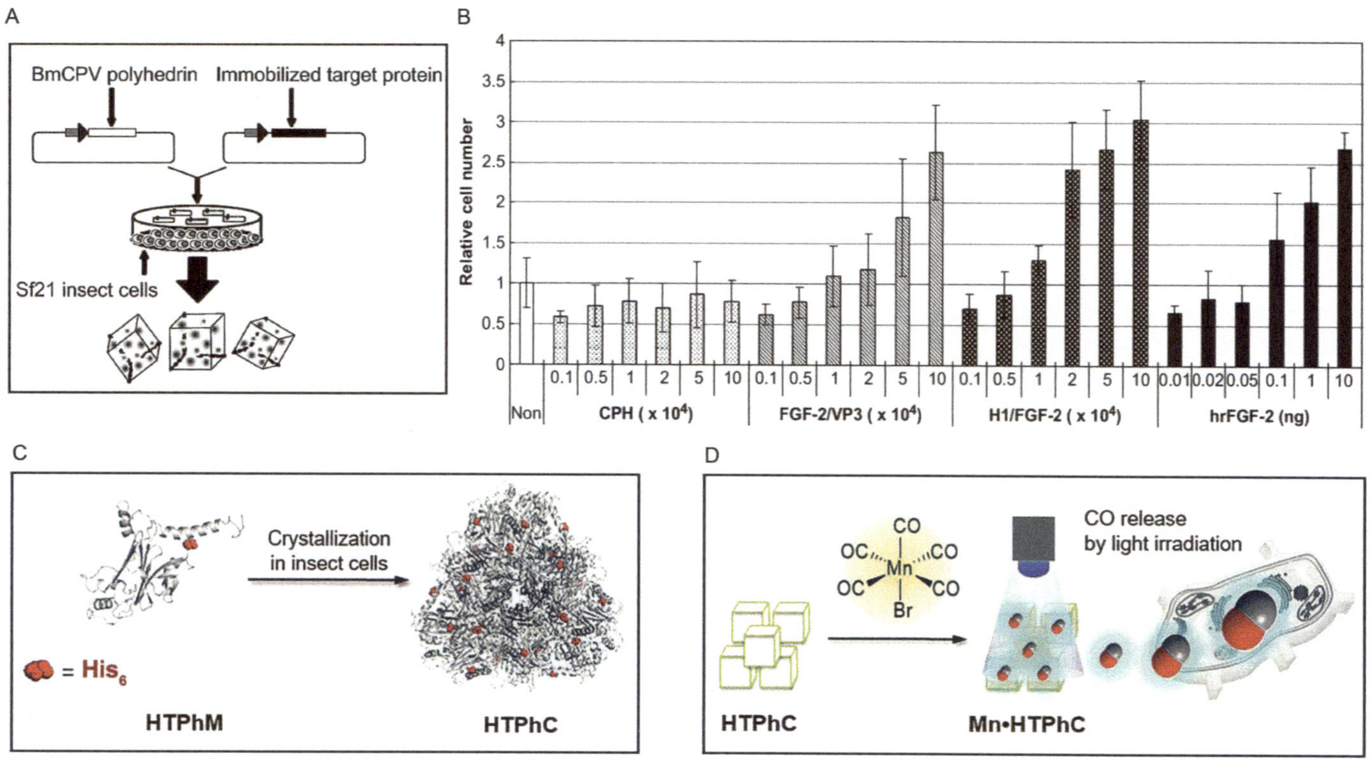

Fig. 8 See legend on next page.

that leukemia inhibitory factor (LIF) immobilized in polyhedra promoted the proliferation of embryonic (ES) and induced pluripotent (iPS) stem cells. Slow-release of the embedded cytokine sustained iPS for 14 days whereas recombinant LIF needed to be replenished every 2–3 days (Nishishita et al., 2011). When applied *in vivo* in atelocollagen sponges, polyhedra containing bone morphogenetic protein 2 facilitated bone repair more efficiently than soluble, recombinant protein. Crystals were observed for up to 10 weeks suggesting long-term slow release of the cargo without evidence of inflammation (Matsumoto et al., 2012, 2015).

Polyhedra have also been engineered to encapsulate functional enzymes such as protein kinase C. Release was achieved at pH 8.5 due to mutation of residue Arg_{13} to an alanine. These crystals preserve the enzymatic activity after air drying for a week at 25 °C (Abe et al., 2015).

4.3 Polyhedra as nanochambers for small molecules

Cypovirus polyhedra have also been engineered for the encapsulation of small molecules for chemical applications. Nanocontainer containing metal carbonyl complexes were developed for storage of CO (Fig. 8C and D) and delivery to living cells, where CO may have anti-inflammatory, anti-apoptotic and antiproliferative activities. While wild-type crystals reacted with the organometallic Ru carbonyl complexes, CO incorporation into the crystals was boosted threefold by addition of a C-terminal histidine-

Fig. 8 Polyhedra as microcontainers for proteins and small molecules. (A) Production of recombinant crystals of CPV1 polyhedrin embedding an immobilized target protein. Two recombinant baculovirus vectors are used to infect Sf21 cells. Crystals are purified from doubly-infected cells by cell lysis and differential centrifugation. (B) NIH/3T3 cells were incubated with CPV polyhedra containing FGF-2 incorporated using the VP3 C-terminal sequence tag (FGF-2/VP3) or H1 N-terminal sequence tag (H1/FGF-2). Proliferation is compared to control crystals of CPV1 polyhedrin (CPH) and recombinant FGF-2 (hrFGF-2, 0.01–10 ng). (B) Schematic representation of polyhedrin crystals (HTPhC) of a CPV polyhedrin modified with a C-terminal His-tag (HTPhM). (C) The HTPhC incorporate the manganese-CO complex and can be activated by light irradiation to release CO as a probe of the cellular effects of CO. *Panel (A): From Ijiri, H., Coulibaly, F., Nishimura, G., Nakai, D., Chiu, E., Takenaka, C., Ikeda, K., Nakazawa, H., Hamada, N., Kotani, E., Metcalf, P., Kawamata, S., Mori, H., 2009. Structure-based targeting of bioactive proteins into cypovirus polyhedra and application to immobilized cytokines for mammalian cell culture. Biomaterials 30 (26), 4297–4308 with permission; panels (B) and (C): From Tabe, H., Shimoi, T., Boudes, M., Abe, S., Coulibaly, F., Kitagawa, S., Mori, H., Ueno, T., 2016. Photoactivatable CO release from engineered protein crystals to modulate NF-κB activation. Chem. Commun. (Camb.) 52 (24), 4545–4548 with permission.*

tag to the polyhedrin sequence (Tabe et al., 2015). Controlled released could be demonstrated with a manganese-CO complex where release of the CO gas is induced by irradiation with visible light (Fig. 8D) (Tabe et al., 2016).

It is anticipated that further biohybrid materials will be developed based on natural protein crystals of known structure. In this perspective, EV spindles are particularly attractive owing to their native metal binding site formed by the histidine brace and their ultra-stable crystalline matrix crosslinked by disulfide bonds.

5. Examples of non-crystalline viral superstructures

5.1 Viral pyramids

Archaeal viruses display an extraordinary diversity in the architectures of their virus particles (Prangishvili, 2013; Prangishvili et al., 2017). *Sulfolobus islandicus* rod-shaped virus 2 (SIRV2) is the first virus shown to package its double-stranded DNA genome in an A-form (DiMaio et al., 2015). Perhaps an even more striking feature of this virus is that it induces the production of pyramid-like structures in infected cells (Fig. 9) (Quax et al., 2011). These structures, also produced by *Sulfolobus* turreted icosahedral virus 1 (STIV1), mediate the release of virions from infected cells.

SIRV2 and STIV1 have completely distinct structures, replication processes and genomic organizations. Accordingly, they have been assigned to the unrelated *Rudiviridae* and *Turriviridae* families, respectively. However, these two viruses share analogous egress strategies involving the formation of specialized pyramidal structures. In the second half of the infectious cycle, these viruses produce virus-associated pyramids (VAPs) at the cell membrane that grow outward, piercing through the S-layer and eventually opening up to form ~150 nm channels through which virions can escape the cell (Fig. 9A–H). Each cell presents about 12 VAPs that open concomitantly, leading to the eventual lysis of the host cell (Daum et al., 2014).

VAPs are composed of a single viral protein of ~10 kDa called PVAP. The protein is predicted to have an N-terminal transmembrane segment, followed by a helical intracellular domain with three predicted alpha-helices (Fig. 9I). Although it is not recognized as a typical signal peptide by sequence analysis, the N-terminus of PVAP directs its insertion into membranes in homologous and heterologous systems (Daum et al., 2014).

The model of VAP assembly proposes that PVAP is produced as a monomeric species that assembles into a heptameric complex upon membrane insertion. This heptameric complex acts as a nucleation point from which

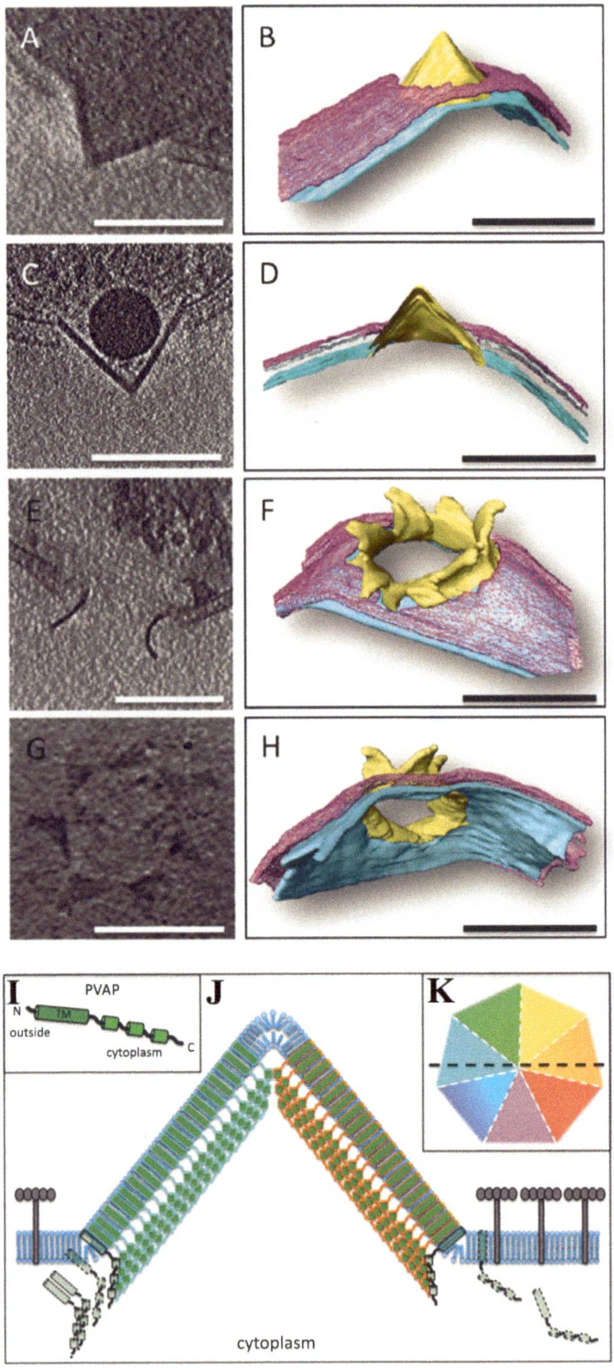
A
B
C
D
E
F
G
H
I
PVAP
N
TM
outside
cytoplasm
C
J
K
cytoplasm

each facet of the pyramid grows. Growth requires the addition of PVAP, which are recruited through at least two types of interactions, in-plane to form the facets and out-of-plane to form the edges of the pyramid (Fig. 9I–K). The nature of the oligomeric species recruited to the growing pyramids is not yet known. Interactions at the edges are labile, allowing opening of the VAP to permit virion egress. VAPs between 20 and 150 nm in diameter have been observed. When VAPs reach ~150 nm they open first from the tip followed by unzipping of the facets. The final structure resembles a flower bloom where each facet has a pronounced anti-clockwise twist when viewed from outside the cell.

Despite differences in size between individual VAP, structural analysis of the top of the pyramids using sub-tomogram averaging produced a 3D map of VAPs at a mean resolution of 4 nm. This structure revealed the unique architecture of VAPs, composed of seven facets forming an opening angle of ~80° at the tip of the pyramid. Each facet has an isosceles triangular shape with an angle of 35° at the tip apex and a slight concavity toward the inside of the hollow pyramid. Sub-tomogram averaging revealed two distinct layers separated by a gap of ~5.8 nm. The outer layer is continuous with the cell membrane and has been assigned to the predicted transmembrane helix of PVAP. The inner layer is placed on the cytoplasmic side and has been proposed to be formed by self-assembly of the C-terminal domain of PVAP. Self-assembly of this domain is suggested to be the driving force deforming the membrane pushed outward due to a staggered arrangement compared to the transmembrane domain.

Interestingly, expression of the VAP protein in heterologous system is sufficient for the formation of pyramids. These structures have thus an

---

Fig. 9 Viral pyramids, elegant egress structures produced by archaeal viruses. (A–H) Electron tomography of SIRV2-infected *S. islandicus* cells. Representation of a tomogram slice (A, C, E, G) and rendered 3D volumes (B, D, F, H) with the S-layer in purple, the cell membrane in blue and the VAP in yellow. Scale bars are 200 nm. (I–K) Schematic model of the PVAP and VAP molecular organization. (I) The PVAP topology is shown with the transmembrane (TM) helix, N- and C-termini labeled. (J) The pyramid presents an external layer continuous with the cell membrane (blue) into which the TM of PVAP inserts. The internal layer of the VAP is proposed to be formed by three predicted helices in the cytoplasmic domain of PVAP. (K) Orthogonal view from the outside of the cell. The doted black line indicates the plane of view shown in (J). *From Daum, B., Quax, T.E.F., Sachse, M., Mills, D.J., Reimann, J., Yildiz, O., Hader, S., Saveanu, C., Forterre, P., Albers, S.V., Kuhlbrandt, W., Prangishvili, D., 2014. Self-assembly of the general membrane-remodeling protein PVAP into sevenfold virus-associated pyramids. Proc. Natl. Acad. Sci. U. S. A. 111 (10), 3829–3834.*

intrinsic membrane remodeling capacity, opening novel avenues to manipulate membranes. However, in a non-native context, these structures are not able to open and do not induce cell lysis (Daum et al., 2014). This functional difference may be due to the absence of other viral factors and suggests that the opening is an active process triggered by a viral protein or specific cellular changes.

This egress system may be more widely distributed than these two related examples, since analogous pyramid structures have been observed in a putative virus of the hyperthermophilic archaeon *Pyrobaculum oguniense* (Quax and Daum, 2018; Rensen et al., 2015). However, the sixfold symmetry differs from VAPs and no homolog of the PVAP has been detected in the viral genome.

5.2 Nucleus-like superstructures

In Section 2, we have mentioned how eukaryotic RNA viruses remodel the cell to segregate proteins and the viral RNA in specialized compartments forming viral factories or viroplasm. Recently, a similar process has been discovered for a DNA virus infecting *Pseudomonas chlororaphis*, bacteriophage 201ϕ2-1 (Chaikeeratisak et al., 2017b). In this case, the analogy with the eukaryotic nucleus is striking. This compartment is roughly spherical and segregates viral and cellular proteins according to their function (Fig. 10A). Cellular and viral DNA and proteins involved in replication and transcription localize to the compartment. The proteins involved in translation of the viral proteins and completion of virion assembly remain outside the compartment together with the bacterial ribosomes. Capsids initially assemble near the bacterial membrane but then relocate to the periphery of the viral compartment. Upon docking to the surface of the shell, the capsid takes up the viral genome. They are subsequently released from the compartment, further matured by addition of the phage tail component, before exiting the cell upon lysis of the host (Fig. 10B).

The compartment is ~500 nm wide surrounded by a ~5 nm protein shell. It is pushed to the center of the cell by bipolar spindles composed of a phage tubulin-like cytoskeletal protein called PhuZ (Kraemer et al., 2012). This structure appears to be in rotation within the cell and to allow post-translational transport into the shell.

To investigate the structure of this compartment, cells were milled using a focused ion beam in cryogenic conditions (cryo-FIB) and imaged by cryo-electron tomography. This approach allowed imaging at higher resolution in

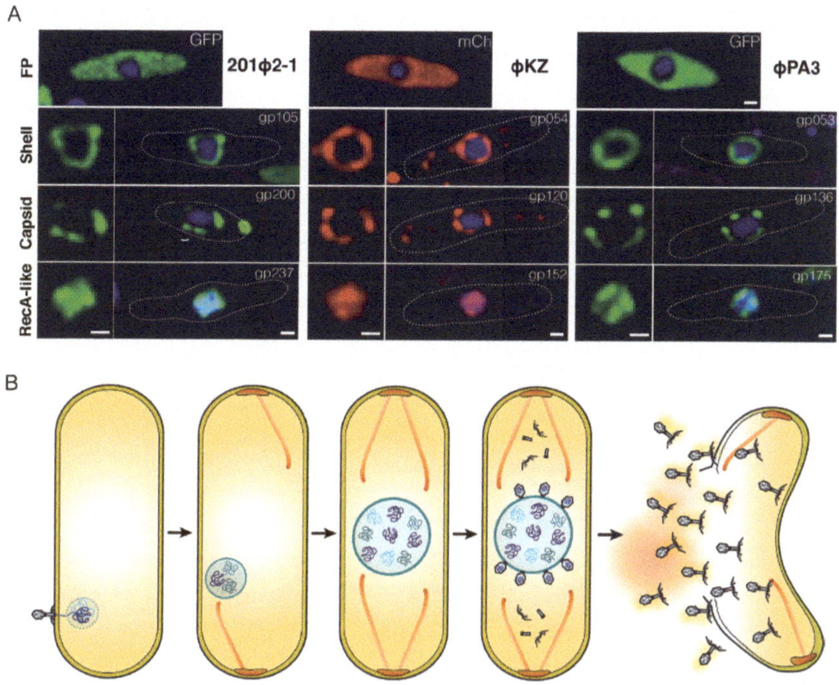

Fig. 10 Nucleus-like superstructures and viral tubulin of large Pseudomonas bacteriophages. (A) Fluorescence microscopy of *P. chlororaphis* (left) and *P. aeruginosa* cells (middle and right) that were infected with phages 201ϕ2-1, ϕKZ and ϕPA3. Cells were expressing GFP or mCherry alone (top row, FP), in fusion with the nuclear shell proteins $gp105_{201\phi2\text{-}1}$, $gp054_{\phi KZ}$ or $gp053_{\phi PA3}$ (second row, shell), in fusion with the major capsid proteins $gp200_{201\phi2\text{-}1}$, $gp120_{\phi KZ}$ or $gp136_{\phi PA3}$ (third row, capsid) or in fusion with the RecA-like proteins $gp237_{201\phi2\text{-}1}$, $gp152_{\phi KZ}$ or $gp175_{\phi PA3}$ (bottom row, RecA-like). (B) Model of the phage nucleus assembly. Upon injection of the phage genome, the viral shell protein is produced and assembles a nucleus-like structure around the DNA. This compartment is pushed toward the center of the cell by the viral tubulin PhuZ, which assembles a bipolar spindle. The nucleus-like compartment grows as replication proceeds until phage assembly begins outside the compartment. Phage heads dock onto the compartment for genome packaging. Eventually, a complete phage assembles and the cell is lysed. *From Chaikeeratisak, V., Nguyen, K., Egan, M.E., Erb, M.L., Vavilina, A., Pogliano, J., 2017. The phage nucleus and tubulin spindle are conserved among large pseudomonas phages. Cell Rep. 20 (7), 1563–1571.*

near native conditions reducing artifacts caused by embedding and sectioning. The compartment appears to be membrane-less and primarily composed of a ~5 nm-thick shell of phage protein gp105. The shell is likely to be continuous since soluble GFP does not diffuse into the compartment (Chaikeeratisak et al., 2017b). How is transport in and out of the compartment regulated? What is the molecular organization of the gp105 shell? New insights into these questions will be undoubtedly provided using the modern arsenal of imaging, biophysics and structural biology approaches.

What are the main functions of this structure? By analogy with eukaryotic viroplasms, it is likely that compartmentalized replication enhances the efficiency of replication by providing a structured environment and concentrating essential viral and cellular factors. It was also proposed that these structures may also help shield the viral DNA from the antiviral defense of the cell (Mendoza et al., 2018).

How widely distributed these structures are remains to be determined. The hallmarks of this structure are conserved in two other large *Pseudomonas* phages, ϕKZ and ϕPA3 (Fig. 10A), including the shell protein and bipolar spindles (Chaikeeratisak et al., 2017a). Replication foci have also been identified in cells infected by the DNA archaeal virus, SIRV2 (Martínez-Alvarez et al., 2017). These foci are located in proximity of the cellular membrane and it is not known whether membrane remodeling is required to form these viral replication centers.

6. Structural biology of viral superstructures

The structural analysis of regular structures induced by viruses in infected cells has benefited from advances in structural biology. In turn, these systems have been used as models to develop and validate innovative structural approaches. In the final sections of this chapter, we provide examples of advances in structural technologies applied to the virus-induced superstructures described in the preceding sections.

6.1 Serial X-ray microcrystallography

The high symmetry of regular structures produced by viruses facilitates their structure determination. However, the fact that they are produced in infected cells means that they are intrinsically limited in size to less than a few microns. Even in the most favorable cases where these structures are highly ordered crystals, technical issues have precluded structure determination until the advent of new approaches in microcrystallography. The advent of powerful microfocus beamlines, new handling methods and serial processing approaches have helped overcome the low signal-to-noise and severe radiation damage associated with these natural microcrystals (Boudes et al., 2014; Giegé et al., 2011; Smith et al., 2012).

The first structures were obtained by microcrystallography of CPV polyhedra where larger recombinant crystals, 5–12 μm in size, could be produced (Coulibaly et al., 2007). These studies took advantage of advanced microfocus beamlines developed at third-generation synchrotron light sources (Boudes et al., 2014; Cusack et al., 1998; Engström et al., 1995; Rossetti et al., 2002) and used micromesh mounts for serial analysis (Thorne et al., 2003). Approaches used for several decades, consisting on crystal harvesting with a nylon loop, are largely unsuitable for microcrystals. The large volume captured by the nylon loop results in the inability to locate microcrystals, overlapping crystals and a strong background of X-ray scattering due to the surrounding layer of vitrified ice. Polyimide micromeshes (MiTegen) have been developed to overcome these limitations. Crystals deposited on these meshes at a suitable dilution will form a single layer of crystals evenly distributed in a thin layer of cryoprotectant once excess liquid has been blotted away (Boudes et al., 2017). The regular pattern and indexing of the meshes allows systematic testing of the whole surface. Automatized workflows analogous to this methodology have been coined Mesh&Collect (Cianci et al., 2019; Zander et al., 2015).

This method has proved essential for subsequent structures, which required the merging of a large number of crystals owing to a weak diffracting power of each individual crystal and severe radiation damage. This step has now been automated at many beamlines to allow rapid rastering of the entire mesh in a few minutes. This grid scan can be used for crystal detection only to focus data collection on diffracting crystals or for X-ray diffraction experiments using shutterless data collection (Axford et al., 2014; Gati et al., 2014). Several fixed-target approaches have been developed as an alternative to micromeshes. For instance, micropatterned silicon chips have allowed structure determination of CPV18 polyhedra at high resolution on a synchrotron source (Roedig et al., 2015) and collect of a full dataset in <10 min at an XFEL source (Roedig et al., 2017). In the case of synchrotron data where multiple incomplete datasets are collected, selection of compatible datasets has also been automated using a hierarchical classification (Giordano et al., 2012; Liu et al., 2012) as implemented in programs like BLEND (Foadi et al., 2013).

Experimental phasing has been and remain a major obstacle to structure determination from microcrystals. The high sensitivity to radiation damage means that the redundancy and accuracy of the diffraction data are often insufficient for structure determination based on anomalous signal. Thus, initial structures of CPV and baculovirus polyhedra were both determined by the isomorphous replacement method using crystals soaked in heavy atom solutions (Coulibaly et al., 2007, 2009). Subsequent structures have been solved by molecular replacement with or without additional phasing information from single isomorphous replacement for polyhedra (Ji et al., 2015), granulin crystals (Gati et al., 2017) and spindles (Chiu et al., 2015). Progress have been made to use anomalous signal from sulfur atoms presents in native crystals (Basu et al., 2019; Liu et al., 2012; Terwilliger et al., 2016) but they are yet to be applied to natural microcrystals induced by viruses.

6.2 *In cellula* X-ray crystallography

Virus-induced crystals are produced naturally in infected cells, which differs from the vast majority of crystals used for structure determination that need to be generated *in vitro* from purified components. The initial success of structure determination from purified crystals highlighted two methodological questions. Can structures be determined in the context of the cell, without isolation of the crystals? Is there a cost to adopting such a direct *in cellula* approach? Using a workflow where crystal-containing cells rather than purified crystals are dispensed onto micromeshes, *in cellula* diffraction data was used to determine the structure of *Operophtera brumata* polyhedra (Fig. 11A–C). The data showed a slightly lower mosaicity and a larger cell for crystals analyzed in the context of the cell (Axford et al., 2014). The higher quality of *in cellula* diffraction data was confirmed in a systematic study that found an improvement of ~0.1 Å on the average resolution of 74 partial datasets (Boudes et al., 2016). The resulting best complete datasets had a resolution of 1.55 Å for data collected *in cellula* vs. 1.90 Å for data collected on purified crystals. A step of flow sorting of virus-containing cells based on side scattering using conventional cell sorter improves the efficiency of the described *in cellula* pipeline (Fig. 12D and E), which outperforms analysis of isolated crystals, even in the case of highly stable polyhedra (Boudes et al., 2016, 2017). Importantly, the pipeline is compatible with experimental phasing using the multiple isomorphous replacement method (MIR) based on heavy metal binding (Fig. 12G). As a proof-of-concept, a *de novo* structure of CPV1 polyhedra was re-determined using this pipeline (PDB ID: 5EXY) in 8 days from initial protein expression to final refinement (Boudes et al., 2016).

In addition to a better preservation of crystals and an improved efficiency, maintaining crystals in a cellular environment has the potential to

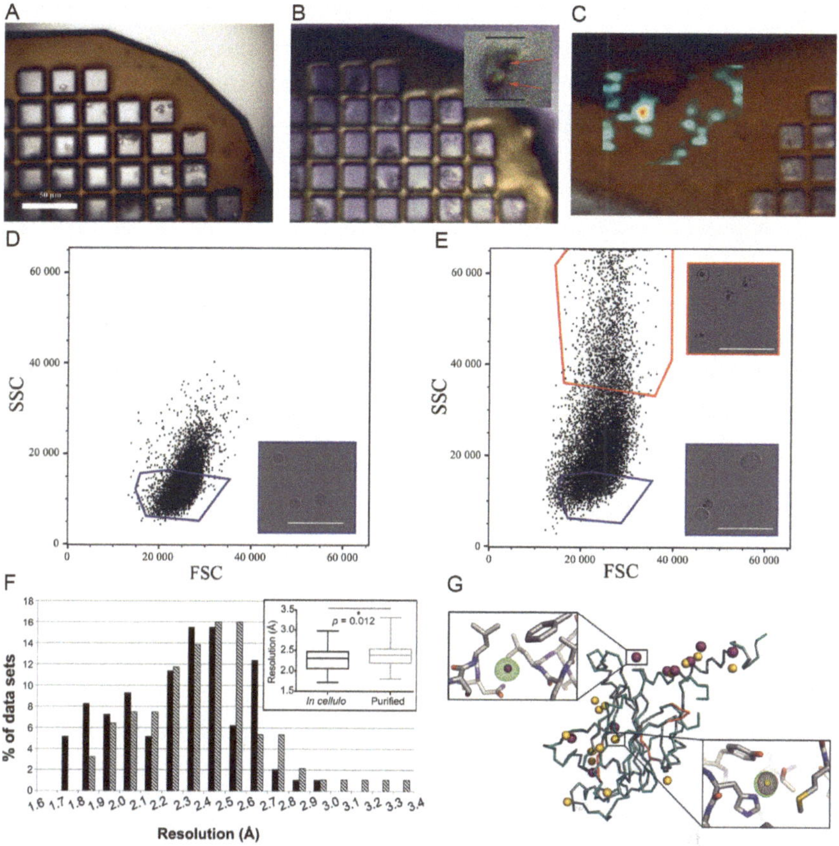

Fig. 11 *In cellula* microcrystallography of viral polyhedra. (A) Polyhedra spread onto a MicroMesh for data collection. Scale bar: 50 μm. (B) CPV polyhedra containing Sf9 cells. Inset: zoom on two polyhedrin crystals inside a single cell. (C) A grid scan of X-ray diffraction was performed over an area of 100 μm × 150 μm on a mesh similar to (B). The gradient blue-red indicates the observed diffracting power. (D-E) Baculovirus-infected cells flow sorted according to the forward (FSC) and side scattering (SSC). Cells that do not express the CPV polyhedrin have low SSC (blue polygon, D). Cells infected with a CPV polyhedrin-expressing baculovirus have an additional population with a range of high SSC. All cells contain at least one visible crystal (red polygon, E). Insets: sorted Sf9 cell. (F) Distribution of the diffracting resolution of polyhedrin crystals analyzed *in cellula* (solid bars) and after purification (striped bars). The inset represents a box-and-whisker plot of data shown in the main panel. (G) Ribbon representation of the polyhedrin structure determined by experimental phasing from polyhedra analyzed *in cellula* (PDB ID: 5EXY). The automatically build model is shown in teal superposed to the final refined model in gray. Red sections were missing from the autobuilt model. Bound iodine (purple spheres) and gold (gold spheres) are shown. The insets show the anomalous and isomorphous maps at 5 sigma as a green meshes, respectively. *Panels (A)–(C): From Axford, D., Ji, X., Stuart, D.I., Sutton, G., 2014. In cellulo structure determination of a novel cypovirus polyhedrin. Acta Crystallogr. D Biol. Crystallogr. 70 (Pt. 5), 1435–1441; panels (D)–(G): From Boudes, M., Garriga, D., Fryga, A., Caradoc-Davies, T., Coulibaly, F., 2016. A pipeline for structure determination of in vivo-grown crystals using in cellulo diffraction. Acta Crystallogr. D Struct. Biol. 72 (Pt. 4), 576–585.*

reveal native interactions with ligands and partners that may be lost during purification. The observed binding of heavy atoms to CPV polyhedra suggests that screening of biologically relevant ligands could be tested in a similar *in cellula* approach.

In cellula diffraction remains limited but has currently been applied to modified polyhedra (Abe et al., 2017) and non-viral proteins (Baskaran et al., 2015; Tsutsui et al., 2015) at synchrotron sources and XFEL sources (Duyvesteyn et al., 2018; Jakobi et al., 2016; Sawaya et al., 2014). Efforts to understand the process of intracellular crystallization are underway (Hasegawa, 2019; Tsutsui et al., 2015) and have been nicely reviewed recently (Schönherr et al., 2018).

6.3 X-ray free electron laser (XFEL)

The limitation owing to the diffraction power of microcrystals has been overcome by illuminating crystals with intense, microfocused X-ray beams at dedicated synchrotron facilities. This approach causes radiation damage mitigated by the merging of several, partial datasets. The advent of free electron laser sources of X-rays has allowed to push this strategy to the extreme. Because of the ultra-intense X-ray pulses, only a single image can be collected before the crystal is destroyed. Merging thousands of frames allows the reconstruction of a complete dataset of sufficient quality to determine a structure. This approach has been applied to the previously undetermined CPV17 polyhedrin crystals (Ginn et al., 2015). Crystals were delivered to the beam using a gas virtual nozzle injecting crystals into a jet of a few μm in diameter presumably at room temperature (similar to Fig. 13A). Crystals were injected at a concentration of 3×10^9 crystals per mL at $30\,\mu L\,min^{-1}$, i.e., ~1.8 billion crystals for a 20 min experiment. Improved processing methods were developed to solve this structure resulting in higher efficiency of indexing and a lower requirement in data redundancy. The structure was solved from 5787 indexable images collected in 20 min, a much lower number of images than previous XFEL experiments, and refined at 1.75 Å.

The structure of virus-containing crystals of granulin was also solved from XFEL diffraction data (PDB ID: 5G3X and 5G0Z) (Fig. 13D–F) and compared with a structure determined using synchrotron radiations (Gati et al., 2017; Oberthuer et al., 2017). The XFEL structure is of higher quality with a better preservation of disulfide bonds suggesting lower radiation damage and the visualization of a protein segment that was not visible in any of the previous structures (Fig. 13F).

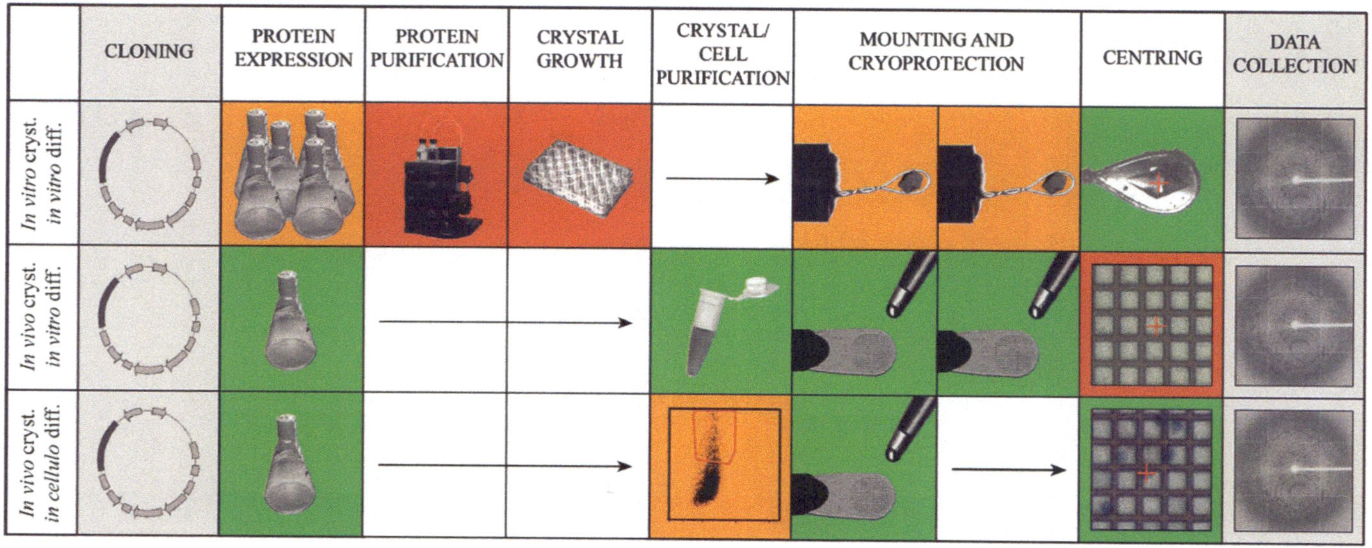

Fig. 12 A pipeline for *in cellula* X-ray microcrystallography. Overview of different structure determination pipelines contrasting the successive steps involved in each workflow and their perceived difficulty or typical success rate (green-orange-red from easy-to-challenging). The workflow for *in cellula* diffraction has the least steps and requires less skills and data collection time than classical microcrystallography in the model case of CPV polyhedra.

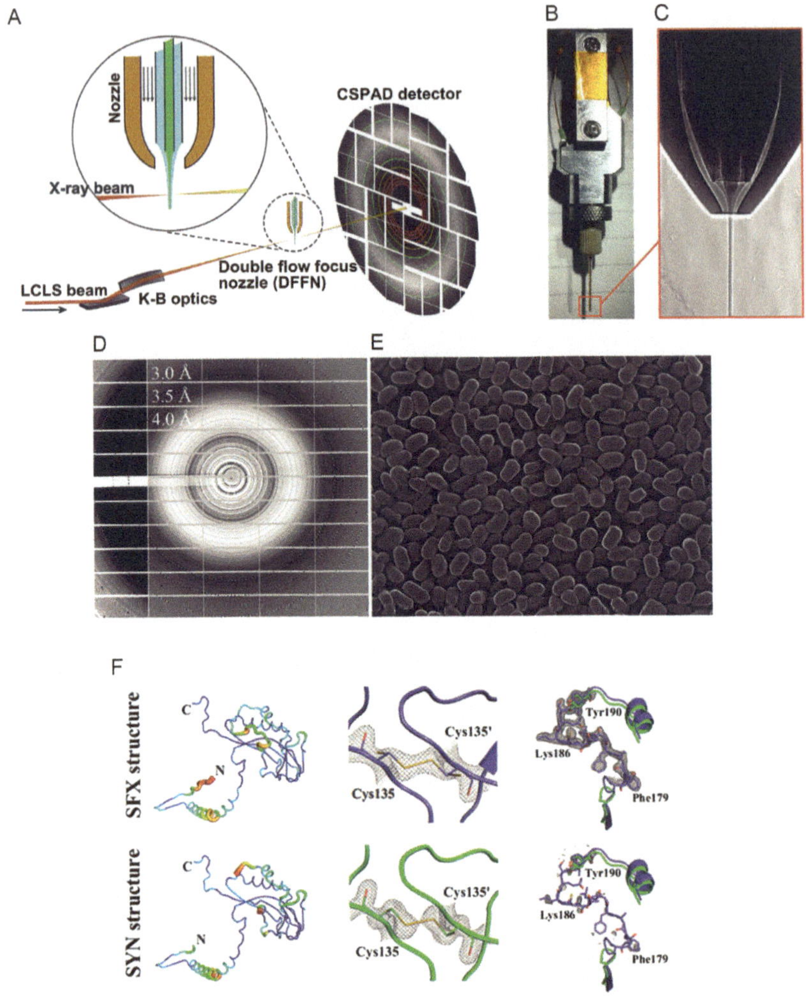

Fig. 13 XFEL analysis of viral granules. (A) Diagram of the experimental set-up for a diffraction before destruction experiment. Crystals are injected in the double flow focus nozzle (DFFN) and intersect with the X-ray free electron laser beam. At LCLS, diffraction is collected on a CSPAD detector. (B, C) DFFN. (D) Powder X-ray diffraction image obtained from GV particles. (E) GV infectious particles. (F) Comparison of the structures obtained from an XFEL source (SFX, top row) and a synchrotron source (SYN, bottom row). Left: a ribbon representation. The blue-red gradient and increasing thickness of the ribbon correspond to the low-to-high B factors. Middle: disulfide bonds have a better density in SFX maps indicating lower radiation damage. Right: residues 176–190 is clearly visible in SFX structure and was not modeled before in a baculovirus polyhedrin structure. *Panels (A)–(C): From Oberthuer, D., Knoška, J., Wiedorn, M.O., Beyerlein, K.R., Bushnell, D.A., Kovaleva, E.G., Heymann, M., Gumprecht, L., Kirian, R.A., Barty, A., Mariani, V., Tolstikova, A., Adriano, L., Awel, S., Barthelmess, M., Dörner, K., Xavier, P.L.,*

6.4 Electron microscopy

Early characterization of ordered structures induced upon viral infection has utilized electron microscopy of serial cell sections. This approach led to the classification of inclusion bodies according to their composition (protein-only vs. virus-containing), intracellular localization and structure (amorphous, paracrystalline or crystalline). However, the molecular organization has remained enigmatic until high-resolution techniques provided 3D visualizations of these objects. For crystalline objects, micro electron diffraction or MicroED is an alternative to X-ray micro-crystallography. In MicroED, sub-micron crystals are illuminated with a low-dose electron beam to collect electron diffraction data. While X-ray experiments at synchrotron still require crystals larger than 1 μm or so, sub-micron crystals are perfectly suitable for MicroED and indeed preferable to limit dynamic scattering. In this respect, MicroED is similar to XFEL but requires minute amounts of crystal compared to a typical XFEL experiments. The structure of granulovirus crystals has been determined by MicroED (PDB ID: 6S2O, unreleased) and it will be very interesting to have a direct comparison of structures determined by synchrotron X-ray crystallography, XFEL and MicroED.

Non-crystalline components in polyhedra require different approaches based on sectioning (e.g., FIB- SEM as described for the nucleus-like compartment of phage 201ϕ2-1) or tomography. A relevant example mentioned earlier in this chapter is the electron tomography of CPV polyhedra, which revealed an exclusion zone around occluded virions (Chen et al., 2011). These virions appeared to contact the crystalline matrix through the A-type fiber rather than the virion surface, in this case contradicting the role

Yefanov, O., James, D.R., Nelson, G., Wang, D., Calvey, G., Chen, Y., Schmidt, A., Szczepek, M., Frielingsdorf, S., Lenz, O., Snell, E., Robinson, P.J., Šarler, B., Belšak, G., Maček, M., Wilde, F., Aquila, A., Boutet, S., Liang, M., Hunter, M.S., Scheerer, P., Lipscomb, J.D., Weierstall, U., Kornberg, R.D., Spence, J.C.H., Pollack, L., Chapman, H.N., Bajt, S., 2017. Double-flow focused liquid injector for efficient serial femtosecond crystallography. Sci. Rep. 7, 44628 with permission; panels (D)–(F): From Gati, C., Oberthuer, D., Yefanov, O., Bunker, R.D., Stellato, F., Chiu, E., Yeh, S.-M., Aquila, A., Basu, S., Bean, R., Beyerlein, K.R., Botha, S., Boutet, S., DePonte, D.P., Doak, R.B., Fromme, R., Galli, L., Grotjohann, I., James, D.R., Kupitz, C., Lomb, L., Messerschmidt, M., Nass, K., Rendek, K., Shoeman, R.L., Wang, D., Weierstall, U., White, T.A., Williams, G.J., Zatsepin, N.A., Fromme, P., Spence, J.C.H., Goldie, K.N., Jehle, J.A., Metcalf, P., Barty, A., Chapman, H.N., 2017. Atomic structure of granulin determined from native nanocrystalline granulovirus using an X-ray free-electron laser. Proc. Natl. Acad. Sci. U. S. A. 114 (9), 2247–2252.

of the turret protein VP3 in virus incorporation. For less symmetrical objects, tomography is a very powerful tool and can be used in cryogenic conditions combined with sub-tomogram averaging to reach higher resolution.

7. Conclusions

Superstructures formed by viruses have been among the most fascinating objects in biology owing to their elegant architectures and, at least at first, mysterious function. Initially, most of these superstructures have been very useful as signatures betraying an interesting underlying viral process. As we progress in their characterization, their architecture, composition, function and molecular structures are elucidated typically requiring a vast arsenal of molecular techniques, imaging, biophysics and structural biology. Some of the systems described herein are now well understood at a molecular level, enabling or facilitating their use for biotechnological applications.

In turn, these systems often constitute ideal targets for the development of new approaches in structural biology and imaging. Their complexity is a challenge, but one that is partly offset by their highly symmetrical nature and the fact that, as most viral components, they have evolved for optimum efficiency, such as high stability and the minimum set of components compared to cellular systems. They offer a fertile ground for inter-disciplinary research bringing together cellular biologists, virologists, biochemists, structural biologists, biophysicists and chemists, as touched upon in this chapter when mentioning biotechnological applications of viral crystals and recent developments in structural biology. The future seems to be toward in situ analysis of these structures using *in cellula* or even *in vivo* analysis, and for most of these systems their dynamic nature remains to be understood. Beyond, one can hope that the techniques tried and tested on these regular superstructures will prefigure the more perilous elucidation of less symmetrical and ordered objects such as membrane-less liquid organelles.

Acknowledgments

My first thanks go to Peter Metcalf and Elaine Chiu from the University of Auckland, with whom I embarked on the initial structure determinations of viral polyhedra. At Monash University, many members of my laboratory have contributed to our research on viral microcrystals, I would like to express particular gratitude to Marcel Hijnen, Marion Boudes, Kaheina Aizel and Damià Garriga. I warmly thank Max Bergoin for sharing with me a fraction of his vast knowledge on insect virology and Leo Chavas for stimulating discussions on in cellula microcrystallography. Research in this field would not have been possible without frequent access to the Swiss Light Source and Australian Synchrotron, and the Ramaciotti Center for cryo-EM at Monash University. The author was funded in part by the Australian Research Council as a recipient of Future Fellowship FT120100893 and Discovery Projects DP1314885 and DP120114039.

References

Abe, S., Ijiri, H., Negishi, H., Yamanaka, H., Sasaki, K., Hirata, K., Mori, H., Ueno, T., 2015. Design of enzyme-encapsulated protein containers by in vivo crystal engineering. Adv. Mater. 27 (48), 7951–7956.

Abe, S., Tabe, H., Ijiri, H., Yamashita, K., Hirata, K., Atsumi, K., Shimoi, T., Akai, M., Mori, H., Kitagawa, S., Ueno, T., 2017. Crystal Engineering of Self-Assembled Porous Protein Materials in Living Cells. ACS nano 11, 2410–2419.

Abrescia, N.G.A., Bamford, D.H., Grimes, J.M., Stuart, D.I., 2012. Structure unifies the viral universe. Annu. Rev. Biochem. 81, 795–822.

Alenquer, M., Vale-Costa, S., Etibor, T.A., Ferreira, F., Sousa, A.L., Amorim, M.J., 2019. Influenza A virus ribonucleoproteins form liquid organelles at endoplasmic reticulum exit sites. Nat. Commun. 10 (1), 1629.

Anduleit, K., Sutton, G., Diprose, J.M., Mertens, P.P.C., Grimes, J.M., Stuart, D.I., 2005. Crystal lattice as biological phenotype for insect viruses. Protein Sci. 14 (10), 2741–2743.

Arif, B.M., 1995. Recent advances in the molecular biology of entomopoxviruses. J. Gen. Virol. 76 (Pt 1), 1–13.

Axford, D., Ji, X., Stuart, D.I., Sutton, G., 2014. In cellulo structure determination of a novel cypovirus polyhedrin. Acta Crystallogr. D Biol. Crystallogr. 70 (Pt 5), 1435–1441.

Banerjee, S., Montaville, P., Chavas, L.M.G., Ramaswamy, S., 2018. The new era of microcrystallography. J. Indian Inst. Sci. 98 (3), 273–281.

Baskaran Y., Ang K.C., Anekal P.V., Chan W.L., Grimes J.M., Manser E., et al., An in cellulo-derived structure of PAK4 in complex with its inhibitor Inka1, Nat. Commun. 6, 2015, 8681.

Basu, S., Finke, A., Vera, L., Wang, M., Olieric, V., 2019. Making routine native SAD a reality: lessons from beamline X06DA at the Swiss Light Source. Acta Crystallogr D Struct Biol 75, 262–271. https://doi.org/10.1107/S2059798319003103. (2019).

Bergoin, M., Devauchelle, G., Vago, C., 1971. Electron microscopy study of Melolontha poxvirus: the fine structure of occluded virions. Virology 43 (2), 453–467.

Boudes, M., Garriga, D., Coulibaly, F., 2014. Reflections on the many facets of protein microcrystallography. Aust. J. Chem. 67 (12), 1793.

Boudes, M., Garriga, D., Fryga, A., Caradoc-Davies, T., Coulibaly, F., 2016. A pipeline for structure determination of in vivo-grown crystals using in cellulo diffraction. Acta Crystallogr. D Struct. Biol. 72 (Pt 4), 576–585.

Boudes, M., Garriga, D., Coulibaly, F., 2017. Microcrystallography of protein crystals and in cellulo diffraction. J. Vis. Exp. (125), 55793.

Boulanger, P.A., Torpier, G., Rimsky, A., 1974. Crystallographic study of intranuclear adenovirus type 5 crystals. Intervirology 2 (1), 56–62.

Carstens, E.B., Krebs, A., Gallerneault, C.E., 1986. Identification of an amino acid essential to the normal assembly of Autographa californica nuclear polyhedrosis virus polyhedra. J. Virol. 58 (2), 684–688.

Caspar, D.L.D., 1956. Structure of bushy stunt virus. Nature 177 (4506), 475–476.

Caspar, D.L., Klug, A., 1962. Physical principles in the construction of regular viruses. Cold Spring Harb. Symp. Quant. Biol. 27, 1–24.
Chaikeeratisak, V., Nguyen, K., Egan, M.E., Erb, M.L., Vavilina, A., Pogliano, J., 2017a. The phage nucleus and tubulin spindle are conserved among large Pseudomonas phages. Cell Rep. 20 (7), 1563–1571.
Chaikeeratisak, V., Nguyen, K., Khanna, K., Brilot, A. F., Erb, M. L., Coker, J. K., Vavilina, A., Newton, G. L., Buschauer, R., Pogliano, K., Villa, E., Agard, D. A. & Pogliano, J. (2017bc) Assembly of a nucleus-like structure during viral replication in bacteria. Science, 355(6321), 194-197.
Chen, J., Sun, J., Atanasov, I., Ryazantsev, S., Zhou, Z.H., 2011. Electron tomography reveals polyhedrin binding and existence of both empty and full cytoplasmic polyhedrosis virus particles inside infectious polyhedra. J. Virol. 85 (12), 6077–6081.
Cheng, X.W., Carner, G.R., Fescemyer, H.W., 1998. Polyhedrin sequence determines the tetrahedral shape of occlusion bodies in Thysanoplusia orichalcea single-nucleocapsid nucleopolyhedrovirus. J. Gen. Virol. 79 (Pt 10), 2549–2556.
Chiu, E., Coulibaly, F., Metcalf, P., 2012. Insect virus polyhedra, infectious protein crystals that contain virus particles. Curr. Opin. Struct. Biol. 22, 234–240.
Chiu, E., Hijnen, M., Bunker, R.D., Boudes, M., Rajendran, C., Aizel, K., Olieric, V., Schulze-Briese, C., Mitsuhashi, W., Young, V., Ward, V.K., Bergoin, M., Metcalf, P., Coulibaly, F., 2015. Structural basis for the enhancement of virulence by viral spindles and their in vivo crystallization. Proc. Natl. Acad. Sci. U. S. A. 112 (13), 3973–3978.
Cianci, M., Nanao, M., Schneider, T.R., 2019. Long-wavelength Mesh&Collect native SAD phasing from microcrystals. Acta Crystallogr. D Struct. Biol. 75, 192–199. (2019). 1-8. https://doi.org/10.1107/S2059798319002031.
Cornalia, E., 1856. Monographia del bombice del gelso (*Bombyx mori Linneo*) Lettere ed Arti, 3-387. Memorie dell' I. R. Instituo Lombardo di Scienze.
Coulibaly, F., Chiu, E., Ikeda, K., Gutmann, S., Haebel, P.W., Schulze-Briese, C., Mori, H., Metcalf, P., 2007. The molecular organization of cypovirus polyhedra. Nature 446 (7131), 97–101.
Coulibaly, F., Chiu, E., Gutmann, S., Rajendran, C., Haebel, P.W., Ikeda, K., Mori, H., Ward, V.K., Schulze-Briese, C., Metcalf, P., 2009. The atomic structure of baculovirus polyhedra reveals the independent emergence of infectious crystals in DNA and RNA viruses. Proc. Natl. Acad. Sci. U. S. A. 106 (52), 22205–22210.
Courtade, G., Aachmann, F.L., 2019. Chitin-active lytic polysaccharide monooxygenases. In: Yang, Q., Fukamizo, T. (Eds.), Targeting Chitin-Containing Organisms. Singapore, Springer Singapore, pp. 115–129.
Crick, F.H.C., Watson, J.D., 1956. Structure of small viruses. Nature 177 (4506), 473–475.
Cusack, S., Belrhali, H., Bram, A., Burghammer, M., Perrakis, A., Riekel, C., 1998. Small is beautiful: protein micro-crystallography. Nat. Struct. Biol. 5, 634.
Dales, S., Eggers, H.J., Tamm, I., Palade, G.E., 1965. Electron microscopic study of the formation of poliovirus. Virology 26 (3), 379–389.
Daum, B., Quax, T.E.F., Sachse, M., Mills, D.J., Reimann, J., Yildiz, O., Hader, S., Saveanu, C., Forterre, P., Albers, S.V., Kuhlbrandt, W., Prangishvili, D., 2014. Self-assembly of the general membrane-remodeling protein PVAP into sevenfold virus-associated pyramids. Proc. Natl. Acad. Sci. U. S. A. 111 (10), 3829–3834.
DiMaio, F., Yu, X., Rensen, E., Krupovic, M., Prangishvili, D., Egelman, E.H., 2015. Virology. A virus that infects a hyperthermophile encapsidates A-form DNA. Science 348 (6237), 914–917.
Dohner, K., Nagel, C.H., Sodeik, B., 2005. Viral stop-and-go along microtubules: taking a ride with dynein and kinesins. Trends Microbiol. 13 (7), 320–327.
Doye, J., Poon, W., 2006. Protein crystallization in vivo. Curr. Opin. Colloid Interface Sci. 11 (1), 40–46.

Duyvesteyn, H.M.E., Ginn, H.M., Pietilä, M.K., Wagner, A., Hattne, J., Grimes, J.M., Hirvonen, E., Evans, G., Parsy, M.-L., Sauter, N.K., Brewster, A.S., Huiskonen, J.T., Stuart, D.I., Sutton, G., Bamford, D.H., 2018. Towards in cellulo virus crystallography. Sci. Rep. 1–7.

Edwardson, J.R., 1966. Cylindrical inclusions in the cytoplasm of leaf cells infected with tobacco etch virus. Science 153 (3738), 883–884.

Engström, P., Fiedler, S., Riekel, C., 1995. Microdiffraction instrumentation and experiments on the microfocus beamline at the ESRF. Rev. Sci. Instrum. 66 (2), 1348.

Fauquet, C.M., Mayo, M.A., 2005. In: Virus Taxonomy, Fauquet, C.M., Mayo, M.A., Maniloff, J., Desselberger, U., Ball, L.A. (Eds.), Academic Press, San Diego, pp. 9–18.

Foadi, J., Aller, P., Alguel, Y., Cameron, A., Axford, D., Owen, R.L., Armour, W., Waterman, D.G., Iwata, S., Evans, G., 2013. Clustering procedures for the optimal selection of data sets from multiple crystals in macromolecular crystallography. Acta Crystallogr. D Biol. Crystallogr. 69 (8), 1617–1632.

Forsberg, Z., Mackenzie, A.K., Sørlie, M., Røhr, A.K., Helland, R., Arvai, A.S., Vaaje-Kolstad, G., Eijsink, V.G.H., 2014. Structural and functional characterization of a conserved pair of bacterial cellulose-oxidizing lytic polysaccharide monooxygenases. Proc. Natl. Acad. Sci. U. S. A. 111 (23), 8446–8451.

Frandsen, K.E., Simmons, T.J., Dupree, P., Poulsen, J.C., Hemsworth, G.R., Ciano, L., Johnston, E.M., Tovborg, M., Johansen, K.S., von Freiesleben, P., Marmuse, L., Fort, S., Cottaz, S., Driguez, H., Henrissat, B., Lenfant, N., Tuna, F., Baldansuren, A., Davies, G.J., Lo Leggio, L., Walton, P.H., 2016. The molecular basis of polysaccharide cleavage by lytic polysaccharide monooxygenases. Nat. Chem. Biol. 12 (4), 298–303.

Franqueville, L., Henning, P., Magnusson, M., Vigne, E., Schoehn, G., Blair-Zajdel, M.E., Habib, N., Lindholm, L., Blair, G.E., Hong, S.S., Boulanger, P., 2008. Protein crystals in Adenovirus type 5-infected cells: requirements for intranuclear crystallogenesis, structural and functional analysis. PLoS One 3 (8), e2894.

Gati, C., Bourenkov, G., Klinge, M., Rehders, D., Stellato, F., Oberthur, D., Yefanov, O., Sommer, B.P., Mogk, S., Duszenko, M., 2014. Serial crystallography on in vivo grown microcrystals using synchrotron radiation. IUCrJ 1 (2), 1–8.

Gati, C., Oberthuer, D., Yefanov, O., Bunker, R.D., Stellato, F., Chiu, E., Yeh, S.-M., Aquila, A., Basu, S., Bean, R., Beyerlein, K.R., Botha, S., Boutet, S., DePonte, D.P., Doak, R.B., Fromme, R., Galli, L., Grotjohann, I., James, D.R., Kupitz, C., Lomb, L., Messerschmidt, M., Nass, K., Rendek, K., Shoeman, R.L., Wang, D., Weierstall, U., White, T.A., Williams, G.J., Zatsepin, N.A., Fromme, P., Spence, J.C.H., Goldie, K.N., Jehle, J.A., Metcalf, P., Barty, A., Chapman, H.N., 2017. Atomic structure of granulin determined from native nanocrystalline granulovirus using an X-ray free-electron laser. Proc. Natl. Acad. Sci. U. S. A. 114 (9), 2247–2252.

Gauthier, L., Cousserans, F., Veyrunes, J.C., Bergoin, M., 1995. The Melolontha melolontha entomopoxvirus (MmEPV) fusolin is related to the fusolins of lepidopteran EPVs and to the 37K baculovirus glycoprotein. Virology 208 (2), 427–436.

Giegé, R., Evans, G., Axford, D., Owen, R.L., 2011. The design of macromolecular crystallography diffraction experiments. Acta Crystallogr. D Biol. Crystallogr. 67, 261–270. (2011). 1-10. https://doi.org/10.1107/S0907444911007608.

Ginn, H.M., Messerschmidt, M., Ji, X., Zhang, H., Axford, D., Gildea, R.J., Winter, G., Brewster, A.S., Hattne, J., Wagner, A., Grimes, J.M., Evans, G., Sauter, N.K., Sutton, G., Stuart, D.I., 2015. Structure of CPV17 polyhedrin determined by the improved analysis of serial femtosecond crystallographic data. Nat. Commun. 6, 6435.

Giordano, R., Leal, R.M., Bourenkov, G.P., McSweeney, S., Popov, A.N., 2012. The application of hierarchical cluster analysis to the selection of isomorphous crystals. Acta Crystallogr. D Biol. Crystallogr. 68 (Pt 6), 649–658.

Hasegawa, H., 2019. Simultaneous induction of distinct protein phase separation events in multiple subcellular compartments of a single cell. Exp Cell Res 379 (1), 92–109.
Hegedus, D., Erlandson, M., Gillott, C., Toprak, U., 2009. New insights into peritrophic matrix synthesis, architecture, and function. Annu. Rev. Entomol. 54 (1), 285–302.
Heinrich, B.S., Maliga, Z., Stein, D.A., Hyman, A.A., Whelan, S.P.J., 2018. Phase transitions drive the formation of vesicular stomatitis virus replication compartments. mBio 9 (5), e02290-17.
Hemsworth, G.R., Davies, G.J., Walton, P.H., 2013. Recent insights into copper-containing lytic polysaccharide mono-oxygenases. Curr. Opin. Struct. Biol. 23 (5), 660–668.
Herniou, E.A., Jehle, J.A., 2007. Baculovirus phylogeny and evolution. Curr. Drug Targets 8 (10), 1043–1050.
Hiebert, E., Purcifull, D.E., Christie, R.G., Maramorosch, K., Koprowski, H., 1984. Purification and immunological analyses of plant viral inclusion bodies. In: Methods in Virology. Elsevier, pp. 225–280.
Hirai, Y., Hirano, Y., Matsuda, A., Hiraoka, Y., Honda, T., Tomonaga, K., 2016. Borna disease virus assembles porous cage-like viral factories in the nucleus. J. Biol. Chem. 291 (50), 25789–25798.
Hukuhara, T., Hayakawa, T., Wijonarko, A., 1999. Increased baculovirus susceptibility of armyworm larvae feeding on transgenic rice plants expressing an entomopoxvirus gene. Nat. Biotechnol. 17 (11), 1122–1124.
Ijiri, H., Coulibaly, F., Nishimura, G., Nakai, D., Chiu, E., Takenaka, C., Ikeda, K., Nakazawa, H., Hamada, N., Kotani, E., Metcalf, P., Kawamata, S., Mori, H., 2009. Structure-based targeting of bioactive proteins into cypovirus polyhedra and application to immobilized cytokines for mammalian cell culture. Biomaterials 30 (26), 4297–4308.
Ikeda, K., Nakazawa, H., Shimo-Oka, A., Ishio, K., Miyata, S., Hosokawa, Y., Matsumura, S., Masuhara, H., Belloncik, S., Alain, R., Goshima, N., Nomura, N., Morigaki, K., Kawai, A., Kuroita, T., Kawakami, B., Endo, Y., Mori, H., 2006. Immobilization of diverse foreign proteins in viral polyhedra and potential application for protein microarrays. Proteomics 6 (1), 54–66.
Iwanowski, D., 1903. Über die Mosaikkrankheit der Tabakspflanze. Z. Pflanzenkrankh. Pflanzenschutz 13 (1), 1–41.
Jakobi AJ, Passon DM, Knoops K, Stellato F, Liang M, White TA, et al. In cellulo serial crystallography of alcohol oxidase crystals inside yeast cells. IUCrJ 2016,3(Pt 2),88–95.
Jarvis, D.L., Bohlmeyer, D.A., Garcia Jr., A., 1991. Requirements for nuclear localization and supramolecular assembly of a baculovirus polyhedrin protein. Virology 185 (2), 795–810.
Jehle, J.A., Blissard, G.W., Bonning, B.C., Cory, J.S., Herniou, E.A., Rohrmann, G.F., Theilmann, D.A., Thiem, S.M., Vlak, J.M., 2006. On the classification and nomenclature of baculoviruses: a proposal for revision. Arch. Virol. 151 (7), 1257–1266.
Ji, X., Sutton, G., Evans, G., Axford, D., Owen, R., Stuart, D.I., 2010. How baculovirus polyhedra fit square pegs into round holes to robustly package viruses. EMBO J. 29 (2), 505–514.
Ji, X., Axford, D., Owen, R., Evans, G., Ginn, H.M., Sutton, G., Stuart, D.I., 2015. Polyhedra structures and the evolution of the insect viruses. J. Struct. Biol. 192 (1), 88–99.
Jin, X., Cao, X., Wang, X., Jiang, J., Wan, J., Laliberté, J.-F., Zhang, Y., 2018. Three-dimensional architecture and biogenesis of membrane structures associated with plant virus replication. Front. Plant Sci. 9, 57.
Juhl, S.B., Chan, E.P., Ha, Y.-H., Maldovan, M., Brunton, J., Ward, V., Dokland, T., Kalmakoff, J., Farmer, B., Thomas, E.L., Vaia, R.A., 2006. Assembly of Wiseana iridovirus: viruses for colloidal photonic crystals. Adv. Funct. Mater. 16 (8), 1086–1094.
Kawakita, H., Miyamoto, K., Wada, S., Mitsuhashi, W., 2016. Analysis of the ultrastructure and formation pattern of the peritrophic membrane in the cupreous chafer, Anomala

cuprea (Coleoptera: Scarabaeidae). Appl. Entomol. Zool. 51, 133. https://doi.org/10.1007/s13355-015-0382-3.

King, A.M., Lefkowitz, E., Adams, M.J., Carstens, E.B. (Eds.), 2011. Virus taxonomy: ninth report of the International Committee on Taxonomy of Viruses, Vol. 9. Elsevier, p. 164.

Kjellen, L., Lagermalm, G., Svedmyr, A., Thorsson, K.G., 1955. Crystalline-like patterns in the nuclei of cells infected with an animal virus. Nature 175 (4455), 505–506.

Koopmann, R., Cupelli, K., Redecke, L., Nass, K., DePonte, D.P., White, T.A., Stellato, F., Rehders, D., Liang, M., Andreasson, J., Aquila, A., Bajt, S., Barthelmess, M., Barty, A., Bogan, M.J., Bostedt, C., Boutet, S., Bozek, J.D., Caleman, C., Coppola, N., Davidsson, J., Doak, R.B., Ekeberg, T., Epp, S.W., Erk, B., Fleckenstein, H., Foucar, L., Graafsma, H., Gumprecht, L., Hajdu, J., Hampton, C.Y., Hartmann, A., Hartmann, R., Hauser, G., Hirsemann, H., Holl, P., Hunter, M.S., Kassemeyer, S., Kirian, R.A., Lomb, L., Maia, F.R.N.C., Kimmel, N., Martin, A.V., Messerschmidt, M., Reich, C., Rolles, D., Rudek, B., Rudenko, A., Schlichting, I., Schulz, J., Seibert, M.M., Shoeman, R.L., Sierra, R.G., Soltau, H., Stern, S., Strüder, L., Timneanu, N., Ullrich, J., Wang, X., Weidenspointner, G., Weierstall, U., Williams, G.J., Wunderer, C.B., Fromme, P., Spence, J.C.H., Stehle, T., Chapman, H.N., Betzel, C., Duszenko, M., 2012. In vivo protein crystallization opens new routes in structural biology. Nat. Methods 9 (3), 259–262.

Kraemer, J.A., Erb, M.L., Waddling, C.A., Montabana, E.A., Zehr, E.A., Wang, H., Nguyen, K., Pham, D.S., Agard, D.A., Pogliano, J., 2012. A phage tubulin assembles dynamic filaments by an atypical mechanism to center viral DNA within the host cell. Cell 149 (7), 1488–1499.

Krupovic, M., Bamford, D.H., 2010. Order to the viral universe. J. Virol. 84 (24), 12476–12479.

Krupovic, M., Bamford, D.H., 2011. Double-stranded DNA viruses: 20 families and only five different architectural principles for virion assembly. Curr. Opin. Virol. 1 (2), 118–124.

Krupovic, M., Koonin, E.V., 2017. Multiple origins of viral capsid proteins from cellular ancestors. Proc. Natl. Acad. Sci. U. S. A. 114 (12), E2401–E2410.

La Scola, B., Audic, S., Robert, C., Jungang, L., de Lamballerie, X., Drancourt, M., Birtles, R., Claverie, J.M., Raoult, D., 2003. A giant virus in amoebae. Science 299 (5615), 2033.

Lacey, L.A., Thomson, D., Vincent, C., Arthurs, S.P., 2008. Codling moth granulovirus: a comprehensive review. Biomed. Sci. Technol. 18 (7), 639–663.

Lacey, L.A., Grzywacz, D., Shapiro-Ilan, D.I., Frutos, R., Brownbridge, M., Goettel, M.S., 2015. Insect pathogens as biological control agents: back to the future. J. Invertebr. Pathol. 132, 1–41.

Lefkowitz, E.J., Dempsey, D.M., Hendrickson, R.C., Orton, R.J., Siddell, S.G., Smith, D.B., 2018. Virus taxonomy: the database of the International Committee on Taxonomy of Viruses (ICTV). Nucleic Acids Res. 46 (D1), D708–D717.

Li, Z., Li, C., Yang, K., Wang, L., Yin, C., Gong, Y., Pang, Y., 2003. Characterization of a chitin-binding protein GP37 of Spodoptera litura multicapsid nucleopolyhedrovirus. Virus Res. 96 (1-2), 113–122.

Lifchitz, A., Rimsky, A., Torpier, G., Boulanger, P.A., 1975. Crystallographic study of an adenovirus-induced protein crystal in KB cells: a structural model. Acta Crystallogr. B 31 (2), 441–445.

Lin, G., Zhong, J., Wang, X., 2000. Abnormal formation of polyhedra resulting from a single mutation in the polyhedrin gene of Autographa californica multicapsid nucleopolyhedrovirus. J. Invertebr. Pathol. 76 (1), 13–19.

Liu, Q., Dahmane, T., Zhang, Z., Assur, Z., Brasch, J., Shapiro, L., Mancia, F., Hendrickson, W.A., 2012. Structures from anomalous diffraction of native biological macromolecules. Science 336 (6084), 1033–1037.

Lopez, M.G., Diez, M., Alfonso, V., Taboga, O., 2018. Biotechnological applications of occlusion bodies of Baculoviruses. Appl. Microbiol. Biotechnol. 102 (16), 6765–6774.
Lunger, P.D., Maddux, T.C., 1972. Fine-structure studies of the avian infectious bursal agent. I. In vivo viral morphogenesis. Avian Dis. 16 (4), 874–893.
Martelli, G.P., Russo, M., 1977. Plant virus inclusion bodies. In: Lauffer, M.A., Bang, F.B., Maramorosch, K., Smith, K.M. (Eds.), Advances in Virus Research. Academic Press, pp. 175–266.
Martínez-Alvarez, L., Deng, L., Peng, X., 2017. Formation of a viral replication focus in Sulfolobus cells infected by the rudivirus Sulfolobus islandicus rod-shaped virus 2. J. Virol. 91 (13), e00486-17.
Matsumoto, G., Ueda, T., Shimoyama, J., Ijiri, H., Omi, Y., Yube, H., Sugita, Y., Kubo, K., Maeda, H., Kinoshita, Y., Arias, D.G., Shimabukuro, J., Kotani, E., Kawamata, S., Mori, H., 2012. Bone regeneration by polyhedral microcrystals from silkworm virus. Sci. Rep. 2, 935.
Matsumoto, G., Ueda, T., Sugita, Y., Kubo, K., Mizoguchi, M., Kotani, E., Oda, N., Kawamata, S., Segami, N., Mori, H., 2015. Polyhedral microcrystals encapsulating bone morphogenetic protein 2 improve healing in the alveolar ridge. J. Biomater. Appl. 30 (2), 193–200.
Mendoza, S.D., Berry, J.D., Nieweglowska, E.S., Leon, L.M., Agard, D., Bondy-Denomy, J., 2018. A nucleus-like compartment shields bacteriophage DNA from CRISPR-Cas and restriction nucleases. bioRxiv 370791. https://doi.org/10.1101/370791.
Mertens, P.P.C., Attoui, H., Sutton, G., 2011. Cypovirus. In: Tidona, C., Darai, G. (Eds.), The Springer Index of Viruses. Springer Science & Business Media, pp. 1580–1587.
Miller, L.K., 2013. The Baculoviruses. Translated from English by Springer Science & Business Media.
Mitsuhashi, W., Miyamoto, K., 2003. Disintegration of the peritrophic membrane of silkworm larvae due to spindles of an entomopoxvirus. J. Invertebr. Pathol. 82 (1), 34–40.
Mitsuhashi, W., Kawakita, H., Murakami, R., Takemoto, Y., Saiki, T., Miyamoto, K., Wada, S., 2007. Spindles of an entomopoxvirus facilitate its infection of the host insect by disrupting the peritrophic membrane. J. Virol. 81 (8), 4235–4243.
Mitsuhashi, W., Murakami, R., Takemoto, Y., Miyamoto, K., Wada, S., 2008. Stability of the viral-enhancing ability of entomopoxvirus spindles exposed to various abiotic factors. Appl. Entomol. Zool. 43 (4), 483–489.
Mitsuhashi, W., Asano, S., Miyamoto, K., Wada, S., 2013. Further research on the biological function of inclusion bodies of Anomala cuprea entomopoxvirus, with special reference to the effect on the insecticidal activity of a Bacillus thuringiensis formulation. Pest Manag. Sci. 70, 46–54.
Mitsuhashi, W., Inoue, T.A., Miyamoto, K., Sugimoto, T.N., 2019. Persistence of perorally inoculated entomopoxvirus spindles in the midgut of Bombyx mori larvae. J. Asia-Pac. Entomol. 22 (3), 723–727.
Morgan, C., Howe, C., Rose, H.M., Moore, D.H., 1956. Structure and development of viruses observed in the electron microscope: IV. Viruses of the RI-APC group. J. Biophys. Biochem. Cytol. 2 (3), 351–360.
Mori, H., Metcalf, P., 2010. Cypoviruses. In: Asgari, S., Johnson, K. (Eds.), Insect Virology. Caister Academic Press, Norfolk, pp. 307–323.
Mori, H., Ito, R., Nakazawa, H., Sumida, M., Matsubara, F., Minobe, Y., 1993. Expression of Bombyx mori cytoplasmic polyhedrosis virus polyhedrin in insect cells by using a baculovirus expression vector, and its assembly into polyhedra. J. Gen. Virol. 74 (Pt 1), 99–102.
Mori, H., Shukunami, C., Furuyama, A., Notsu, H., Nishizaki, Y., Hiraki, Y., 2007. Immobilization of bioactive fibroblast growth factor-2 into cubic proteinous microcrystals (Bombyx mori cypovirus polyhedra) that are insoluble in a physiological cellular environment. J. Biol. Chem. 282 (23), 17289–17296.

Moshe, A., Gorovits, R., 2012. Virus-induced aggregates in infected cells. Viruses 4 (10), 2218–2232.
Netherton, C.L., Wileman, T., 2011. Virus factories, double membrane vesicles and viroplasm generated in animal cells. Curr. Opin. Virol. 1 (5), 381–387.
Netherton, C., Moffat, K., Brooks, E., Wileman, T., 2007. A guide to viral inclusions, membrane rearrangements, factories, and viroplasm produced during virus replication. Adv. Virus Res. 70, 101–182.
Nikolic, J., Le Bars, R., Lama, Z., Scrima, N., Lagaudrière-Gesbert, C., Gaudin, Y., Blondel, D., 2017. Negri bodies are viral factories with properties of liquid organelles. Nat. Commun. 8 (1), 1–12.
Nishishita, N., Ijiri, H., Takenaka, C., Kobayashi, K., Goto, K., Kotani, E., Itoh, T., Mori, H., Kawamata, S., 2011. The use of leukemia inhibitory factor immobilized on virus-derived polyhedra to support the proliferation of mouse embryonic and induced pluripotent stem cells. Biomaterials 32 (14), 3555–3563.
Nysten, P.H., 1808. Recherches sur les Maladies des vers à Soie. Impr. Impériale, Paris, France.
Oberthuer, D., Knoška, J., Wiedorn, M.O., Beyerlein, K.R., Bushnell, D.A., Kovaleva, E.G., Heymann, M., Gumprecht, L., Kirian, R.A., Barty, A., Mariani, V., Tolstikova, A., Adriano, L., Awel, S., Barthelmess, M., Dörner, K., Xavier, P.L., Yefanov, O., James, D.R., Nelson, G., Wang, D., Calvey, G., Chen, Y., Schmidt, A., Szczepek, M., Frielingsdorf, S., Lenz, O., Snell, E., Robinson, P.J., Šarler, B., Belšak, G., Maček, M., Wilde, F., Aquila, A., Boutet, S., Liang, M., Hunter, M.S., Scheerer, P., Lipscomb, J.D., Weierstall, U., Kornberg, R.D., Spence, J.C.H., Pollack, L., Chapman, H.N., Bajt, S., 2017. Double-flow focused liquid injector for efficient serial femtosecond crystallography. Sci. Rep. 7, 44628.
Paez-Espino, D., Eloe-Fadrosh, E.A., Pavlopoulos, G.A., Thomas, A.D., Huntemann, M., Mikhailova, N., Rubin, E., Ivanova, N.N., Kyrpides, N.C., 2016. Uncovering Earth's virome. Nature 536, 425.
Prangishvili, D., 2013. The wonderful world of archaeal viruses. Annu. Rev. Microbiol. 67, 565–585.
Prangishvili, D., Bamford, D.H., Forterre, P., Iranzo, J., Koonin, E.V., Krupovic, M., 2017. The enigmatic archaeal virosphere. Nat. Rev. Microbiol. 15 (12), 724–739.
Prasad, B.V.V., Schmid, M.F., 2012. Principles of virus structural organization. Adv. Exp. Med. Biol. 726, 17–47.
Quax, T.E.F., Daum, B., 2018. Structure and assembly mechanism of virus-associated pyramids. Biophys. Rev. 10, 1–7.
Quax, T.E.F., Lucas, S., Reimann, J., Péhau-Arnaudet, G., Prevost, M.-C., Forterre, P., Albers, S.-V., Prangishvili, D., 2011. Simple and elegant design of a virion egress structure in Archaea. Proc. Natl. Acad. Sci. U. S. A. 108 (8), 3354–3359.
Rees, J.S., Jarrett, P., Ellar, D.J., 2009. Peritrophic membrane contribution to Bt Cry delta-endotoxin susceptibility in Lepidoptera and the effect of Calcofluor. J. Invertebr. Pathol. 100 (3), 139–146.
Rensen, E., Krupovic, M., Prangishvili, D., 2015. Mysterious hexagonal pyramids on the surface of Pyrobaculum cells. Biochimie 118, 365–367.
Roedig, P., Vartiainen, I., Duman, R., Panneerselvam, S., Stübe, N., Lorbeer, O., Warmer, M., Sutton, G., Stuart, D.I., Weckert, E., David, C., Wagner, A., Meents, A., 2015. A micro-patterned silicon chip as sample holder for macromolecular crystallography experiments with minimal background scattering. Sci. Rep. 5, 10451.
Roedig, P., Ginn, H.M., Pakendorf, T., Sutton, G., Harlos, K., Walter, T.S., Meyer, J., Fischer, P., Duman, R., Vartiainen, I., Reime, B., Warmer, M., Brewster, A.S., Young, I.D., Michels-Clark, T., Sauter, N.K., Kotecha, A., Kelly, J., Rowlands, D.J., Sikorsky, M., Nelson, S., Damiani, D.S., Alonso-Mori, R., Ren, J., Fry, E.E., David, C., Stuart, D.I., Wagner, A., Meents, A., 2017. High-speed fixed-target serial virus crystallography. Nat. Methods 14 (8), 805–810.

Rohrmann, G.F., 1986. Polyhedrin structure. J. Gen. Virol. 67 (Pt 8), 1499–1513.
Rohrmann, G.F., 2019. Baculovirus molecular biology, Translated from Englishfourth ed. National Center for Biotechnology Information (US), Bethesda, MD[eBook].
Rossetti, D., Lienert, U., Pradervand, C., Schneider, R., Shi, M., Zelenika, S., Rossat, M., Hignette, O., Rommeveaux, A., Schulze-Briese, C., 2002. Design and performance of the flexural-hinge-based mirror bender at the SLS protein crystallography beamline X06SA. Proc. SPIE 4782, X-Ray Mirrors, Crystals, and Multilayers II 86–93.
Rossmann, M.G., 2013. Structure of viruses: a short history. Q. Rev. Biophys. 46 (2), 133–180.
Rubio Huertos, M., 1954. Rapid extraction of intact crystalline inclusions from the cells of plants infected with Tobacco Mosaic Virus. Nature 174 (4424), 313.
Salvador, R., Ferrelli, M.L., Berretta, M.F., Mitsuhashi, W., Biedma, M.E., Romanowski, V., Sciocco-Cap, A., 2012. Analysis of EpapGV gp37 gene reveals a close relationship between granulovirus and entomopoxvirus. Virus Genes 45 (3), 610–613.
Sawaya, M.R., Cascio, D., Gingery, M., Rodriguez, J., Goldschmidt, L., Colletier, J.-P., et al., 2014. Protein crystal structure obtained at 2.9 Å resolution from injecting bacterial cells into an X-ray free-electron laser beam. Proc Natl Acad Sci USA 111 (35), 12769–12774.
Schönherr, R., Rudolph, J.M., Redecke, L., 2018. Protein crystallization in living cells. Biol. Chem. 399 (7), 751–772.
Shi, M., Lin, X.-D., Tian, J.-H., Chen, L.-J., Chen, X., Li, C.-X., Qin, X.-C., Li, J., Cao, J.-P., Eden, J.-S., Buchmann, J., Wang, W., Xu, J., Holmes, E.C., Zhang, Y.-Z., 2016. Redefining the invertebrate RNA virosphere. Nature 540, 539–543.
Shi, M., Lin, X.-D., Chen, X., Tian, J.-H., Chen, L.-J., Li, K., Wang, W., Eden, J.-S., Shen, J.-J., Liu, L., Holmes, E.C., Zhang, Y.-Z., 2018. The evolutionary history of vertebrate RNA viruses. Nature 556 (7700), 197–202.
Simmonds, P., Adams, M.J., Benkő, M., Breitbart, M., Brister, J.R., Carstens, E.B., Davison, A.J., Delwart, E., Gorbalenya, A.E., Harrach, B., Hull, R., King, A.M.Q., Koonin, E.V., Krupovic, M., Kuhn, J.H., Lefkowitz, E.J., Nibert, M.L., Orton, R., Roossinck, M.J., Sabanadzovic, S., Sullivan, M.B., Suttle, C.A., Tesh, R.B., van der Vlugt, R.A., Varsani, A., Zerbini, F.M., 2017. Consensus statement: virus taxonomy in the age of metagenomics. Nat. Rev. Microbiol. 15, 161–168.
Smith, J.L., Fischetti, R.F., Yamamoto, M., 2012. Micro-crystallography comes of age. Curr. Opin. Struct. Biol. 22 (5), 602–612.
Steere, R.L., Williams, R.C., 1953. Identification of crystalline inclusion bodies extracted intact from plant cells infected with tobacco mosaic virus. Am. J. Bot. 40 (2), 81–84.
Summers, M.D., 1971. Electron microscopic observations on granulosis virus entry, uncoating and replication processes during infection of the midgut cells of Trichoplusia ni. J. Ultrastruct. Res. 35 (5), 606–625.
Summers, M.D., 2006. Milestones leading to the genetic engineering of baculoviruses as expression vector systems and viral pesticides. Adv. Virus Res. 68, 3–73.
Summers, M.D., Arnott, H.J., 1969. Ultrastructural studies on inclusion formation and virus occlusion in nuclear polyhedrosis and granulosis virus-infected cells of Trichoplusia ni (Hübner). J. Ultrastruct. Res. 28 (5), 462–480.
Tabe, H., Shimoi, T., Fujita, K., Abe, S., Ijiri, H., Tsujimoto, M., Kuchimaru, T., Kizaka-Kondo, S., Mori, H., Kitagawa, S., Ueno, T., 2015. Design of a CO-releasing extracellular scaffold using in vivo protein crystals. Chem. Lett. 44 (3), 342–344.
Tabe, H., Shimoi, T., Boudes, M., Abe, S., Coulibaly, F., Kitagawa, S., Mori, H., Ueno, T., 2016. Photoactivatable CO release from engineered protein crystals to modulate NF-κB activation. Chem. Commun. (Camb.) 52 (24), 4545–4548.
Takemoto, Y., Mitsuhashi, W., Murakami, R., Konishi, H., Miyamoto, K., 2008. The N-terminal region of an entomopoxvirus fusolin is essential for the enhancement of peroral infection, whereas the C-terminal region is eliminated in digestive juice. J. Virol. 82 (24), 12406–12415.

Terwilliger, T.C., Bunkoczi, G., Hung, L.W., Zwart, P.H., Smith, J.L., Akey, D.L., Adams, P.D., 2016. Can I solve my structure by SAD phasing? Planning an experiment, scaling data and evaluating the useful anomalous correlation and anomalous signal. Acta Crystallogr. D Struct. Biol. 72 (Pt 3), 359–374.

Thorne, R., Stum, Z., Kmetko, J., O'Neill, K., Gillian, R., 2003. Microfabricated mounts for high-throughput macromolecular cryocrystallography. J. Appl. Cryst. 36 (6), 1455–1460.

Tsutsui, H., Jinno, Y., Shoda, K., Tomita, A., Matsuda, M., Yamashita, E., et al., 2015. A diffraction-quality protein crystal processed as an autophagic cargo. Mol Cell 58 (1), 186–193.

Vaaje-Kolstad, G., Westereng, B., Horn, S.J., Liu, Z., Zhai, H., Sørlie, M., Eijsink, V.G.H., 2010. An oxidative enzyme boosting the enzymatic conversion of recalcitrant polysaccharides. Science 330 (6001), 219–222.

Vaaje-Kolstad, G., Forsberg, Z., Loose, J.S., Bissaro, B., Eijsink, V.G., 2017. Structural diversity of lytic polysaccharide monooxygenases. Curr. Opin. Struct. Biol. 44, 67–76.

Vago, C., Croissant, O., 1959. Recherches sur la pathogènese des viroses d'insectes. La libération des virus dans le tube digestif de l'insecte à partir des corps d'inclusion ingérés. Ann. Epiphyt. 1, 5–18.

Vida, M.H., 1527. De bombice (poem). Rome.

Wang, Q., Bosch, B.-J., Vlak, J.M., van Oers, M.M., Rottier, P.J., van Lent, J.W.M., 2016. Budded baculovirus particle structure revisited. J. Invertebr. Pathol. 134, 15–22.

Weber, J., Liao, S.K., 1969. Light and electron microscopy of virus-associated, intranuclear paracrystals in cultured cells infected with types 2,4,6, and 18 human adenoviruses. Can. J. Microbiol. 15 (8), 841–845.

Wileman, T., 2006. Aggresomes and autophagy generate sites for virus replication. Science 312 (5775), 875–878.

Williams, T., Ward, V., 2010. Iridoviruses. In: Assgari, S., Johnson, K. (Eds.), Insect Virology. Caister Academic Press, UK, pp. 123–152.

Wodak, S.J., Janin, J., 2002. Structural basis of macromolecular recognition. Adv. Protein Chem. 61, 9–73.

Xeros, N., 1954. A second virus disease of the Leatherjacket, Tipula paludosa. Nature 174 (4429), 562–563.

Xu, J., Hukuhara, T., 1992. Enhanced infection of a nuclear polyhedrosis virus in larvae of the armyworm *Pseudaletia separata* by a factor in the spheroids of an entomopoxvirus. J. Invertebr. Pathol. 60 (3), 259–264.

Yu, X., Ge, P., Jiang, J., Atanasov, I., Zhou, Z.H., 2011. Atomic model of CPV reveals the mechanism used by this single-shelled virus to economically carry out functions conserved in multishelled reoviruses. Structure 19 (5), 652–661.

Zander, U., Bourenkov, G., Popov, A.N., de Sanctis, D., Svensson, O., McCarthy, A.A., Round, E., Gordeliy, V., Mueller-Dieckmann, C., Leonard, G.A., 2015. MeshAndCollect: an automated multi-crystal data-collection workflow for synchrotron macromolecular crystallography beamlines. Acta Crystallogr. D Biol. Crystallogr. 71, 2328–2343. https://doi.org/10.1107/S1399004715017927.

Kolibri 360 GmbH
Ferdinand-Jühlke-Straße 7
99095 Erfurt, Deutschland
produktsicherheit@kolibri360.de